SOLID STATE LASER

Edited by **Amin H. Al-Khursan**

Solid State Laser
Edited by Amin H. Al-Khursan

Published by InTech

Janeza Trdine 9, 51000 Rijeka, Croatia

Publishing Process Manager Iva Simcic

Technical Editor Teodora Smiljanic

Cover Designer InTech Design Team

Additional hard copies can be obtained from orders@intechopen.com

Solid State Laser, Edited by Amin H. Al-Khursan
p. cm.
ISBN 978-953-51-0086-7

For EU product safety concerns:
IN TECH d.o.o., Prolaz Marije Krucifikse Kozulić 3, 51000 Rijeka, Croatia,
info@intechopen.com or visit our website at intechopen.com.

Meet the editor

Dr Amin H. Al-Khursan graduated from the Science College, Basrah University, where he obtained an MSc in Nuclear Track Detectors in 1992, working at the Science College, Thi-Qar University, Nassiriya in 1992, as assistant lecturer. In 1997, he returned to the Science College, Basrah University as a PhD student and obtained a PhD in physics of "Quantum Dot Semiconductor Lasers" in October 2001, under the supervision of Professor R.S. Fyath, an electrical engineer, he became the first Iraqi researcher to work in the nanostructure field. Dr Al-Khursan worked as a lecturer in 2001, as an assistant professor in 2005, and then as a professor in 2010, all at the Science College, Thi-Qar University. Between 2007 and 2008, he received a fellowship from the Politecnico di Torino, Italy, with Professor Montrosset and M. Gioannini. Dr Al-Khursan founded the "Nassiriya Nanotechnology Research Laboratory (NNRL)" in 2010, where MSc and PhD groups from different Iraqi universities work. He has published 30 papers and has been a member of the editorial board of "Recent Patents in Electrical Engineering", since 2007, and "Physics Express", since 2010, as well as three other Iraqi journals. He is also a member of IEEE Photonics Society and OSA.

Contents

Preface

Even though the first laser light was emitted from a solid-state ruby crystal more than half a century ago, a wide range of applications still use solid-state lasers today. This is because they comprise a variety of materials and cover a wide range of the spectrum. Many new materials have already been discovered, but the door to add new ones is still open. Accordingly, this book covers some recent developments and reviews, to a certain extent, the work done in the past years.

This book consists of eight chapters divided into four sections. The first section deals with the waveguide optimization in solid-state lasers where, in the first chapter, the optimum design of the optical coupling system and the mode matching to achieve the desired gain and thermal properties of an end-pumped laser are derived analytically. Chapter two discusses the pump uniformity, beam distortion and cooling non-uniformities in the planar waveguide guides to obtain high efficiency and good beam quality.

In the second section, the rare-earth-doped lasers are discussed in chapters three and four. The recent study on the rare-earth double borate family laser crystals, in the growth, thermal, optical and spectrum characteristics are reviewed in chapter three. Then, in chapter four, the rare-earth-doped chloride and bromide laser crystals operating in bands II and III of the atmosphere transmission window are discussed.

Section three (chapters five, six, and seven) deals with the nonlinearity in solid-state lasers. First, high-energy solid-state oscillators, where the levels of intensity are in the order of 10^{14} W/cm^2, are important from a practical point of view. To exert it, the femtosecond pulse technology must be used. Such oscillators suffer from some impediment results from nonlinear effects, and when they work at the normal dispersion region, they are called "chirped pulse oscillators". Chapter five deals with these oscillators from both theoretical, as well as experimental standpoints. In chapter six, the use of solid-state lasers in intracavity nonlinear frequency generation has been explored. However, the ability of Cr^{3+}-doped colquiriites lasers in nonlinear conversion is discussed. A description of the physics involving second-order nonlinear interactions is presented, where the allowed processes for three-wave mixing are detailed. A review of techniques for controlling the phase matching among interacting waves is made. Another important nonlinear phenomenon is the upconversion frequency where the absorption of two or more photons is followed by the emission of a high-energy photon. The upconversion using lanthanides doped host matrices and different excitation processes, mainly the energy transfer and excitation state absorption process responsible for the luminescence in the rare earth ions, are reviewed in chapter seven.

Section four deals with the nanostructure quantum dot semiconductor lasers through chapter eight, which discusses the feedback in quantum dot lasers. The delay differential equations system is used to elucidate the behavior of different states in the quantum dot laser.

Finally, I wish to acknowledge the publishing process manager, Iva Simcic, for her effort and patience through the book process, and InTech for giving me this opportunity. I would also like to thank the researchers who contributed to this book.

Amin H. Al-Khursan
Nassiriya Nanotechnology Research Laboratory (NNRL)
Physics Department, College of Science
Thi-Qar University, Nassiriya
Iraq

Part 1

Waveguide Optimization in Solid-State Lasers

1

Optimum Design of End-Pumped Solid-State Lasers

Gholamreza Shayeganrad
Institute of Photonics Technologies, Department of Electrical Engineering,
National Tsinghua University,
Taiwan

1. Introduction

The principle of laser action was first experimentally demonstrated in 1960 by T. Maiman (Maiman, 1960). This first system was a solid-state laser in which a ruby crystal and a flashlamp served as gain medium and pump source, respectively. Soon after this first laser experiment, it was realized that solid-state lasers are highly attractive sources for various scientific and industrial applications such as laser marking, material processing, holography, spectroscopy, remote sensing, lidar, optical nonlinear frequency conversion, THz frequency generation (Koechner, 2006; Hering et al., 2003; Ferguson & Zhang, 2003; Sennaroghlu, 2007). Since the early 1980's with the development of reliable high power laser diode and the replacement of traditional flashlamp pumping by laser-diode pumping, the diode-pumped solid-state lasers, DPSSL's, have received much attention and shown the significant improvements of laser performance such as optical efficiency, output power, frequency stability, operational lifetime, linewidth, and spatial beam quality.

Nd:YAG and Nd:YVO_4 crystals have been extensively used as a gain medium in commercial laser products with high efficiency and good beam quality. The active ion of Nd^{3+} has three main transitions of $^4F_{3/2}\rightarrow{}^4I_{9/2}$, $^4F_{3/2}\rightarrow{}^4I_{11/2}$, and $^4F_{3/2}\rightarrow{}^4I_{13/2}$ with the respective emission lines of 0.94, 1.06 and 1.3 μm. The emission wavelengths of DPSSL's associated with nonlinear crystals cover a wide spectral region from ultraviolet to the mid-infrared range and very often terahertz range by difference frequency mixing process in simultaneous multi-wavelength solid-state lasers (Saha et al., 2006; Guo etal., 2010).

DPSSL's are conventionally categorized as being either end-pumped or side pumped lasers. End-pumping configuration is very popular because of higher efficiency, excellent transverse beam quality, compactness, and output stability which make it more useful for pumping tunable dye and Ti: Sapphire laser, optical parametric oscillator/amplifier, and Raman gain medium. The better beam quality is due to the high degree of spatial overlap between pump and laser modes while the high efficiency is dependent on good spatial mode-matching between the volume of pump and laser modes or nondissipating of pump energy over pumping regions that are not used by laser mode. In addition, end pumping allows the possibility of pumping a thin gain medium such as disk, slab, and microchip lasers that are not be accessed from the side-pumping (Fan &. Byer, 1998; Alfrey, 1989; Carkson & Hanna,1988; Sipes, 1985; Berger et al. 1987).

The side-pumping geometry allows scaling to high-power operation by increasing the number of pump sources placed around the gain medium before occurring thermal fracture. In this arrangement the pump power is uniformly distributed and absorbed over a large volume of the crystal which leads to reduce the thermal effects such as thermal lensing and thermal induced stress. However, the power scaling of end-pumped lasers is limited due to the physically couple of many diode-lasers into a small pumped volume and the thermal distortion inside the laser crystal. To improve power scaling of an end-pumped laser, a fiber-coupled laser-diode array with circular beam profile and high-output power and a crystal with better thermal properties can be employed as a pump source and gain medium, respectively (Hemmeti & Lesh, 1994; Fan & Sanchez, 1989; Mukhopadhyay, 2003; Hanson, 1995; Weber, 1998; Zhuo, 2007; Sulc, 2002; MacDonald, 2000).

Laser performance is characterized by threshold and slope efficiency. The influence of pump and laser mode sizes on the laser threshold and slope efficiency has been well investigated (Hall et al. 1980; Hall, 1981; Risk, 1988; Laporta &. Brussard, 1991; Fan & Sanchez, 1990; Clarkson & Hanna, 1989; Xiea et al., 1999). It is known a smaller value of the pump radius leads to a lower threshold and a higher slope efficiency. However, in the case of fiber-coupled end-pumped lasers, due to pump beam quality, finite transverse dimension, diffraction, absorption and finite length of the gain medium, the pump size can be decreased only to a certain value.

It is worthwhile to mention, that for both longitudinal and transverse pumping, the pump radius varies within the crystal mainly because of absorption and diffraction. It is possible to consider a constant pump radius within the crystal when the crystal length is much smaller than the Rayleigh range of the pump beam and also than the focal length of thermal lens. However, in the case of longitudinal pumping, the pump intensity is still a function of distance from the input end even this circumstance is also satisfied. Meanwhile, the lower brightness of the laser-diodes than the laser beam makes the Rayleigh distance of the pump beam considerably be shorter than the crystal length.

The effect of pump beam quality on the laser threshold and slope efficiency of fiber-coupled end-pumped lasers has been previously investigated (Chen et al., 1996, 1997). The model is developed based on the space-dependent rate-equations and the approximations of paraxial propagation on pump beam and gain medium length much larger than absorption length. Further development was made by removing the approximation on gain medium length (Chen, 1999), while for a complete description, rigorous analysis is required.

In this chapter, we initially reviewed the space-dependent rate equation for an ideal four-level end-pumped laser. Based on the space-dependent rate equation and minimized root-mean-square of pump beam radius within the gain medium, a more comprehensive and accurate analytical model for optimal design an end-pumped solid-state laser has been presented. The root-mean-square of the pump radius is developed generally by taking a circular–symmetric Gaussian pump beam including the M^2 factor. It is dependent on pump beam properties (waist location, M^2 factor, waist radius, Rayleigh range) and gain medium characteristics (absorption coefficient at pump wavelength and gain medium length). The optimum mode-matching is imposed by minimizing the root-mean-square of pump beam radius within the crystal. Under this condition, the optimum design key parameters of the optical coupling system have been analytically derived. Using these parameters and the linear approximate relation of output power versus input power, the parameters for optimum design of laser cavity are also derived. The requirements on the pump beam to achieve the desired gain at the optimum condition of mode-matching are also investigated. Since thermal effects are the final limit for scaling end-pumped solid-state lasers, a relation for thermal focal length at this condition is developed as a function of pump power, pump beam M^2 factor, and physical and thermal-optics of gain medium properties. The present model provides a straightforward procedure to design the optimum laser resonator and the optical coupling system.

2. Space-dependent rate equation

The rate equation is a common approach for dynamically analyzing the performance of a laser. For a more accurate analysis of characteristics of an end-pumped laser, particularly the influence of the pump to laser mode sizes, it is desirable to consider the spatial distribution of inversion density and the pump and laser modes in the rate equation. The space-dependent rate equation based on single mode operation for an ideal four level laser is developed by Laporta and. Brussard (Laporta &. Brussard, 1991):

$$\frac{dN(x,y,z)}{dt} = R(x,y,z) - \sigma_e c_0 \frac{S(x,y,z)}{h\nu_l} N(x,y,z) - \frac{N(x,y,z)}{\tau} \tag{1}$$

$$\frac{dq}{dt} = \sigma_e c_0 \iiint \frac{S(x,y,z)}{h\nu_l} N(x,y,z) dV - \frac{q}{\tau_c} \tag{2}$$

where z is the propagating direction, N is the upper energy level population density, R is the total pumping rate into the upper level per unit volume, S is the cavity mode energy density, σ_e is the cross section of laser transition, c_0 is the light velocity in the vacuum, h is the Plank's constant, ν_l is the frequency of the laser photon , q is the total number of photons in the cavity mode, τ is the upper-level life-time, and τ_c is the photon lifetime. In Eq. (2) the integral is calculated over the entire volume of the active medium. The photon lifetime can be expressed as $\tau_c=2l_e/\delta c_0$, where $l_e=l_{ca}+(n-1)l$ is the effective length of the resonator, n is refractive index of the active material, l_{ca} and l are the geometrical length of the resonator and the active medium, respectively, and $\delta=2\alpha_i l-\ln(R_1R_2)+\delta_c+\delta_d\approx 2\delta_i+T+\delta_c+\delta_d$ is the total logarithmic round-trip cavity-loss of the fundamental intensity, T is the power transmission of the output coupler, δ_i represents the loss proportional to the gain medium length per pass such as impurity absorption and bulk scattering, δ_c is the non-diffraction internal loss such as scattering at interfaces and Fresnel reflections, and δ_d is the diffraction losses due to thermally induced spherical aberration. The approximation is valid for the small values of T.

Note that to write Eq. (2) the assumption of a small difference between gain and logarithmic loss has been assumed which maintains when intracavity intensity is a weak function of z. For a continuous-wave (CW) laser this situation always holds while for a pulsed laser, it is valid only when the laser is not driven far above the threshold. It follows that this analysis is appropriated to describe the behavior of low gain diode-pumped lasers, but is not adequate for gain-switched or Q-switched lasers and in general for high gain lasers. It is also assumed that the transverse mode profile considered for the unloaded resonator is not substantially modified by the optical material inside the cavity.

The pumping rate R can be related to the input pump power P_{in}

$$\iiint R(x,y,z)dV = \eta_p \frac{P_{in}}{h\nu_l} \tag{3}$$

where $\eta_p=\eta_t\eta_a(\nu_l/\nu_p)$ is the pumping efficiency, η_t is the optical transfer efficiency (ratio between optical power incident on the active medium and that of emitted by the pump source), and $\eta_a\approx 1-exp(-al)$ is the absorption efficiency (ratio between power absorbed in the active medium and that of entering the gain medium), α is the absorption coefficient at pump wavelength, l is the crystal length, ν_p is the frequency of the pump photon, and the integral extends again over the volume of the active material. Under the stationary condition, a relationship between the energy density in the cavity and the pumping rate can be easily derived. We define a normalized pump distribution within the gain medium as

$$r_p(x,y,z) = \frac{R(x,y,z)}{R_0} \tag{4}$$

where $\iiint r_p(x,y,z)dV = 1$ and R_0 represents therefore the total number of photons absorbed per unit time in the active medium. We also define a normalized mode distribution as

$$s_l(x,y,z) = \frac{S(x,y,z)}{S_0} \tag{5}$$

where $\int_1 ns_l(x,y,z)dV + \int_2 s_l(x,y,z)dV = 1$, and S_0 is the total energy of cavity mode corresponding to the total number of photons $q=S_0/h\nu_l$ The first integral is taken over the whole field distribution in the region of the active medium and the second in the remaining volume of the resonator.

Substituting Eq. (1) into (2), and considering Eqs. (3)-(5), under the steady-state condition, we have

$$P_{in} = \frac{\delta h\nu_l}{2\eta_p l_e \sigma_e \tau}\left[\iiint \frac{s_l(x,y,z)r_p(x,y,z)}{c_0 S_0 s_l(x,y,z)/I_{sat}+1}dV\right]^{-1} \tag{6}$$

where $I_{sat}= h\nu_l/\sigma_e\tau$ is the saturation intensity. In the threshold limit ($S_0\approx 0$) we obtain the following formula for the threshold pump power:

$$P_{th} = \frac{\delta\, h\nu_l}{2\sigma_e \tau \eta_p l_e} V_{eff} \tag{7}$$

where

$$V_{eff} = \left[\iiint s_l(x,y,z).r_p(x,y,z)dV\right]^{-1} \tag{8}$$

introduces the effective volume of spatial overlap between pump and cavity modes.

In the approximation of intracavity intensity much less than saturation intensity, the argument of the integral in (6) can be expanded around zero based on Taylor series and keep the first term as

$$\frac{s_l r_p}{c_0 S_0 s_l / I_{sat} + 1} \cong (1 - c_0 S_0 s_l / I_{sat}) s_l r_p \tag{9}$$

Inserting Eq. (9) into Eq. (6) and developing the integral with assuming the plane wave approximation, $c_0S_0/l_e=2P$, where $P=P_{out}/T$ is the intracavity power of one of the two circulating waves in the resonator, yields

$$P_{out} = \eta_s \left[P_{in} - P_{th} \right] \tag{10}$$

where

$$\eta_s = \frac{T}{\delta} \eta_p V_{slope} \tag{11}$$

is the slope efficiency and

$$V_{slope} = \frac{\left(\iiint s_l(x,y,z).r_p(x,y,z)dV \right)^2}{\iiint s_l^2(x,y,z).r_p(x,y,z)dV} \tag{12}$$

represents the mode-matching efficiency. The slope efficiency η_s can be defined as the product of the pumping efficiency η_p, the output coupling efficiency $\eta_c=T/\delta$, and the spatial overlap efficiency V_{slope}. The slope efficiency measures the increase of the output power as the pump power increases. It is generally somewhat larger than the total power conversion efficiency. For high slope efficiency, one wants high η_p, and low δ. It can also be achieved by increasing T if other losses are not low, but this is undesirable because it increases threshold pump power.

We should note that for mode-to-pump size ratio greater than unity, the linear approximation in Eq. (10) is valid also when the intracavity intensity is comparable with the saturation intensity. It should be also noted that for simplicity we have considered the plane wave approximation, but the formalism can be easily expanded for non-plane wave, such as a Gaussian beam profile.

From Eq. (7), threshold pump power depends linearly on the effective mode volume, and inversely on the product of the effective stimulated-emission cross-section and the lifetime of laser transition. Thus, if laser transition lifetime be the only variable, it seems the longer lifetime results in a lower pump threshold for CW laser operation. However, the stimulated-emission cross section is also inversely proportional to the lifetime of laser transition. Offsetting this is the relation between the laser transition lifetime and the stimulated emission cross-section. In many instances, the product of these two factors is approximately constant for a particular active ion. Consequently, threshold is roughly and inversely proportional to the product of the effective stimulated emission cross-section and the lifetime of the laser transition. Notice that larger stimulated-emission cross section is useful in a lower pump threshold for CW laser operation and a smaller cross section has advantages in Q-switch operation. On the other hand, slope efficiency depends on the overlap or mode-matching efficiency and losses as well. Overlap efficiency is dependent on the particular laser design but generally it is easier to achieve when laser pumping is used rather than flashlamp pumping.

The total round-trip internal loss, $L_i=2\delta_i+\delta_c+\delta_d$, in the system can be determined experimentally by the Findlay-Clay analysis. This was done by measuring the different pumping input power at the threshold versus the transmission of output coupling mirror as (Findlay & Clay 1966)

$$T = K_p P_{th} - L_i \tag{13}$$

where $K_p=(2\eta_p l_e/I_{sat} V_{eff})$ is the pumping coefficient.

According to Koechner (Koechner, 2006) the optimum output coupler transmission T_{opt} can be calculated using the following standard formula:

$$T_{opt} = (\sqrt{g_0 l_e / L_i} - 1)L_i \tag{14}$$

where g_0 is the small signal round-trip gain coefficient. The small-signal round-trip gain coefficient for an ideal four-level end-pumped laser is often expressed as:

$$g_0 = 2\sigma_e \iiint s_l(x,y,z)N(x,y,z)dV \tag{15}$$

According to Eqs. (1), (2) and (15), the small-signal round-trip gain coefficient which under the steady-state condition can be found as

$$g_0 = \frac{2\eta_p}{I_{sat} V_{eff}} P_{in} \tag{16}$$

As can be seen in (16), for an ideal four-level laser, the small-signal gain coefficient g_0 varies linearly with pump power and inversely with effective mode volume.

3. Optimum pumping system

A common configuration of a fiber-coupled laser diode end-pumped laser is shown in Fig. 1. In this arrangement, the coupled pump energy from a laser-diode into a fiber is strongly focused by a lens onto the gain medium. The w_{po} and w_{l0} are the pump and beam waists, respectively, l is the gain medium length, and z_0 is the location of pump beam waist.

Assuming a single transverse Gaussian fundamental mode (TEM_{00}) propagates in the cavity and neglecting from diffraction over the length of the gain medium, s_l can be expressed as:

$$s_l(x,y,z) = \frac{2}{\pi w_l^2(z) l_e} \exp\left(-2\frac{x^2+y^2}{w_l^2(z)}\right) \tag{17}$$

where

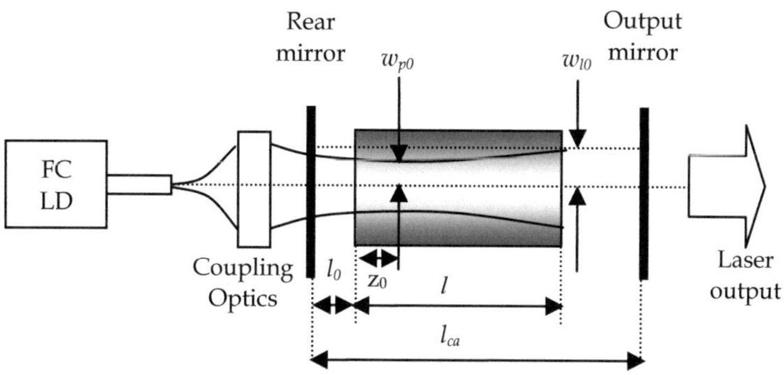

Fig. 1. Schematic diagram of a fiber-coupled laser-diode (FC LD) end-pumped solid-state laser.

$$w_l^2(z) = w_{l0}^2\left[1+\left(\frac{z\lambda_l}{n\pi w_{l0}^2}\right)^2\right] \approx w_{l0}^2 \quad (18)$$

represents the spot size at a distance z, where w_{l0} is the waist of the Gaussian beam, n is the refraction index of the crystal, λ_l is the fundamental laser wavelength in free space. The neglect of diffraction is justified if $n\pi w_{l0}^2 / \lambda_l$ is much larger than physical length l of the gain medium.

The intensity of the output beam comes out from a fiber-coupled laser-diode, r_p, may be described by a circular Gaussian function (Gong et al., 2008; Mukhopadhyay, 2003)

$$r_p(x,y,z) = \frac{2\alpha}{\pi w_p^2(z)[1-\exp(-\alpha l)]}\exp\left(-2\frac{x^2+y^2}{w_p^2(z)} - \alpha z\right) \quad (19)$$

Here, α is the absorption coefficient at pump wavelength, l is the gain medium length, and $w_p(z)$ is the pump beam spot size given by:

$$w_p^2(z) = w_{p0}^2\left[1+\left(\frac{z-z_0}{z_R}\right)^2\right] \quad (20)$$

where w_{p0} is the waist of the pump beam, $z=z_0$ is the beam waist location and z_R is the Rayleigh range:

$$z_R = \frac{n\pi w_{p0}^2}{\lambda_p M^2} \quad (21)$$

Note in the above equations z =0 is taken at the incidence surface of the gain medium. In Eq. (21) M^2 is the times diffraction limited factor which indicates how close a laser is to being a single TEM_{00} beam. An increasing value of M^2 represents a mode structure with more and more transversal modes. Beam M^2 factor is a key parameter which defines also how small a spot of a laser can be focused and the ability of the laser to propagate as a narrow thereby in some literatures it is called beam focusability factor. An important related quantity is the confocal parameter or depth of focus of the Gaussian beam b=2z_R. It is a measure of the longitudinal extent of the focal region of the Gaussian beam or the distance that the Gaussian beams remains well collimated. In other word, over the focal region, the laser field, called the near field, stays roughly constant with a radius varying from w_{p0} to $\sqrt{2}\, w_{p0}$. We see from (21), Rayleigh range is directly proportional to the beam waist w_{p0} and inversely proportional to the pump wavelength λ_p. Thus, when a beam is focused to a small spot size, the confocal beam parameter is short and the focal plane must be located with greater accuracy. A small spot size and a long depth of focus cannot be obtained simultaneously unless the wavelength of the light is short.

Inherent property of the laser beam is the relationship between beam waist w_0, far-field angle θ, and the index of refraction n. Based on the brightness theorem (Born & Wolf, 1999)

$$n\theta w_0 = C \geq M^2 \lambda_p / \pi \tag{22}$$

where C is a conserved parameter during focusing associated to the beam quality. For a fiber-coupled laser-diode, the value of C can simply be calculated from the product of fiber core radius and beam divergence angle. From Eq. (22), focusing a laser beam to a small spot size increases the beam divergence to reduce the intensity outside the Rayleigh range.

Putting Eqs. (17) and (19) into Eqs. (7) and (11), we obtain

$$P_{th} = \frac{\pi \delta w_{l0}^2 I_{sat}}{2\eta_p F(\alpha, l, w_{l0}, w_{p0}, z_0, z_R)} \tag{23}$$

$$\eta_s = \frac{T\eta_p}{\delta} \frac{F(\alpha, l, w_{l0}, w_{p0}, z_0, z_R)^2}{F(\alpha, l, w_{l0} / \sqrt{2}, w_{p0}, z_0, z_R)} \tag{24}$$

where

$$F(\alpha, l, \omega_{l0}, \omega_{p0}, z_0, z_R) = \frac{1}{1-\exp(-\alpha l)} \times \int_0^l \frac{\alpha w_{l0}^2 \exp(-\alpha z)}{w_{l0}^2 + w_{p0}^2[1+(z-z_0)^2 / z_R^2]} dz \tag{25}$$

is the mode-matching function describes the spatial-overlap of pump beam and resonator mode. The maximum value of the mode-matching function leads to the lowest threshold and the highest slope efficiency (Laporta & Brussard, 1991; Fan & Sanchez, 1990). Thereby the mode-matching function is the most important parameter to improve the laser performance. In general, this function cannot be solved analytically and to obtain the optimum pump focusing, Eq. (25) should be numerically solved. A closed form solution can be found by defining a suitable average pump spot size inside the active medium:

$$F(\alpha, l, \omega_{l0}, \omega_{p0}, z_0, z_R) = \frac{1}{1-\exp(-\alpha l)} \times \int_0^l \frac{\alpha w_{l0}^2 \exp(-\alpha z)}{w_{l0}^2 + \overline{w_p^2(z)}} dz = \frac{w_{l0}^2}{w_{l0}^2 + w_{p,rms}^2} \tag{26}$$

In this equation, $w_{p,rms}$ is the root-mean-square (RMS) of pump beam spot size within the active medium:

$$w_{p,rms} = \sqrt{\overline{w_p^2(z)}} \tag{27}$$

where $\overline{w_p^2(z)}$ is the mean of square pump beam spot size along the active medium given by (Shayeganrad & Mashhadi, 2008):

$$\overline{w_p^2(z)} = \frac{\int_0^l w_p^2(z)\exp(-\alpha z)dz}{\int_0^l \exp(-\alpha z)dz} \tag{28}$$

The exp(-αz) is the weighting function comes from the absorption of the pump beam along z direction. After putting Eq. (20) into Eq. (28) and performing the integrations we can obtain:

$$\overline{w_p^2(z)} = w_{p0}^2\left(1 + \frac{Z_0^2 - 2Z_0 f(L) + 2 - L(L+2)/[\exp(L)-1]}{Z_R^2}\right) \tag{29}$$

where $Z_0 = \alpha z_0$, $Z_R = \alpha z_R$,, and $L=\alpha l$ are dimensionless waist location, Rayleigh range and crystal length, respectively. In Eq. (29) $f(L) = 1 - \exp(-L)L / L_{eff}$ where $L_{eff} = 1 - \exp(-L)$ is the dimensionless parameter which defines the effective interaction length. Note that $L_{eff} \to L$ for $L<<1$, and $l_{eff} \to 1$ for $L>>1$. Thus for a strongly absorbing optical material ($l \gg 1/\alpha$) the effective interaction length is much shorter than physical length of the medium. This configuration can be useful for designing the disk or microchip laser with high absorption coefficient and short length gain medium.

We see from (26) that the maximum value of mode-matching can be raised by minimizing the RMS of beam spot size at a constant mode size. A minimum value of $w_{p,rms}$ can occur when $\partial w_{p,rms}/\partial Z_0$ is equal to zero at a fixed L, w_{p0} and Z_R. The solution is

$$Z_{0,opt} = f(L) \Rightarrow L/2 \tag{30}$$

After substituting Eq. (30) into (29) we obtain

$$\overline{w_{p,rms}^2(z)} = w_{p0}^2\left(1 + \frac{g^2(L)}{Z_R^2}\right) = \beta\left(Z_R + \frac{g^2(L)}{Z_R}\right) \tag{31}$$

where $\beta = C/n\alpha$ is pump beam quality which is often quoted in square millimeter and $g^2(L) = 1 - \exp(-L)\left(L/L_{eff}\right)^2$. The value of parameter β can be calculated by substituting the value of C and the properties of the active medium, n, and α.

Differentiating (31) respect to Z_R and put it equal to zero, we find

$$Z_{R,opt} = g(L) \Rightarrow L/2\sqrt{3} \tag{32}$$

In each expression the last form gives the asymptotic value for small L compared to unity. One sees that asymptotically the optimum waist location and Rayleigh range depend only on crystal length. While in the case of L≫1 or strong absorbing gain medium, they both tend to absorption lenght 1/α and are much shorter than physical length of the gain medium.

Fig. (2) shows the dimensionless optimum waist location and Rayleigh range of the pump beam. We can see, when absorption coefficient α increases, the optimum waist location and optimum Rayleigh range move closer to the incident surface of the active medium and they both increase with increasing active medium length *l* at a fixed α. These results were expected; because for large value of α, the pump beam is absorbed in a short length of the active medium. It can be also seen that the optimum waist location is larger than optimum Rayleigh range for $1<L<8$ and for large *L* ($L \geq 8$) they both tend to the absorption length $1/\alpha$. For L=1.89 and L=1.26 optimum Rayleigh range and optimum waist location are equal to half of the absorption length independently on the gain medium length that is considered as the optimum range in several papers (Laporta & Brussard, 1991; Berger et al., 1987).

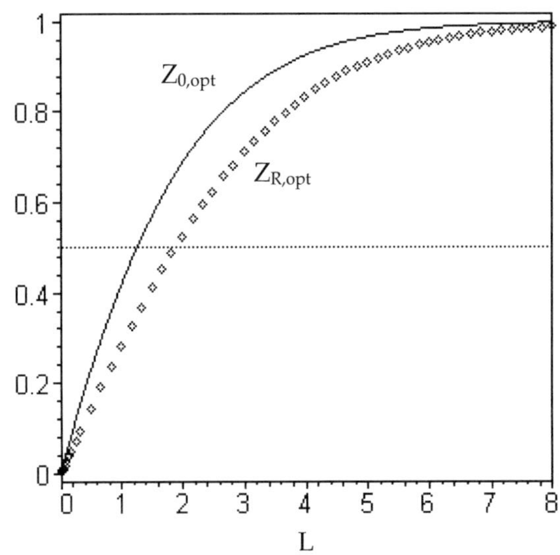

Fig. 2. The optimum dimensionless waist location and Rayleigh rang of the pump beam as a function of *L*.

Substitution (32) into (21) gives

$$w_{p0,opt} = \sqrt{\beta g(L)} \Rightarrow \sqrt{\beta L / 2\sqrt{3}} \tag{33}$$

Hence $w_{p,opt}$ becomes, after replacing Z_R by its value $g(L)$ into Eq. (31):

$$w_{p,opt} = \sqrt{2\beta g(L)} \tag{34}$$

From this equation, minimum pump spot size is a function of M^2 factor and two characteristic lengths: crystal length and absorption length $1/\alpha$. For $L \ll 1$, the minimum pump beam spot size $w_{p,opt}$ can be expressed in the following form:

$$w_{p,opt} \approx \sqrt{\beta L / \sqrt{3}} \tag{35}$$

As a result, optimum mode-matching function depends on the pump beam quality and gain medium characteristics as well. In practice, the experimentally measured optimum pump beam spot size $w_{p,opt}$, usually differs from that of calculated based on Eq. (34) because of the diffraction and thermal effects in a realistic laser gain medium. Nevertheless, this formula can provide a very good estimate for the $w_{p,opt}$.

Putting Eq. (34) into (22), optimum far-field-angle of pump beam is given by:

$$\theta_{p,opt} = \alpha\sqrt{\frac{\beta}{g(L)}} \tag{36}$$

It can be seen from Eqs. (34) and (36), optimum pump spot size and optimum pump beam divergence angle increase with increasing β to obtain maximum mode-matching efficiency. Equations (30), (33) and (36) provide a good guideline to design an optimum optical-coupling system. Again, these parameters are governed by the absorption coefficient, the gain medium length and the pump beam M^2 factor.

To reach the optimal-coupling, the incident Gaussian beam should be fitted to the aperture of the focusing lens with the largest possible extent without severe loss of pump power due to the finite aperture of the focusing lens and also serious edge diffraction. As one reasonable criterion for practical design, we might adapt the diameter of the focusing lens to πw_p, where w_p is the pump spot size of the Gaussian beam at the focusing lens. The waist and waist location for a Gaussian beam after passing through a thin lens of focal length f can be calculated with the ABCD Matrix method. For a collimated beam with radius w_p, they can be respectively described as

$$w_{p0} = w_p \frac{f / z'_p}{\sqrt{1+(f / z'_p)^2}} \tag{37.a}$$

$$z'_0 = \frac{f}{f^2 / z'^2_p + 1} \tag{37.b}$$

where $z'_p = n\pi w_p^2 / M^2\lambda_p$ is the Rayleigh range of the incoming beam. In these two equations, for simplicity, z=0 is considered the location of the lens. If we assume $z'_p >> f$, which is usually satisfied for fiber-coupled end-pumped lasers, Eq. (37) are reduced to:

$$w_{po} = \frac{f\lambda_p M^2}{n\pi w_p} \tag{38.a}$$

$$z'_0 = f \tag{38.b}$$

From Eqs. (33) and (38.a) we obtain

$$F_{opt} = w_p\sqrt{\frac{g(L)}{\beta}} \tag{39}$$

where $F_{opt}=af_{opt}$ is dimensionless optimal focal length of the focusing lens is plotted in Fig. 3 as a function of L for $w_p=1$ and several pump beam quality factors β. At a fixed β, optimal focal length of the focusing lens is an increasing function of L and is not very sensitive to L when pump beam quality is poor. It can be also seen, for a specific active medium, when pump beam quality increases by increasing divergence angle and/or core diameter of the fiber a lens with a small focal length satisfies in Eq. (39) is needed to achieve an optimal focusing and consequently a higher mode-matching efficiency. At a fixed l and β, if absorption coefficient α increases the optimal focal length decreases because of the moving pump beam waist location closer to the incident surface of the active medium. Putting the values of β, L and w_p into (30) and (39), optimal focal length lens and optimal location of the focusing lens can be determined.

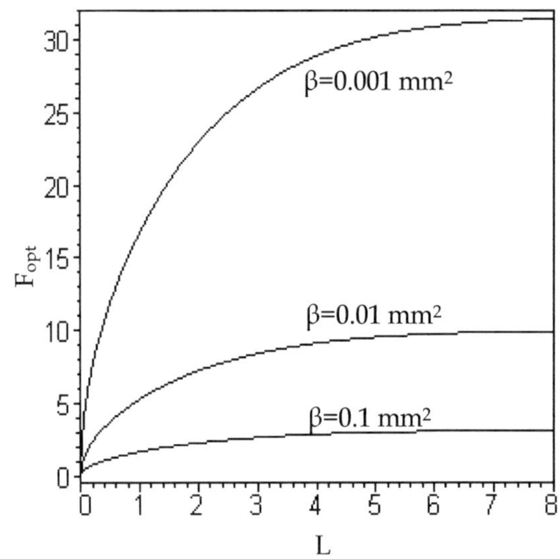

Fig. 3. Dimensionless optimized focal length of the focusing lens, af_{opt}., as a function of L for $\omega_p=1$ and several pump beam quality factors β.

On the other hand, based on the paraxial approximation, pump spot size $w_p(z)$ may be given by (Fan &. Sanchez, 1990)

$$w_p(z) = w_{p0} + \theta_p |z - z_0| \tag{40}$$

Several authors (Fan & Sanchez, 1990; Laporta & Brussard, 1991; Chen, 1999; Chen et al., 1996, 1997) have considered Eq. (40) to describe the evolution of pump beam radius within the gain medium in their model. This functional dependence is appropriate for beams with partial-spatial coherence (Fan & Sanchez, 1990). Also, if one is focusing the beam to a small spot size, the paraxial approximation is not justified and making questionable using Eq. (40) which is derived under the paraxial approximation. Using this function to describe the evolution of pump beam radius, the optimum pump spot size is defined as (Chen, 1999):

$$w_{p,opt} = 2\sqrt{\beta\left[ln\left(\frac{2}{1+\exp(-L)}\right)\coth\left(\frac{L}{2}\right) - \frac{L}{\exp(L)-1}\right]} \tag{41}$$

For L≫1, this equation yields (Chen et al., 1997)

$$w_{p,opt} = 2\sqrt{\beta ln(2)} \tag{42}$$

Fig. (4) shows comparison of the optimum pump spot size using Eqs. (34) and (41). It can be seen, at a fixed β, minimum pump size is an increasing function of L. For the case of poor pump beam quality, it initially increases rapidly and then this trend becomes saturate and is not significantly sensitive to L, while for the case of a good pump beam quality, it varies smoothly with increasing L. Further, for a specific active medium with a defined L, a poorer pump beam quality leads to a higher $w_{p,opt}$ to maximize the mode matching because of governing focusability with beam quality. Note that a good agreement between the Chen's model (Chen, 1999) and present model is obtained only when the pump beam has a good quality and the deviation increases with increasing pump beam quality β.

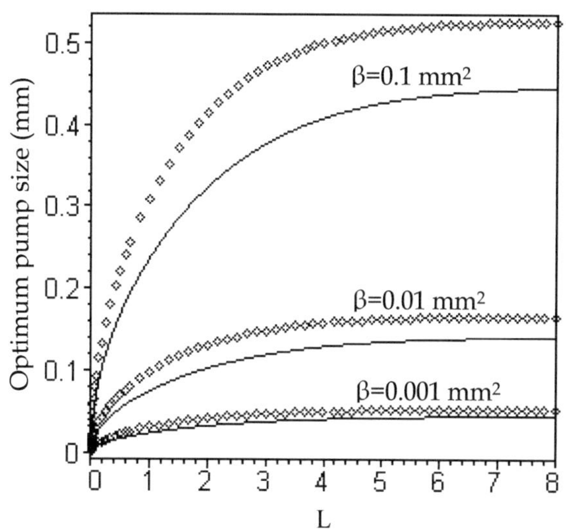

Fig. 4. Comparsion of optimum pump spot size, $w_{p,opt}$, as a function of L for values of β=0.1 and β=0.01 mm². Solid and pointed curves are calculated from Eqs. (36) and (43), respectively.

The saturation of the minimum pump spot size and hence the optimum mode-matching efficiency is due to the limit of interaction length which causes by the finite overlap distance of the beams in space. When crystal length becomes larger than the beams-overlap length in the crystal, an increasing in crystal length no larger contributing to generate the laser. To achieve the maximum mode-matching efficiency for a given crystal length and a pump beam M² factor, when absorption coefficient increases the optimum pump size should decrease. Hence, in the case of poor pump beam quality, the mode-overlapping could not be maintained through the length of the crystal and slow saturation prevented us from using a short crystal with a high absorption coefficient to improve the overlap.

To examine the accuracy of the present model, we compared Eq. (34) with the results determined by Laporta and Brussard (Laporta and Brussard, 1991). They have found that the average pump size

$$\bar{w}_p = \left[\frac{1}{l'} \int_0^{l'} w_p^2(z) dz \right]^{1/2} \tag{43}$$

with $l' = (-2.3\theta_p + 1.8)(1/\alpha)$ for $\theta_p \leq 0.2$ rad or $\alpha l' \leq 1.34$ can give a fairly accurate estimate of the overlap integral. l' is the effective length related to the absorption length $1/\alpha$ of the pump radiation and the divergence angle θ_p of the pump beam inside the crystal.

Fig. 5 shows a comparison of the minimum average pump size within the active medium using. Eqs. (34) and (41)-(43). It can be seen that the results calculated from (34) are in a good agreement with the results evaluated by Laporta and Brussard. Again, it can be also seen that, the optimum pump spot size in the active medium is an increasing function of β.

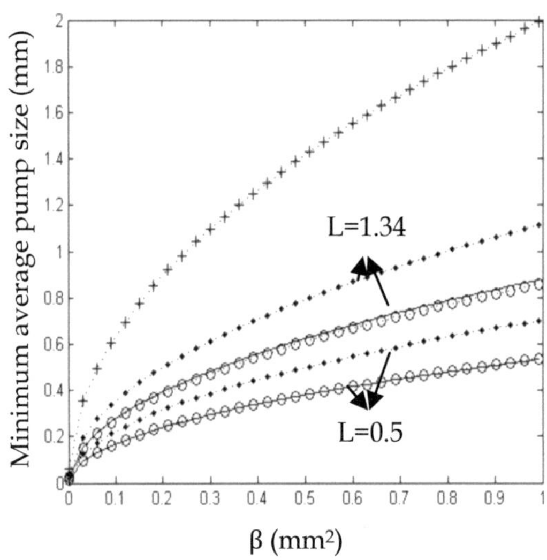

Fig. 5. Minimum pump spot size as a function of pump beam quality β for L=1.34 and L=0.5. Solid, solid diamond, open circle, and plus curves is calculated from Eqs. (34), (41), (42) and (43), respectively.

4. Optimum laser resonator

According to Eqs. (23)-(26) and (10), the output power at the condition of optimum pumping is given by

$$P_{out} = \frac{T\eta_p}{\delta} \frac{\omega_{l0}^2(\omega_{l0}^2 + 4\beta g(L))}{(\omega_{l0}^2 + 2\beta g(L))^2} \left[P_{in} - \frac{\pi \delta I_{sat}}{2\eta_p} (\omega_{l0}^2 + 2\beta g(L)) \right] \tag{44}$$

Now, we define, for generality, the normalized output efficiency as

$$\sigma_{out} = \left(\frac{P_{out}}{P_{in}} \right) \Big/ \left(\frac{T\eta_p}{\delta} \right) = \frac{w_{l0}^2(w_{l0}^2 + 4\beta g(L))}{(w_{l0}^2 + 2\beta g(L))^2} \left[1 - \frac{(w_{l0}^2 + 2\beta g(L))}{\chi} \right] \tag{45}$$

where the input power is normalized as

$$\chi = \frac{p_{in}}{(\pi\delta I_{sat} / 2\eta_p)} \tag{46}$$

It is often quoted in square millimeter. At a fixed β and P_{in}, the optimum mode size, $w_{l0,opt}$ for the maximum output power can be obtained by using the condition

$$\partial P_{out} / \partial w_{l0} = 0 \tag{47}$$

This equation yields the solution

$$w_{l0,opt} = \sqrt{2\beta g(L)\left[h_+(\chi,\beta,L) - h_-(\chi,\beta,L) - 1\right]} \tag{48}$$

where

$$h_{\pm}(\chi,\beta,L) = \left[\frac{\sqrt{3}\sqrt{27(\chi / 2\beta g(L))^2 + 1}}{9} \pm \frac{\chi}{2\beta g(L)}\right]^{1/3} \tag{49}$$

Fig. 6 shows a plot of optimum mode size, $w_{l0,opt}$ as a function of dimensionless crystal length, L, for several values of χ and β. One sees, at a fixed χ and β, $w_{l0,opt}$ is initially a rapidly increasing function of L, and then its dependence on L becomes weak. Also, at a fixed L, the poorer pump beam quality and larger χ leads to a larger mode size to reach a higher slope efficiency and a lower pump threshold. Increasing optimum mode size with increasing pump beam quality is attributed to the increasing optimum pump beam spot size with increasing its beam quality and maintaining the optimum mode-matching.

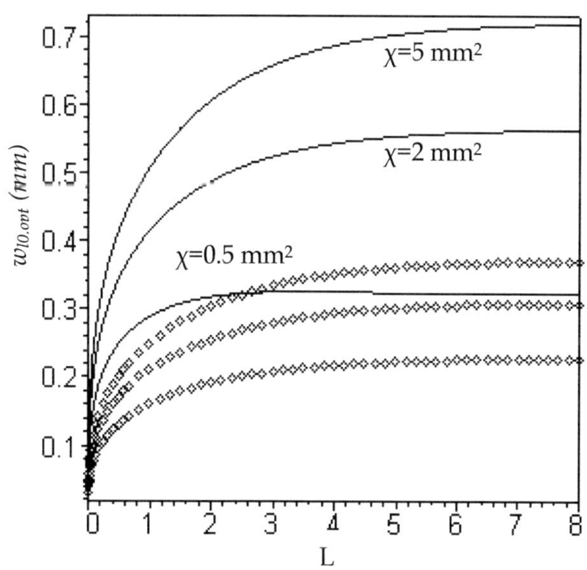

Fig. 6. Optimum mode size, $w_{l0,opt}$, as a function of dimensionless active medium length, L, for several values of χ and β. Pointed lines are for β=0.01 mm² and solid lines are for β=0.1 mm².

Equation (35) can be used as a guideline to design the laser resonator. First, the value of parameter β is calculated by considering values of C, n, and α. Then, for a given P_{in} and gain medium, the value of χ is determined from Eq. (46). Putting β, L and χ into (48), the optimum mode size can be determined and subsequently substituting calculated optimum mode size into (45), the maximum output efficiency $\sigma_{out,max}$ can be also determined.

Fig.7 shows the maximum output efficiency as a function of L for several values of χ and β. It is clear, the maximum output efficiency rapidly decreases with increasing L particularly when the available input power is not sufficiently large and beam quality is poor. It results because of the spatial-mismatch of the pump and laser beam with increasing L. Further, the influence of dimensionless gain medium length is reduced for high input power and better pump beam quality. For a poor pump beam quality, the maximum attainable output power strongly depends on the input power. This can be readily understood in the following way: the increasing pump power leads to increase the gain linearly while the better pump beam quality leads to the better pump and signal beams overlapping regardless of the value of the gain which continues to increase with increasing pump power. The large overlapping of the pump and signal beams in the crystal ensures a more efficient interaction and higher output efficiency. Note that the laser pump power limited by the damage threshold of the crystal, then χ can be an important consideration in the choice of a medium. It looks like, in the case of high pump power, the pump beam quality is a significant factor limiting to scale end-pumped solid state lasers.

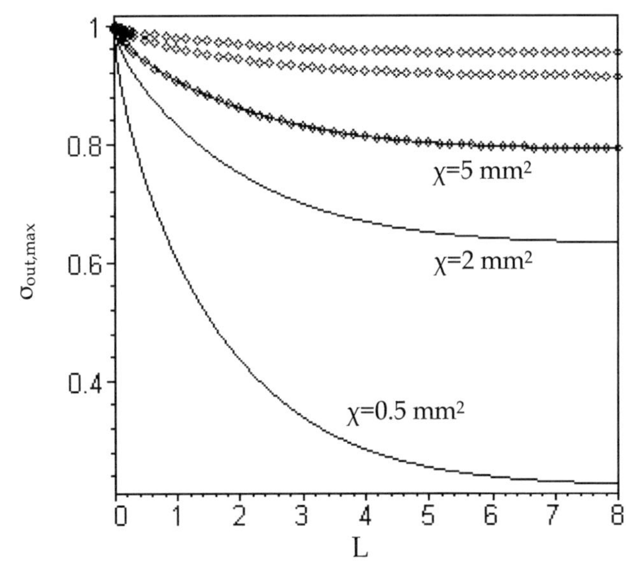

Fig. 7. Maximum output efficiency, $\sigma_{out,max}$, as a function of dimensionless active medium length, L, for several values of χ and β. Pointed and solid curves are for β=0.01 mm² and β=0.1mm², respectively.

In comparison, Fig. 8 shows the maximum output efficiency calculated from Eq. (45), and determined by Chen (Chen, 1999). One sees the Chen's model make a difference compared to the present model. The difference increases for low input power χ and poor pump beam quality with increasing L. The present model shows a higher output efficiency in each value of β, χ and L. Typically, the maximum output efficiency calculated using this model is ~5%, 16%, 12% and 15% higher than those obtained from the Chen's model for sets of (L=8, χ=0.5 mm², β=0.001 mm²), (L=8, χ=0.5 mm², β=0.1 mm²), (L=1, χ=0.5 mm², β=0.1 mm²) and (L=5, χ=0.5 mm², β=0.1 mm²), respectively.

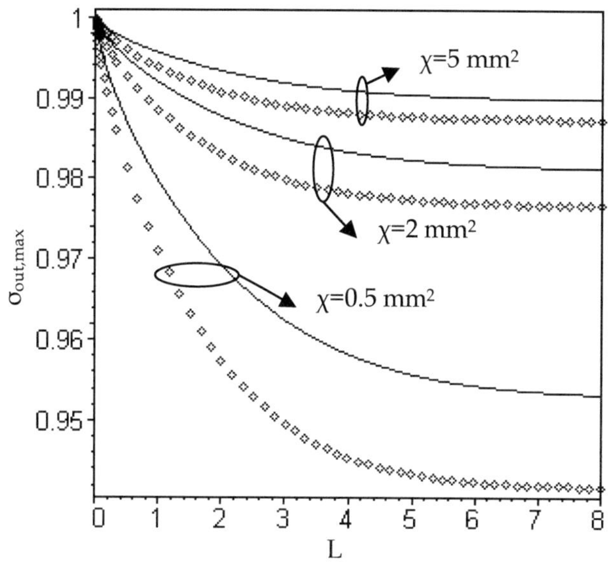

(a)

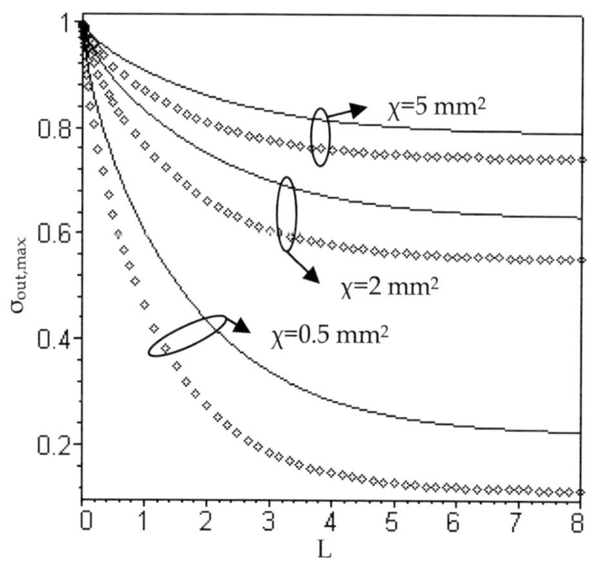

(b)

Fig. 8. Comparison of the maximum output efficiency, $\sigma_{out,max}$, as a function of L for several values of χ and (a) β=0.001 mm², (b) β=0.1 mm². Solid curves calculated from Eq. (45) and pointed curves are determined from Chen's model (Chen, 1999).

Note that we assumed the pump beam distribution comes out from the fiber-coupled laser-diodes is a Gaussian profile. Nevertheless, a more practical distribution for the output beam from the fiber-coupled laser-diodes may be closer to a Top-Hat or super-Gaussian distribution:

$$r_p(x,y,z) = \frac{\alpha}{\pi\varpi_p^2(z)[1-\exp(-\alpha l)]}\Theta(\varpi_p - \sqrt{x^2+y^2})\exp(-\alpha z) \tag{50}$$

where $\Theta(\varpi_p - \sqrt{x^2+y^2})$ is the Heaviside step function and ϖ_p is the average pump-beam radius inside the gain medium. Solving Eqs. (7) and (11) with considering (50) we can obtain mode-match-efficiency as follow:

$$F(\omega_{l0},\varpi_p) = \frac{\omega_{l0}^2}{2\varpi_p^2}\left[1-\exp\left(-2\frac{\varpi_p^2}{\omega_{l0}^2}\right)\right] \tag{51}$$

Fig. 9 shows the mode-match-efficiency calculated from (51) and (26) versus $w_{l0}/w_{p,opt}$. One sees the differences is small, especially for $w_{l0}/w_{p,opt}<1$. Therefore the Gaussian distribution can be considered a reasonable approximation for analysis the optical pump conditions.

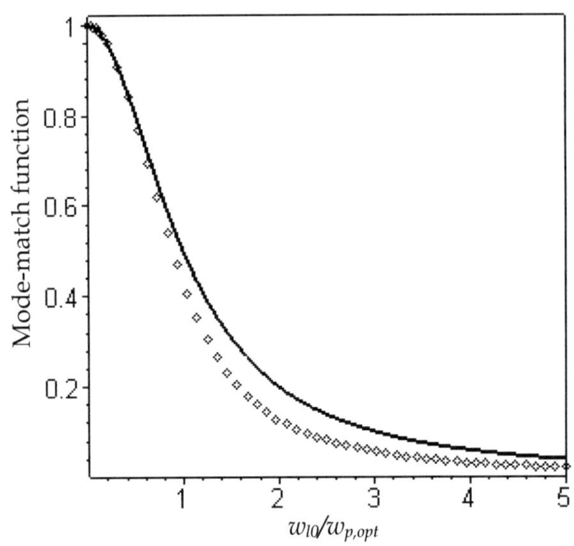

Fig. 9. Mode-match efficiency as a function of $w_{l0}/w_{p,opt}$. Solid and pointed curves represent the results for Gaussian distribution and Top-Hat profile, respectively.

5. Pump source requirements

In an end-pumped laser, the brightness of the pump source may be a critical factor for optimizing the laser performance. For instance, tight focusing of the pump beam is required to enhance the nonlinear effect for mode-locking of a femtosecond laser while the long collimation of a tight-focused pump beam is crucial for mode-matching of the laser beam along the gain medium. According to Eq. (16), the desired exponential unsaturated gain at optimal design can be determined by the optimum mode size and the optimum average pump beam spot size:

$$\Gamma = \frac{2P_{in}\eta_p}{\pi I_{sat}} \frac{1}{w_{l0,opt}^2 + w_{p,opt}^2} \tag{52}$$

Putting Eq. (34) into (52), the small-signal round-trip gain at the optimum condition is expressed as:

$$\Gamma = \frac{2P_{in}\eta_p}{\pi I_{sat}} \frac{1}{w_{l0,opt}^2 + 2\beta g(L)} \tag{53}$$

Note that, if the volume of pump beam stays well within the volume of the fundamental cavity mode, TEM_{00} operation with diffraction-limited beam quality is often possible. It is because of, at this condition, gain of the high-order modes is too small to balance the losses and start to oscillate. Therefore, for oscillating laser in TEM_{00}, we have

$$w_{l0,opt} \geq w_{p0,opt} \tag{54a}$$

or

$$w_{l0}^2 \geq 2\beta g(L) \tag{54b}$$

The constraint condition in (54) can be rewritten as:

$$n w_{p0}\theta_p \leq \frac{w_{l0}^2 n\alpha}{2g(L)} \tag{55}$$

The brightness of the beam in the air B can be defined as (Born & Wolf, 1999):

$$B = \frac{P_{in}}{(\pi n w_{p0}\theta_p)^2} \tag{56}$$

Substituting the constraint of (55) into (56), we find

$$B \geq 4P_{in} \frac{g^2(L)}{n^2\pi^2\alpha^2 w_{l0,opt}^2} \tag{57}$$

Now, we can obtain a relation between the required brightness of the pump beam in air, desired gain Γ, properties of the gain medium (I_{sat}, l, and n) at a given pump power P_i and pump beam quality β:

$$B \geq 4P_{in} \frac{g^2(L)}{n^2\pi^2\alpha^2} \left[\frac{2\eta_p P_{in}}{\Gamma\pi I_{sat}} - 2\beta g(L) \right]^{-2} \tag{58}$$

This is a requirement on B to achieve a gain value of Γ at a pump power P_i. If the inequality in (58) is not satisfied, the laser will not work at the design point. It can be simply shown, in the limit case of $l=2z_R=2g(L)/\alpha$ and $\beta\to0$, Eq. (58) reduces to that of developed by Fan and Sanchez (Fan & Sanchez, 1990):

$$B \geq \frac{1}{P_{in}}\left[\frac{\Gamma l I_{sat}}{2n\eta_p}\right]^2 \tag{59}$$

The limit of $\beta\to0$ is justified when the M^2 factor is small and the absorption coefficient is high in which the effective interaction length is much shorter than physical length of the medium.

Notice the power scaling with maintaining operation in the TEM_{00} mode has been limited by the formation of an aberrated thermal lens within the active medium. Besides the thermal lens, the maximum incident pump power is restricted by thermal fracture of the laser crystal. Therefore, it is of primary importance for the laser design to avoid thermally induced fracture and control the thermal effects.

6. Thermal effects in end-pumped lasers

The thermal lens generated within a gain medium may hinder the power scaling of such lasers by affecting the mode size of the laser inside the resonator and reducing the overlap between the pump and cavity modes. Efficient design consideration usually is dominated by heat removal and the reduction of thermal effects for high-power solid-state lasers. In end-pumped solid-state lasers the requirement for small focusing of pump beam size leads to a very high pump deposition density and further exacerbate thermal effects such as (a) thermal lensing and aberration, (b) birefringence and depolarization caused by thermal stress, (c) fracture and damaging the laser crystal by thermal expansion which limit the power-scaling of end-pumped lasers. The thermal lens in the gain medium will act as another focusing element which should be taken into account in order to optimize the matching between the cavity and the pump beams in the gain medium.

One of the main problems encountered in end-pumped lasers is beam distortion due to the highly aberrated thermal lens, making it extremely difficult to simultaneously achieve high efficiency and good beam quality. Many methods such as low quantum defect level, double-end pumping, composite crystal and low doping concentration have been proposed to reduce the thermal effects and increase output power (Koechner, 2006). The lower doping concentration and longer length crystal decrease the thermal lens effect greatly and also are preferable for the efficient conversion.

In the case of longitudinal pumping, most of the pump energy absorbed close to the end surface of the rod. This means the gradient-index lens is strongest near the pump face and the end effect localized at this first face of the rod. Since the generated lenses are located inside the crystal and the thickness of the gain medium is very small compared the cavity length, the separation between these lenses can be neglected and the combination can be well approximated by an effective single thin lens located at the end of the laser rod with the effective thermal focal length (distance from the end of the rod to the focal point) as (Shayeganrad, 2012)

$$\frac{1}{f_{th}} = \frac{\eta_h P_{in}\xi}{2\pi k_c}\frac{\alpha}{1-e^{-\alpha l}}\int_0^l \frac{e^{-\alpha z}dz}{w_p^2(z)} \tag{60}$$

where η_h is the heat conversion coefficient resulting from fluorescence efficiency, upconversion, and quantum defect, $\xi = \frac{\partial n}{\partial T} + (n-1)(1+\nu)\alpha_T + 2C_{r,t}n^3\alpha_T$ is the-averaged thermo-optic coefficient, $\partial n/\partial T$ is the thermal-optics coefficient, α_T is the thermal expansion coefficient, $C_{r,t}$ is the photo-elastic coefficient of the material, n is refractive index of the gain medium, and v is Poisson's ratio. Note that under efficient laser operation and low-loss cavity $\eta_h=1-\lambda_p/\lambda_l$. Also, the factor (n-1) has to be replaced by n in the case of end-pumped resonators with a high reflectivity coating on end surface of the crystal. The first term results from the thermal dispersion, the second term is caused by the axial mechanical strain, and the third term represents the strain-induced birefringence. Though for most cases contribution of thermal stress effect is small. Further, the Gaussian pump beam leads to a much more highly aberrated thermal lens, which is a factor of two stronger on axis for the same pump spot size and pump dissipation (Fan et al., 2006).

Equation (60) cannot be solved analytically. To obtain the focal length, this equation must be solved numerically. Again, similar solving mode-matching function, an average pump spot size inside the active medium is considered. Then, we have

$$\frac{1}{f_{th}} = \frac{\eta_h P_{in}\xi}{2\pi k_c}\frac{1}{\omega_{p,opt}^2} \tag{61}$$

This function shows that when pump beam spot size increases the thermal focal length increases, while a smaller pump radius needs to achieve a lower threshold and higher slope efficiency. Most importantly, when the pump beam waist is decreased, the temperature and temperature gradient in the laser rod would be very high due to the resulting heat.

Substituting Eq. (34) into Eq. (61), we can obtain thermal focal power D_{th} as:

$$D_{th} = \frac{\eta_h P_{in}\xi}{4\pi k_c \beta g(L)} \tag{62}$$

It is clear that the focal power of the thermal lens depends on the gain medium characteristics, pump beam properties and is independent on the crystal radius. It increases directly with pump power and inversely with pump beam M^2 factor. Increasing P_{in} leads to increase the deposited heat and hence increase the thempratute gradiant and D_{th}. Decreasing M^2, leads to increase focusability of the beam and hence increasing D_{th}. In addition, utilizing a material with small value of ξ, and high values of thermal conductivity k_c can help to reduce the thermal effect.

Eq. (62) has the following properties: for small values of L (L<<1):

$$D_{th} = \frac{\sqrt{3}\eta_h P_{in}\xi}{2\pi k_c \beta} \tag{63}$$

and for large values of L (L>>1):

$$D_{th} = \frac{\eta_h P_{in} \xi}{4\pi k_c \beta} \tag{64}$$

It can be seen in the asymptotic, values are independent of the active medium length.

7. References

Alfrey, A. J. (1989). Modeling of longitudinally pumped CW Ti: Sapphire laser oscillators. *IEEE J. Quant. Electron*, Vol. 25 PP. 760-765.

Berger, J., Welch, D. F., Sciferes, D.R., streifer, W. and Cross, P. (1987). High power, high efficient neodymium: yttrium aluminum garnet laser end-pumped by a laser diode array. *Appl. Phys. Lett.*, Vol. 51 PP. 1212-1214.

Bollig, C., Jacobs, C., Daniel Esser, M. J., Bernhardi, Edward H., and Bergmann, Hubertus M. von. (2010). Power and energy scaling of a diode-end-pumped Nd:YLF laser through gain optimization. *Opt. Express*, Vol. 18, PP. 13993-14003.

Born M., and Wolf E. (1999). Principles of Optics. 7th extended edition, Pergoman Press Ltd. Oxford, England.

Carkson, W. A., and Hanna,D.C. (1988). Effects of transverse mode profile on slop efficiency and relaxation oscillations in a longitudinal pumped laser. *J. Modm Opt.*, Vol. 36 PP. 483-486.

Chen, Y. F. (1999). Design Criteria for Concentration Optimization in Scaling Diode End-Pumped Lasers to High Powers: Influence of Thermal Fracture. *IEEE J. Quantum Electron.*, Vol. 35, PP. 234-239.

Chen, Y. F., Kao, C. F., and Wang, C. S. (1997). Analytical model for the design of fiber-coupled laser-diode end-pumped lasers. *IEEE J. Quantum Electron.*, Vol. 133 PP. 517-524.

Chen, Y. F., Liao, T. S., Kao, C. F., Huang, T. M., Lin, K. H., and Wang, S. C. (1996). Optimization of fiber-coupled laser-diode end-pumped lasers: influence of pump-beam quality. *IEEE J. Quantum Electron.*, Vol. 32, PP. 517-524.

Clarkson, W. A., and Hanna, D. C. (1989). Effects of transverse-mode profile on slope efficiency and relaxation oscillations in a longitudinally-pumped laser. *J. Mod. Opt.* Vol. 27 PP. 483-498.

Fan, S., Zhang, X., Wang, Q., Li, S., Ding, S., and Su, F. (2006). More precise determination of thermal lens focal length for end-pumped solid-state lasers, *Opt. Commun.* Vol. 266, PP. 620-626.

Fan, T. Y., and Byer, B. L. (1988). Diode-pumped solid-state lasers. *IEEE J. Quantum Electron.*, Vol. 24, PP. 895-942.

Fan, T. Y., and Sanchez, A. (1990). Pump source requirements for end pumped lasers. *IEEE J. Quantum Electron.*, Vol. 26 PP. 311-316.

Fan, T. Y., Sanchez, A., and Defeo, W. G. (1989). Scalable end-pumped, Diode-laser pumped lasers. *Opt. Lett.*, Vol. 14 PP. 1057-1060.

Ferguson, B., and Zhang, X. C. (2002). Materials for terahertz science and technology. *Nature Materials*, Vol. 1, PP. 26-33.

Findlay, D., and Clay, R. A. (1966). The measurement of internal losses in 4-level lasers. *Physics Letters*, Vol. 20, PP. 277-278

Gong, Lu, M., Yan, C., P., and Wang, Y. (2008) Investigations on Transverse-Mode Competition and Beam Quality Modeling in End-Pumped Lasers. *IEEE J. Quantum Electron.*, Vol. 44, PP. 1009-1019.

Guo, L., Lan, R., Liu, H., Yu, H., Zhang, H., Wang, J., Hu, D., Zhuang, S., Chen,L., Zhao,Y., Xu, X., and Wang, Z. (2010). 1319 nm and 1338 nm dual-wavelength operation of LD end-pumped Nd:YAG ceramic laser. *Opt. Express*, Vol. 18, PP. 9098–9106.

Hall, D. G. (1981). Optimum mode size criterion for low gain lasers. *Appl. Opt.*, Vol. 20 PP. 1579-1583.

Hall, D. G., Smith, R. J., and Rice, R. R. (1980). Pump size effects in Nd:YAG lasers. *Appl. Opt.*, Vol. 19 PP. 3041-3043.

Hanson, F. (1995). Improved laser performance at 946 and 473 nm from a composite $Nd:Y_3Al_5O_{12}$ rod. *Appl. Phys. Lett.* Vol. 66, PP. 3549-3551.

Hemmeti, H., and Lesh, Jr. (1994). 3.5 w Q-switch 532-nm Nd:YAG laser pumped with fiber-coupled diode lasers. *Opt. Lett.* Vol. 19 PP. 1322-1324.

Hering, P., Lay, J. P., Stry, S., (Eds). (2003). Laser in environmental and life sciences: Modern analytical method. Springer, Heidelberg-Berlin.

Koechner, W. (2006). Solid state laser engineering. Sixth Revised and Updated Edition, Springer-Verlog, Berlin.

Laporta, P., and Brussard, M. (1991). Design criteria for mode size optimization in diode pumped solid state lasers. *IEEE J. Quantum Electron.*, Vol. 27 PP. 2319-1326.

MacDonald, M. P., Graf, Th., Balmer, J. E., and Weber, H. P. (2000). Reducing thermal lensing in diode-pumped laser rods. *Opt. Commun.* Vol. 178, PP. 383-393.

Maiman, T. H. (1960). Stimulated Optical Radiation in Ruby. *Nature*, Vol. 187, PP. 493-494.

Mukhopadhyay, P. K., Ranganthan, K., George, J., Sharma, S. K. and Nathan, T. P. S. (2003). 1.6 w of TEM_{00} cw output at 1.06 μm from Nd:CNGG laser end-pumped by a fiber-coupled diode laser array. *Optics & Laser Technology*, Vol. 35 PP. 173-180.

Risk, W. P. (1988). Modeling of longitudinally pumped solid state lasers exhibiting reabsorption losses. *J. Opt. Amer. B*, Vol. 5PP. 1412-1423.

Saha, A., Ray, A., Mukhopadhyay, S., Sinha, N., Datta, P. K., and Dutta, P. K. (2006). Simultaneous multi-wavelength oscillation of Nd laser around 1.3 μm: A potential source for coherent terahertz generation. *Opt. Express* , Vol. 14, PP. 4721-4726.

Sennaroghlu, A. (Ed.). (2007). Solid atate lasers and applications. CRC Press, Taylor & Francis Group.

Shayeganrad, G. (2012). Efficient design considerations for end-pumped solid-state-lasers. Optics and laser Technology, *Optics & Laser Technology*, Vol. 44, PP. 987–994.

Shayeganrad, G., and Mashhadi, L. (2008). Efficient analytic model to optimum design laser resonator and optical coupling system of diode-end-pumped solid-state lasers: influence of gain medium length and pump beam M^2 factor. *Appl. Opt.*, Vol. 47, PP. 619-627.

Sipes, D. L. (1985). Highly efficient neodymium: yttrium aluminum garnet lasers end-pumped by a semiconductor laser array. *Appl. Phys. Lett.*, Vol. 47 PP. 74-76.

Šulc, J., Jelínková, H., Kubeček, V., Nejezchleb, K.. and Blažek, K.. (2002). Comparison of different composite Nd:YAG rods thermal properties under diode pumping. *Proc. SPIE* Vol. 4630, PP. 128-134.

Weber, R., Neuenschwander, B., Donald, M. M., Roos, M. B., and Weber, H. P. (1998). Cooling schemes for longitudinally diode laser-pumped Nd:YAG rods. *IEEE J. Quantum Electron.* Vol. 34, PP. 1046-1053.

Xiea, W., Tama, S. C., Lama, Y. L., Yanga, H., Gua, J., Zhao, G., and Tanb, W. (1999). Influence of pump beam size on laser diode end-pumped solid state lasers. *Optics & Laser Technology* Vol. 31 PP. 555-558.

Zhuo, Z., Li, T., Li, X. and Yang, H. (2007). Investigation of Nd:YVO_4/YVO4 composite crystal and its laser performance pumped by a fiber coupled diode laser, *Opt. Commun.* Vol. 274, PP. 176-181.

2

Diode Pumped Planar Waveguide/Thin Slab Solid-State Lasers

Jianqiu Xu
Key Laboratory for Laser Plasmas (Ministry of Education) and Department of Physics, Shanghai Jiaotong University, Shanghai China

1. Introduction

Diode-pumped solid-state lasers (DPSSLs) with high power and good output beam quality are widely used in material processing, communication, remote sensing, and medical treatment. A preferable diode-pumped solid-state laser has the characteristics, such as high efficiency, long lifetime, high reliability, compact size and flexibility. Planar waveguide lasers can be formed by transition metal diffusion method, optical bonding technique, or ion-etching technique. Planar waveguides used for high-power lasers are usually the simplest one-dimensional waveguides, of which the width (the dimension along the y-direction) is much large than the laser wavelength. The laser beam is guided only in the x-direction as shown in Fig. 1.1. The behavior of beam along the y-direction is similar as the beam propagating in the free space. We consider a planar isotropic optical waveguide, where the active core $x<\pm d/2$ is occupied by the homogeneous gain medium, and the claddings $x>\pm d/2$ consist of the semi-infinite substrate. The z-axis is taken in the direction of beam propagation. The refractive index of core and cladding are n_0 and n_1, respectively.

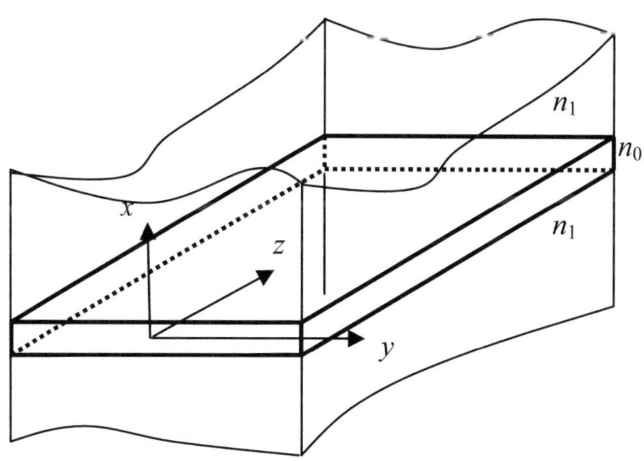

Fig. 1.1. Structure of planar optical waveguide.

Usually, the aspect ratio of the planar waveguide/thin slab lasers is very large. The dimension in the width direction is much larger than beam widths of fundamental modes of stable resonators, and the diffraction-limited beam from a stable resonator is difficult to be achieved. The unstable resonator, with large fundamental mode volume and much higher-order mode discriminations, has been employed successfully in the planar waveguide gas lasers. Unlike gas lasers, thermal distortion and pumping non-uniformity in solid-state lasers are serious limitations for achieving good beam quality in using unstable resonators. These distortion and non-uniformity should be considered in the design of unstable resonators.

1.1 Pump uniformity for planar waveguide lasers

The pump uniformity for planar waveguide lasers is discussed for the absorption of pump power in the active waveguide. Using special composite waveguide design and controlling the incident angle of the pump light would reduce the pump non-uniformity. A typical configuration of planar waveguide lasers pumped by laser diodes from double edges is shown in Fig. 1.2(a). To achieve large beam diameters along the width direction, off-axis unstable resonator is applied. The resonator is constructed by a high reflective mirror and a hard-edge output coupler. Both are concave mirrors. Although negative branch confocal unstable resonator is dispatched in the figure, positive branch confocal unstable resonators can also be used. In the thickness direction the beam characteristics are determined by the waveguide structure, and in the width direction the beam quality is controlled by the cavity design. Because the beam propagation in these two directions is independent [1], we can concentrate our investigations on the width direction.

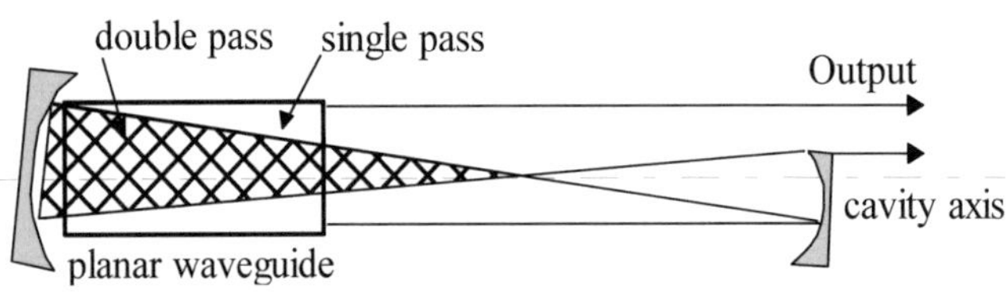

(a)

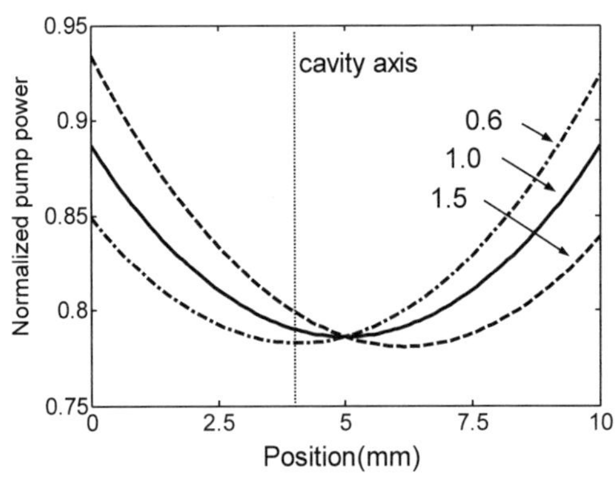

(b)

Fig. 1.2. (a) Edge pumped planar waveguide lasers with negative branch confocal unstable resonator. (b) Normalized pump power along the width direction of the planar waveguide for asymmetric factors $b=0.6$, 1.0, and 1.5. The absorption coefficient $\alpha=0.3$ cm^{-1}.

Because of absorption of pump power in the active waveguide, the pumping distribution along the width direction is non-uniform as shown in Fig. 1.2(b). If identical laser diodes are used on the two sides, a minimum of the pumping density is located at the center of the waveguide [2]. Notice that only a part of the crystal waveguide is double-pass covered by the laser beam in the resonator. The effective fill-factor is defined as [1]

$$F_{eff} = \frac{\iint_{S+} P_{abs}(x,z)\ dxdz + \iint_{S-} P_{abs}(x,z)\ dxdz}{2\iint_{S0} P_{abs}(x,z)\ dxdz} \tag{1.1}$$

where S_0 is area of the crystal waveguide, S_+ and S_- are the area covered by the forward (from left hand side to right hand side in the Fig. 1.2a) and the backward laser beam, respectively. Obviously $S_+ \neq S_-$. Clearly, a unit effective fill-factor means full covering of the laser beam on the waveguide. The distribution of pumping density is calculated by

$$P(x) = C_0 \left\{ \cosh[\frac{\alpha w}{2} - \ln\sqrt{b} - \alpha x] + \mathrm{R}e^{-\alpha w} \cosh[\frac{\alpha w}{2} + \ln\sqrt{b} - \alpha x] \right\} \tag{1.2}$$

with the normalized coefficient

$$C_0 = \frac{P_{in}}{tl} \frac{2(1-R)\sqrt{b}e^{-\frac{\alpha w}{2}}}{(1+b)(1-R^2 e^{-2\alpha w})} \tag{1.3}$$

where α is the absorption coefficient of the active medium; P is launched power from LD in the upper side (Watt/cm); t, l and w are the thickness, length and width of the planar waveguide, respectively; R is the reflectivity of side surface; b is the pumping asymmetric factor, defined by the pump power ratio of down to upper sides as shown in Fig. 1.2(a). The pumping minimum offsets roughly by $\ln\sqrt{b}/\alpha$ from the center with a symmetric factor of b. By choice of different laser diodes in the down and upper sides, we can adjust the asymmetric factor so that the pumping distribution overlaps mostly with the laser beam, and consequently the pumping efficiency is increased. Figure 1.3(a) shows the calculated effective fill-factors for various asymmetrical pumping. The parameters used in the calculations are described in Table 1.1. Fill-factors reach larger values with larger asymmetric pumping. Since high absorption coefficients give rise to large pumping non-uniformities, improvement of fill-factor by asymmetric pumping is more efficient for heavy doping concentration. Lower doping concentration and smaller absorption coefficients give higher fill-factors, but wide waveguide is required to achieve high pump absorption efficiency. An optimum doping concentration is a trade-off of these effects. Because the beam converges more hard in the short cavity, the effective fill-factor is enlarged with increasing cavity length (see Fig. 1.3b).

Radius of curvature rear mirror	Radius of curvature front mirror	Waveguide dimension	Cavity length	Cavity magnification
285.6	200	55(l)×10(w)	242.8	1.428

Table 1.1. Cavity and waveguide parameters (mm)

High efficient absorption and perfectly uniform pumping are very difficult to be achieved simultaneously in the edge-pumped slab lasers, and the pump intensities near edges are always higher than that at the center of slab. The difference between the maximum and minimum pump intensity can be written as a function of the pump absorption efficiency η_{abs} in the form [1]

$$\Delta I = \frac{I_{\max-\min}}{I_{\max}} = 1 - \frac{2\sqrt{1-\eta_{abs}}}{2-\eta_{abs}} \tag{1.4}$$

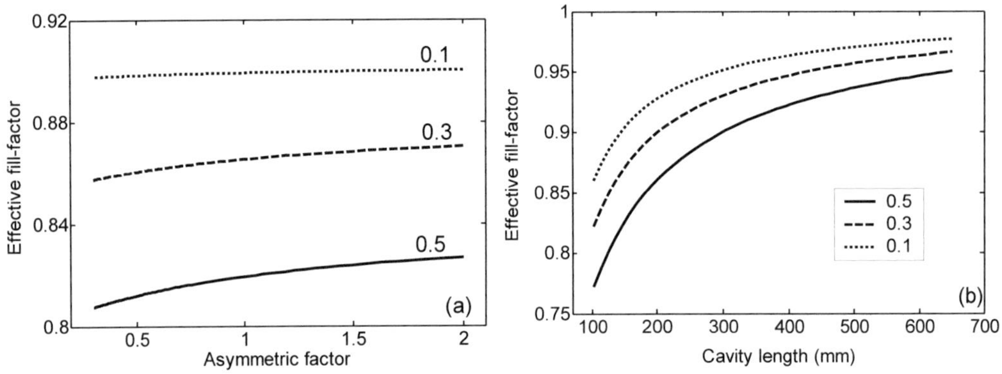

(a) (b)

Fig. 1.3. (a) Effective fill-factor with varying asymmetric factor. (b) Effective fill-factor vs. cavity length with symmetric pumping for absorption coefficient α=0.1, 0.3 and 0.5 cm^{-1}.

The more the pump power is absorbed, the higher the pump non-uniformity occurs. The figure of merit of pump performance is defined as [5],

$$F = \Delta I + \frac{2\sqrt{1-\eta_{abs}}}{2-\eta_{abs}} \tag{1.5}$$

where F=1 for the traditional edge-pumped slab lasers, and F=0 for the case with total absorption and perfect uniformity.

The pump non-uniformity comes mainly from the preferential absorption near the edge. We can control the incident angle of the pump light entering the waveguide to prevent a part of them from being absorbed near the edge, or use special composite waveguide design to improve the pump intensity near the center. The ideals using composite waveguide and oblique pumping are illustrated in Fig. 1.4. The composite YAG/Yb:YAG slab crystal was taken for example. The total width of the slab is 1 mm and the doping section is in the center of the slab with the Yb^{3+}-ion doping concentration of 4 at.%, and its thickness is 0.4 mm.

The method of ray tracing was used to simulate the pump behavior in the waveguide. In figure 1.4, the waveguide with plane cladding and concave cladding is irradiated by the LD at an inclination angle. In our simulation, the pump power of each LD was 100 W. The pump light with the beam divergence angle of b=7.5° enters the waveguide at the inclination angle of a=5°.

Figure 1.4(a) described the pump intensity *versus* width of waveguide with the plane cladding. It was seen that the ratio between the maximum and the minimum intensity of the plane slab is $\Delta I = 33.3\%$ and the absorption efficiency is η_{abs}=90%. The figure of merit is $F = 0.9$ which is better than the result when the absorption η_{abs}=90% is inserted into Eq. (1.4). To make further improvement, the upper cladding is made to a concave profile as shown in Fig. 1.4. (a). Figure 1.4(b) depicted the pump intensity through the width of the concave slab. The difference between the maximum and minimum intensity is $\Delta I = 10.8\%$, corresponding the figure of merit $F = 0.68$, which is in the same level as that in face pumped solid-state slab lasers.

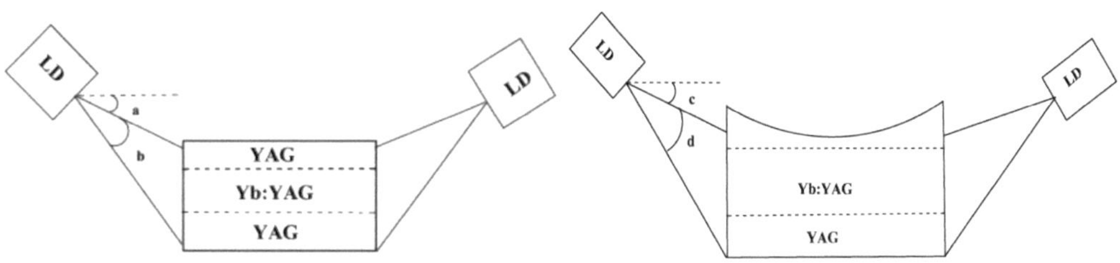

Fig. 1.4. (a). Schematic diagram of edge-pumped slab laser with plane cladding (left), concave cladding (right).

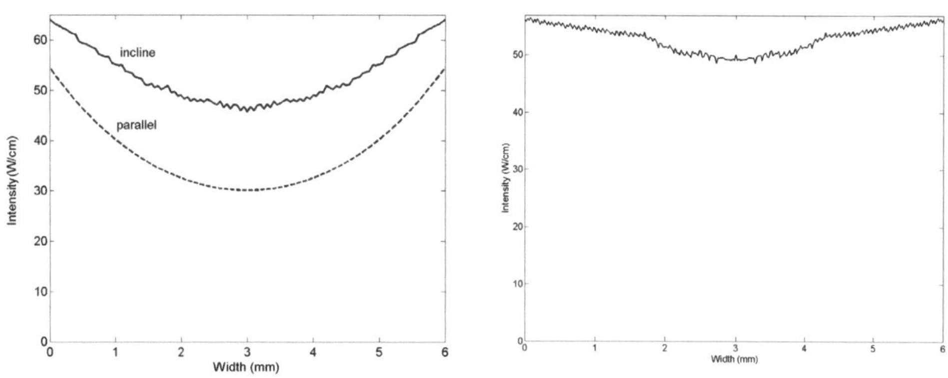

Fig. 1.4. (b). Pump power intensity versus width for plane cladding (left), concave cladding (right).

1.2 Unstable resonator for edge-pumped planar waveguide lasers

One of the most popular unstable resonators is confocal unstable resonators, which have been generally used in the high-power planar waveguide gas lasers [3]. Confocal unstable resonators can provide diffraction-limited output beam in very short cavities with large beam diameters. Detailed investigations on effects of non-uniform pumping with confocal unstable resonators are important in the design of edge pumped solid-state planar waveguide lasers. The characteristics of output laser beam of edge-pumped planar waveguide solid-state lasers with confocal unstable resonators have been investigated numerically considering pumping non-uniformity, gain saturation and beam interaction. A novel quasi-self-imaging planar waveguide is designed for single-mode output.

While the ray matrix technique gives good descriptions to the beam characteristics in stable resonators, diffraction effects in unstable resonators play an important role and make the analytic solutions unavailable. There are a number of methods, which are mainly extensions of the Fox-Li method for simulation of unstable resonators [4~7]. However, no method can treat all types of unstable resonator due to efficiency, accuracy and divergence of iteration. For an confocal unstable resonator with a large Fresnel number, which is commonly used in planar waveguide solid-state lasers, we explore fast Fourier transform (FFT) techniques [8, 9] with equivalent coordinate transform and saturable gain sheets to investigate the output beam quality of edge pumped planar waveguide solid-state lasers. Optimization of pumping uniformity for good output beam quality and high pumping efficiency is analyzed. The influences of doping concentration, cavity length and effective Fresnel number are also discussed. We find that good beam quality and high efficiency can be obtained with asymmetric pumping and negative branch confocal unstable resonators. The results are useful in design of diode pumped solid-state lasers with unstable resonators.

When the laser beam travels back and forth in the resonator, the laser gain is saturated by the double passed laser beam. As we mentioned above, the forward and backward beams go through the crystal waveguide in different passes. Thus, we apply a linearly expanding coordinate transform [4] in the calculation. For the forward laser beam, the primed coordinate is adopted. The gain medium is divided equally into several slices (gain sheets) with constant small signal gain coefficients. Between two adjacent slices, the wave propagates freely and is described with aid of FFT technique [10]. For the backward laser beam, the coordinate is expanded according to the resonator magnificent factor, but the gain on the axis is considered as constant. Therefore, the gain is saturated from the sum of the forward and backward laser beam, varying the distribution in different slices. The gain saturation is described by [11]

$$u_2(x,z) = u_1(x,z)\sqrt{\exp\frac{g_0\Delta z}{1+[I^+(x,z)+I^-(x,z)]/I_s}} \tag{1.6}$$

where $u_2(x, z)$ and $u_1(x, z)$ are the complex amplitudes before and after gain slices; I_s is the saturation intensity, Δz is the propagation length in the gain medium, $I^+(x, z)$ and $I^-(x, z)$ are laser intensity for the forward and the backward beam, respectively. The small signal gain coefficient $g_0=\eta_0 P_{abs}(x)$, where η_0 is a coefficient determined by the quantum efficiency, the stimulated emission cross-section, and the upper level energy transform coefficient. Notice that the x-coordinate for the backward beam should be transformed to the expanding coordinate system.

Without gain saturation, the output beam is well collimated plane waves with nearly constant phase fronts (see Fig. 1.5a). Using edge-pumping, the part of beam near the waveguide edge obtains larger gain than the center part does. The laser beam is pulled to the edge of the waveguide. The output beam diameter is gain-narrowed as shown in Fig.

1.5(b). The beam profile is modulated with an increased phase angle. In a negative branch unstable resonator, the laser beam is reversed when it passes through the intra-cavity focus, i.e., the upper part of the beam swaps with the lower part. In such a way, gain non-uniformity can be averaged by asymmetric pumping when the beam bounces back. As can be seen in Fig. 1.5(c), the main ripple in the output beam profile becomes smoother with asymmetric pumping than that with symmetric pumping. Moreover, the phase aberration is reduced to about 75%. In order to evaluate these effects in only one parameter, we calculated the beam propagation factor M^2 of the output beam.

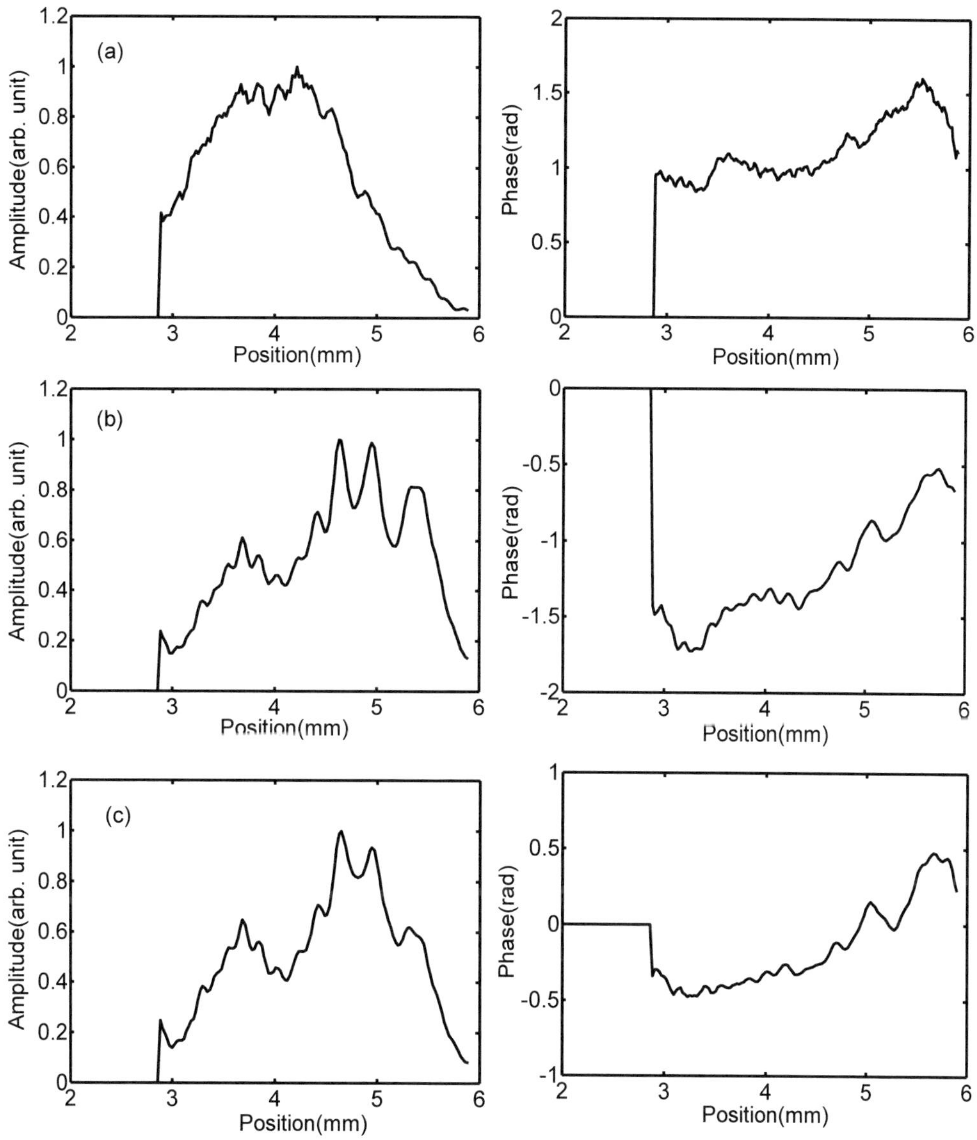

Fig. 1.5. Output beam profiles for (a) bare cavity, (b) symmetric pumping, (c) asymmetric pumping with asymmetric factor b=1.5.

The beam propagation factor M^2 is calculated by propagating the output beam through a long focal length lens. The beam width is defined by the second moment intensity [4].

$$d = 4\sigma_x \tag{1.7}$$

$$\sigma_x^2 = \frac{\int I(x,z)(x-\overline{x})^2 dx}{\int I(x,z)dx} \tag{1.8}$$

where $I(x,z)$ is the laser intensity at position z; $\overline{x}$ is the center of the laser beam given as

$$\overline{x} = \frac{\int I(x,z)xdx}{\int I(x,z)dx} \tag{1.9}$$

The beam waist is calculated by a hyperbolic fit of beam widths at different positions. The beam propagation factors M^2 of the output beam with varying cavity length are shown in Fig. 1.6. With increasing the cavity length, the beam propagation factor displays resonance behavior, which arises from the diffraction nature of hard-edge aperture [10]. The best beam quality can be obtained with an optimum cavity length. This optimum cavity length is the result of combination of pumping non-uniformity, beam control of unstable resonator, and gain saturation.

As we described above, the pumping non-uniformity is enlarged with large absorption coefficients, resulting in bad beam qualities. Using asymmetric pumping could compensate the pumping non-uniformity somewhat by the off-axis resonator, and improve the beam quality then. With an asymmetric factor b=1.5, the beam propagation factor $M^2 \approx 1.25$ is obtained even for the absorption coefficient α=0.5 cm^{-1} (see Fig. 1.6b). Notice that the optimum cavity length is almost independent of the asymmetric factor. In comparison, the beam propagation factor shows $M^2 \approx 1.43$ with an inversed asymmetric factor b=0.6, confirming asymmetric pumping really improving the output beam quality.

The characteristics of output laser beam of edge-pumped planar waveguide solid-state lasers with confocal unstable resonators have been investigated numerically considering pumping non-uniformity, gain saturation, and beam interaction. Optimized methods for cavity length and doping concentration have been given to offer good beam quality and high pumping efficiency. It has been found that good beam quality and high pumping efficient can be achieved by asymmetric pumping and negative branch confocal unstable resonators. These results provide reference for designs of high-power planar waveguide solid-state lasers.

1.3 Single-mode oscillation from quasi-self-imaging multi-mode waveguide

The input beam can reproduces itself at the exit of a self-imaging waveguide [12,13]. Hence, beam quality can be preserved in the laser amplifier made by a self-imaging planar waveguide [14]. Self-imaging planar waveguides associating with self-imaging resonator have also been used in laser oscillators [15]. Because the input and exit beams have the same

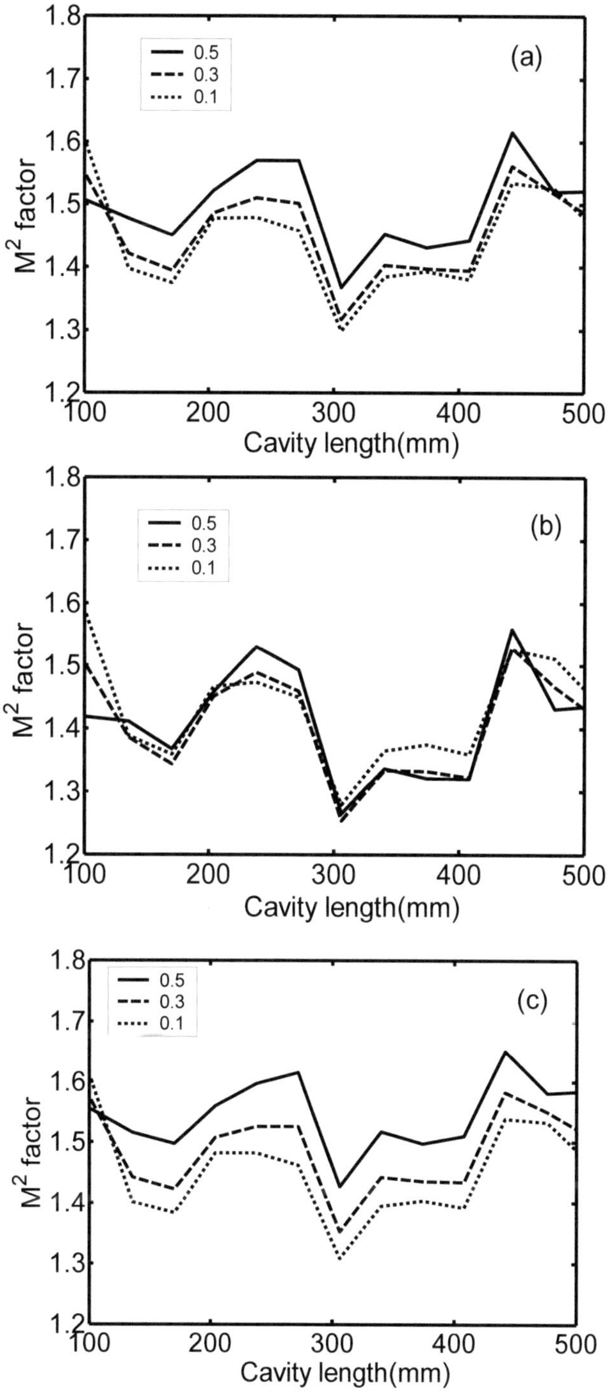

Fig. 1.6. Beam propagation factor M^2 values with symmetric pumping (a) and with asymmetric pumping of $b = 1.5$ (b) and $b = 0.6$ (c).

field profile, single-mode propagation of the laser beam is not necessary in the waveguide. High beam quality is obtained by controlling the beam width at the entrance of the waveguide through optimal resonator design [16].

In this context, we present a novel planar waveguide design, in which self-imaging is incomplete and each guided mode suffers different coupling losses. Single-mode output can be achieved with this quasi-self-imaging planar waveguide. The influence of thermal lens effects on the single guided mode operation is discussed.

The quasi-self-imaging planar waveguide as shown in Fig. 1.7 consists of two parts with different core thicknesses D and d, respectively. Both cores are made by the same rare-earth doped medium surrounded with non-doped cladding. The resonator can be of the plane-plane type fabricated by direct coating on the waveguide facets, or of the hybrid/unstable type with concave mirrors attached to the waveguide ends [2]. The pump light can be coupled from the edges or surfaces into the waveguide. Both the thin and thick waveguide parts support more than one guided mode in the core. The thick waveguide part is constructed with the self-imaging properties while the thin part is a conventional waveguide.

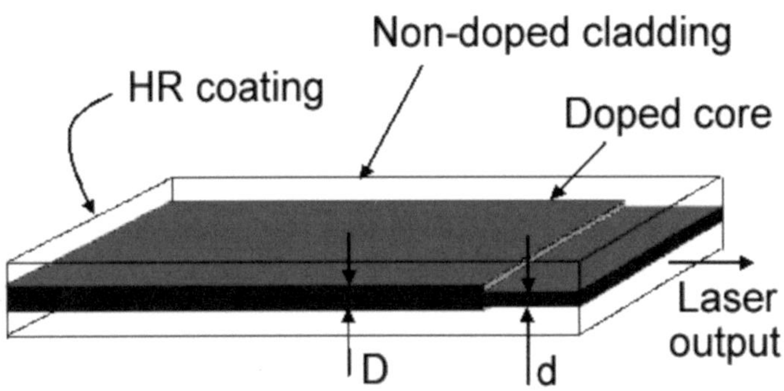

Fig. 1.7. Structure of quasi-self-imaging planar waveguide laser.

For an input field profile Ψ_{in}, the profile Ψ_E at the exit of a multi-mode waveguide of length L is given by the superposition of eigen-modes E_m, [1]

$$\Psi_E = \exp(-jn_{core}k_0L)\sum_{m=0} E_m \exp(j\Delta\beta_m L)\int E_m \Psi_{in} dx \tag{1.10}$$

where n_{core} is the refractive index of the core and k_0 is the wave vector in vacuum. The eigen-mode E_m and the propagation constant β_m can be obtained from the scalar wave equation together with the boundary condition from Maxwell's equations [17]. In the case that the waveguide supports a sufficiently large number of eigen-modes, the field at the exit of the waveguide will reproduce the input field precisely according to Eq. (1.10). However, if a waveguide can support only a few guided modes, the field at the exit can not reproduce the input field precisely. In addition, the propagation constant $\Delta\beta_m$ will be out of proportion to m^2 in a weakly guided waveguide. Thus, the self-imaging in such a waveguide is no longer perfect.

The difference between the exit and input field can be regarded as a kind of coupling loss. Because the quasi-self-imaging waveguide can guide a few modes, the coupling loss is determined by how completely the input field is represented by the guided modes. Thus the coupling loss depends on the input field profile. In an appropriately designed waveguide, the fundamental mode can be the one with lowest loss, so lasers based on such a type of waveguide can provide single fundamental mode output. We note that some fractions of Λ have special self-imaging properties. For instance, as shown in Table 1.2, even modes have the same $\pi/4$ phase shifts, while odd modes show a π phase difference at a propagation distance of $5\Lambda/8$. Actually, the modes have similar phase shift properties at distances equal to an odd multiple of $\pi/8$. In such a way, a symmetric field in a $(2l+1)\pi/8$ long waveguide will still be self-imaged, but an asymmetric field will not.

Mode	TE_0	TE_1	TE_2	TE_3	TE_4	TE_5
Phase shift	$\pi/4$	π	$\pi/4$	2π	$\pi/4$	π

Table 1.2. Phase shifts for several lowest-order modes at a propagation distance of $5\Lambda/8$ [14].

With the symmetry of eigen-mode in mind, the thick waveguide part of the quasi-self-imaging waveguide structure, as described in Fig. 1.7, is designed to be $5\Lambda/16$ long, and a round trip in the thick waveguide part will be $5\Lambda/8$. The eigen-modes of the thin waveguide part are then the input field for the thick waveguide part, i.e., $E_i^{thin} = \Psi_{in}$, where E_i^{thin} is denoted as the ith order mode of the thin waveguide part. When the field travels a round trip in the thick waveguide part, the coupling loss for the ith order mode at the interface between the thick and thin parts is given by [17]

$$\alpha_i = 1 - \int \left| \Psi_E E_i^{thin} \right| dx \tag{1.11}$$

As described above, odd i modes with asymmetric field profiles have large coupling losses if the thick waveguide part is $5\Lambda/16$ long. To achieve low coupling loss for the fundamental mode we choose the core diameter d close to D, i.e., $d \sim D$, and consequently $E_0^{thin} \sim E_0^{thick}$. When E_0^{thin} enters the thick waveguide part, E_0^{thick} carries most of the power of E_0^{thin}. Because the waveguide length is designed for $2\pi / \Delta\beta_0^{thick}$, the coupling loss of the fundamental mode E_0^{thin} is small, and the laser can be operated with a low threshold. Further, the waveguide with $d \sim D$ has a large active mode volume.

As an example, we calculate the coupling loss for a waveguide composed of a 1.0 at.% Nd-doped YAG (n=1.82) core surrounded with non-doped YAG (n=1.8192) cladding. The core in the thin waveguide part is chosen to be 70 μm thick and 10 mm long, while that in the thick part is 90 μm thick and 38.8 mm long. The total thickness is 1 mm. In this waveguide, the thin and thick parts can guide 7 and 10 TE modes, respectively. In Fig. 1.8, field profiles at propagation distances $5\Lambda/16$ and $5\Lambda/8$ are compared with the initial field profiles for the three lowest eigen-modes E_0^{thin} of the thin waveguide part. Clearly, self-imaging of the fundamental mode at a distance of $5\Lambda/8$ is better than that of higher-order modes. The coupling losses are calculated according to Eq. (1.11) for a set of eigen-modes E_i^{thin} and shown in Fig. 1.8(a). The coupling loss for the fundamental mode is about 0.026. In comparison, the losses for higher-order modes are all larger than 0.3.

Due to smaller beam width, the coupling loss of the fundamental mode increases more slowly than that of higher order modes. Thus, mode discrimination can be improved by using a little longer waveguide. For example, the coupling loss for a waveguide with a 42-mm-long thick part is shown in Fig. 1.8(a) (filled dots). The coupling loss of the fundamental mode is about 0.04 while that of the second order mode is 0.7.

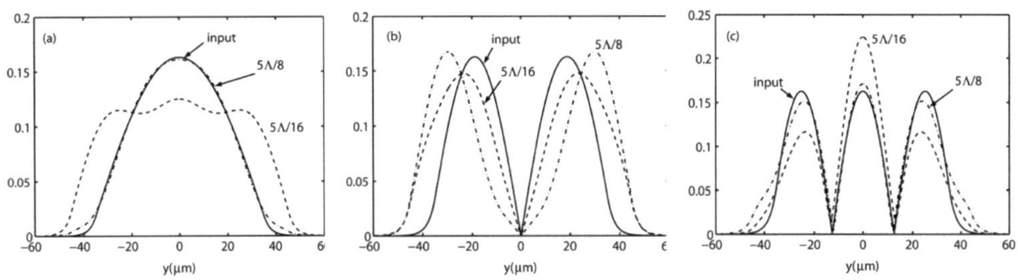

Fig. 1.8. Field profiles of eigen-modes compared with the field profiles at a distance $5\Lambda/16$ and $5\Lambda/8$ for (a) mode i=0, (b) i=1, (c) i=2.

In the multi-mode waveguide lasers, all modes compete through the gain saturation. The photon density of ith order mode is given by the multi-mode rate equations [17]

$$\frac{dI_i^{thin}}{dz} = \left[\frac{g_0}{1 + I / I_s} - \alpha_i^{thin} \right] + I_{spon} \tag{1.12a}$$

$$I = \sum_i \left[I_{i+}^{thin} + I_{i-}^{thin} \right] \tag{1.12b}$$

where I_+ and I_- are the photon intensity of laser beam propagating in forward and backward direction, respectively; α_i^{thin} is the coupling losses of ith order mode, g_0 is the small signal gain, I_s is the saturation intensity, and I_{spon} is the contribution of spontaneous emission. Substitute the coupling losses into Eq. (1.12), we obtain evolution of photon density of each mode with varying the pump power. In the calculation, a uniform gain is assumed in both thin and thick waveguide parts. The gain saturation is included by insert 10 gain sheets in the middle of the propagation. As can be seen in Fig. 1.9, more than 95% of laser power has been caught by the fundamental mode after a few round trips. Even for the Q-switched planar waveguide lasers, it is sufficient to achieve single mode operation.

2. High-power Nd:YAG planar waveguide lasers

Rapid progress has been made in fiber lasers, thin-disc lasers and slab lasers in recent years. A variety of medium have been used to fabricate planar waveguide. Crystal or ceramic

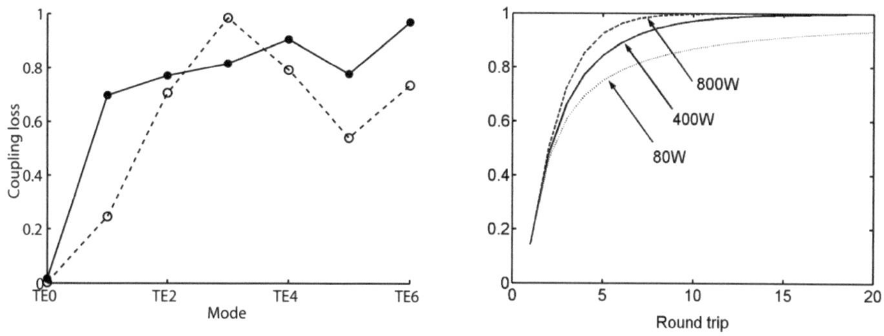

(a) (b)

Fig. 1.9. (a) Coupling losses of eigen-modes for 38.8 mm (=5Λ/16) long (empty dots) and 42 mm long (filled dots) thick waveguide part. (b) Laser power percent carried by the fundamental mode as a function of time for various pump power.

Nd:YAG is one of the most popular active material used for high-power planar waveguide lasers. Compared with fiber and thin-disc lasers, planar waveguide lasers are advantageous such as compact, two-dimensional power scalable and easily integrated. The potential of planar waveguide lasers has been exploited broadly. Double clad concept, sandwich structure and quasi self-imaging structure have been applied to high power planar waveguide lasers [13, 18, 19]. The output beam quality of high power solid-state laser usually suffers from the thermal disturbance in the laser host. The temperature gradient along the width of slab gives rise to optical distortions. How to remove the waste heat and reduce the thermal effects efficiently usually are main considerations in the design of high average power systems. In related articles [20], consequences of thermal and thermo-mechanical effects have been discussed for slab lasers and amplifiers. The analyses of nonlinear thermal effects in slab lasers are also presented in ref. [21]. In this context, we describe some experimental investigations on the high average power Nd:YAG lasers.

2.1 High-power Nd:YAG planar waveguide lasers with YAG and Al_2O_3 coatings

Thermal distortion and pumping non-uniformity in the high-average-power solid-state lasers are serious limitations for achieving good beam quality in using unstable resonators. Using thin slab geometry with large aspect ratio, heat removal can be very efficient. The thermal lenses, thermal birefringence, and their aberration are compensated by zig-zag path of the laser beam. As a type of thin slab lasers, the planar waveguide laser guides the laser beam by refractive index step. The beam distortion induced by thermal effects is resisted by guidance condition. Single-mode waveguides and suitable waveguide resonators can hold near-diffraction limited beam quality in the transverse waveguide axis, whilst unstable resonators are required to match the high Fresnel number in the lateral (width) axis. Pump and cooling non-uniformities exacerbate the optical distortions in the solid-state lasers. Compensation for these non-uniformities is important to obtain high efficiency and good beam quality output.

The planar waveguide laser uses a very thin rare-earth doped core as the active layer, where the heat dissipated in the waveguide can be removed efficiently from its large area surfaces.

The planar waveguide lasers are generally fabricated in symmetrical structures [10, 22~24], where an active core is in the middle of the waveguide and the claddings are symmetrically bonded to the core from two sides. Non-symmetrical structures have also been applied for the planar waveguide lasers [25, 26]. In this section, a planar waveguide laser with simple non-symmetrical structure is proposed to reduce the thermal effects. With non-symmetrical design, heat removal can be improved by directly contacting the active core to the heat sink. In order to improve pump uniformity and absorption efficiency, the pump light is injected at a small oblique angle to the edge of planar waveguide, and then the light follows a zigzag path along the slab. The thermal effects of different cladding materials are discussed. With appropriate cladding materials, the thermal and thermal-mechanic influences can be alleviated.

The structure of non-symmetrical planar waveguide is shown in Fig. 2.1. The ion-doped active core is directly contacted to the heat sink. To preserve the total internal reflection at the bottom surface, a very thin coating is applied between the active core and the heat sink. The pump light is injected to the planar waveguide at a small oblique angle. Since the pump light projects on different position through the un-doped cladding, the exponential absorption of pump power along the width direction is flatted, resulting in uniform pumping and high absorption efficiency. The oblique angle can be estimated from $a = \arctan(T/W)$, where T and W are the thickness and width of waveguide, respectively. In the experiments, the oblique incident angle is calculated around 5 degree. When the pump light is focused into the waveguide, the convergence angle also influences the path of ray inside the waveguide. In Fig. 2.2, the pump intensity along the width direction is plotted for various convergence angle of the focused pump light. When the convergence angle is less than 20 degree, the pump distribution is quite uniform. Compared with the exponential absorption where the maximum pump intensity is at the edge of waveguide [10], the maximum pump intensity moves to the middle of waveguide by the oblique edge-pumping. As it is well known, the edge of slab bears the largest thermal stress. Thus, the thermal stress at the edge is released to some extent with the oblique edge-pumping.

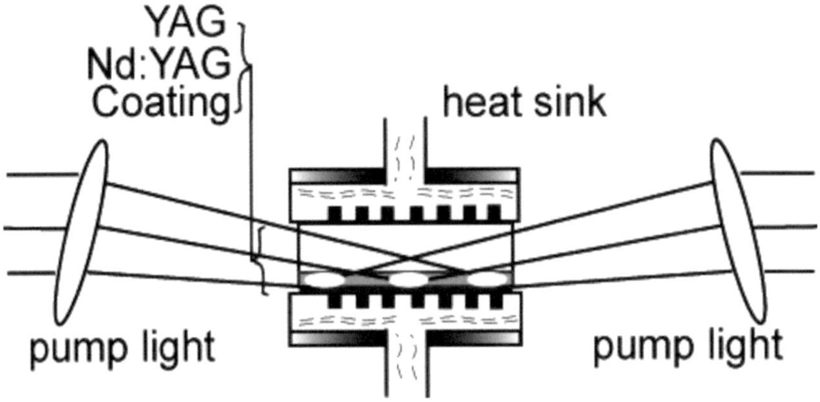

Fig. 2.1. Oblique edge pumping of planar waveguide.

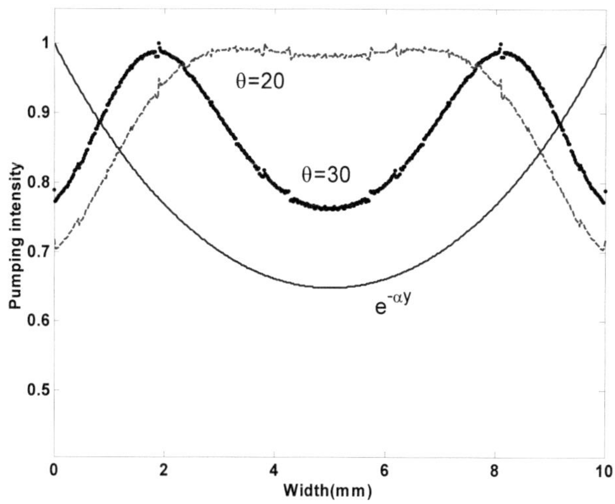

Fig. 2.2. Pump intensity along the width direction when the focused pump light is launched into the waveguide with the oblique angle of 5 degree. The convergence angles of the focused pump light are 20 and 30 degree, respectively. The solid line is for exponential absorption of the pump light.

With oblique edge-pumped technique, the thermal effect along the thickness is considered. We assume the y-axis direction along the thickness and the origin is taken as the center of the core. The thickness of waveguide core is $2h$, and the thickness of upper and bottom claddings are d_1 and d_2, respectively. The corresponding thermal conductivity and the surface heat transfer coefficient are k_1, λ_1 and k_2, λ_2, respectively. The planar waveguide is clipped by the copper micro-channel heat sink from the upper and bottom surfaces. The temperatures on the surfaces of upper and bottom claddings are supposed equal to that of the coolant. Heat flow in the waveguide is then a one-dimensional function. The thermal equation is [17]

$$\nabla^2 T = \frac{\partial^2 T}{\partial y^2} = -\frac{Q}{k} \tag{2.1}$$

where k is the thermal conductivity of the core, and Q is the heat generated per unit volume.

The boundary conditions between the core and the cladding are [27]

$$\left.\frac{dT}{dy}\right|_{y=\pm h} = \pm\frac{\lambda_{(1,2)}}{k}(T_l - T_s) \tag{2.2}$$

where T_s is the temperature at the core boundary, and T_l is the temperature at the cladding. We define the effective thickness $L=(\lambda_1 k_2/\lambda_2 k_1)d_2+d_1+2h$ and the difference in the cladding thickness $S=d_1-(\lambda_1 k_2/\lambda_2 k_1)d_2$. The analytical solutions is given by

$$T_b(y) = \frac{\lambda_1 Qh}{\lambda_2 k_1}(1+S/L)(y+h+d_2)+T_c \tag{2.3a}$$

$$T_c(y) = -\frac{Q}{2k_2}y^2 + \frac{QhS}{k_2L}(y + h + \frac{\lambda_1 k_2}{\lambda_2 k_1}d_2) + \frac{Qh}{k_2}(2h + \frac{\lambda_1 k_2}{\lambda_2 k_1}d_2) + T_c \quad (2.3b)$$

$$T_u(y) = Qh(\frac{S}{k_2L} - \frac{1}{k_2})(y - h - d_1) + T_c \quad (2.3c)$$

where $T_b(y)$, $T_c(y)$ and $T_u(y)$ are the temperature distributions in the bottom cladding, the active core and the upper cladding, respectively. The temperatures along the thickness direction in the non-symmetrical and symmetrical waveguide lasers are shown in Fig. 2.3. The temperature distribution in the active core is quadratic and linear in the claddings. The highest temperature is given as

$$(T_{\max} - T_c)\frac{k_2}{2Qh} = h(\frac{d_1 + h}{L})^2 + \frac{d_1 + h}{L}\frac{\lambda_1}{\lambda_2}\frac{k_2}{k_1}d_2 \quad (2.4)$$

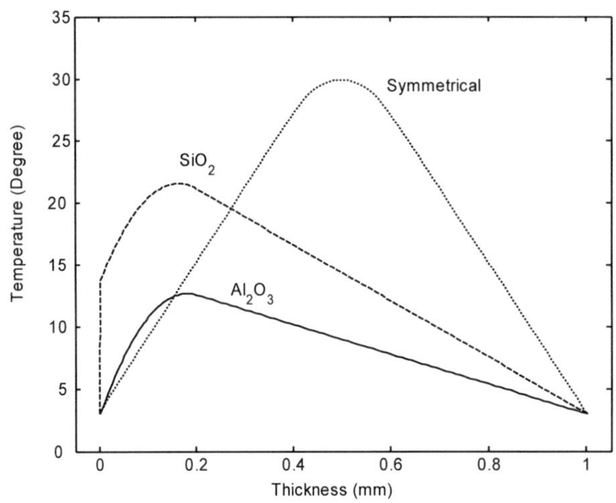

Fig. 2.3. Temperature distributions in non-symmetrical waveguides with Al_2O_3 and SiO_2 coatings, respectively, in comparison with that of symmetrical waveguide when pump power is 1 kW.

in which T_c is the temperature of coolant. The maximum temperature is located at $y = hS/L$. The active core made by 1.0 at.% Nd-doped YAG is 0.2 mm in thickness, while the upper cladding is a 0.8 mm un-doped YAG layer. The bottom cladding is a coating of about 3-μm thick. The condition about SiO_2 bottom coating has been discussed in the work [26]. The material SiO_2 has great differences in the thermal conductivity and the surface heat transfer coefficient with YAG. The material SiO_2 is not a good material for the planar waveguide laser with non-symmetrical structure. Changing bottom cladding material as YAG or Al_2O_3, it is found that the maximum temperature decreases rapidly and the location of the maximum temperature keeps almost the same. Considering maturity in the vacuum coating technology, Al_2O_3 is a prefer choice. The temperature with Al_2O_3 coating is only a half of that with SiO_2 coating (see Fig. 2.4).

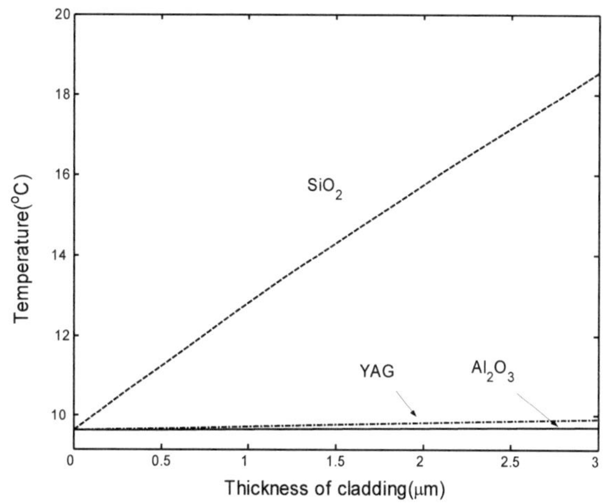

Fig. 2.4. The highest temperature in the waveguide verses coating thickness for Al_2O_3, YAG and SiO_2 coatings when pump power is 1 kW.

Due to the rectilinear pumping and the cooling of the waveguide, the thermal gradient and thermally induced stress are present only in the *y*-axis direction. The thermal stress caused by the thermal gradient ΔT is given by [17]

$$\sigma = \frac{\beta E}{(1-\nu)} \Delta T \tag{2.5}$$

where β is thermal expansion coefficient; E is Young's modulus; and ν is Poisson's ratio. The values of the material parameters for Al_2O_3, YAG, or SiO_2 are listed in Table 2.1. The thermal stress distributions for non-symmetrical waveguide with Al_2O_3, YAG, or SiO_2 coatings are shown in Fig. 2.5. The stress in the center of waveguide is positive (compressive stress) while it is negative (tensile stress) near the surfaces. In the Nd:YAG planar waveguide with SiO_2 coating, there is much stronger stress of 630 kg/cm^2 at the edge due to higher temperature in the waveguide. While in the waveguide with Al_2O_3 or YAG coating the value is 300 kg/cm^2, only a half of that in the waveguide with SiO_2 coating. This is only one-sixth of fracture strength for YAG crystal of 1800~2100 kg/cm^2. Furthermore, although the non-symmetrical structure results in the non-symmetrical temperature distribution, the waveguide with Al_2O_3 and YAG coating have nearly the same tension at the edges with the value about 300 kg/cm^2, whereas the waveguide with SiO_2 coating shows a large non-balanced stress. The thermal deformation with the SiO_2 coating is more severe than that with the Al_2O_3 and YAG coatings.

	thermal expansion coefficient α (10-6/K)	Young's modulus E (106 kg/cm2)	Poisson's ratio ν
Al_2O_3	5.6	4.26	0.309
YAG	6.1	3.17	0.25
SiO_2	0.55	0.73	0.17

Table 2.1. thermal properties of Al_2O_3, YAG, and SiO_2 coatings [17]

As shown in Table 2.1, the thermal expansion coefficients of Al_2O_3 and YAG material are one order of magnitude larger than that of SiO_2. Thus, the tension inside the SiO_2 coating is much smaller than that inside Al_2O_3 and YAG coatings (see Fig. 2.5b). However, difference in the thermal expansion coefficient between SiO_2 and YAG results in large shear stress at the core boundary. Obviously, the Al_2O_3 and YAG coatings are better matched to the Nd:YAG core.

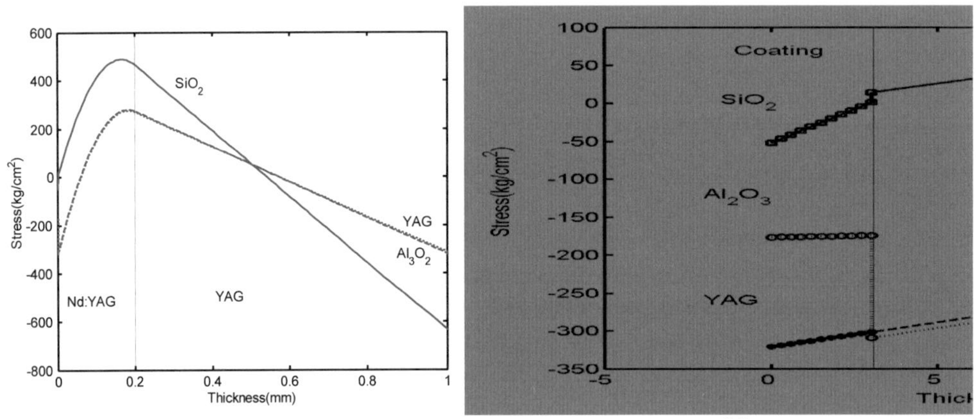

(a) (b)

Fig. 2.5. (a) Stress distributions inside non-symmetrical waveguides with Al_2O_3, YAG and SiO_2 coatings. The curves for YAG and Al_2O_3 coatings are almost overlapped. (b) Stress near the bottom surface for waveguides with the Al_2O_3, YAG and SiO_2 coatings when pump power is 1 kW. Vertical lines refer to the interface of Nd: YAG and YAG.

The non-symmetrical Nd:YAG planar waveguide laser was installed as shown in Fig. 2.6. The planar waveguide is 58×10×1 mm in dimension with 3-μm thick Al_2O_3 coating on the bottom. Two laser diodes (LD) with emission area about 10×48 mm can deliver 640 W average power with 200-μs duration and 1 kHz repetition rate after fast-axis collimation. Two cylindrical lens (f_1=50 mm and f_2=20 mm) were used to launch the pump light into the waveguide. The divergence angle of the pump light is less than 20 degree and the incident angle to the x-axis is around 5 degree. About 90% pump power can be coupled into the waveguide. A 100-μm thick indium foil is sealed between the heat sink and waveguide to ensure a uniform thermal contact with the heat sink. Water flux in the micro-channel heat sink is about 4 l/min and the temperature of coolant is set to 18 °C.

A plane-parallel cavity was used in the first place. Two flat mirrors were placed 1-mm apart from waveguide facets. The experimental output power versus the pump power for various output couplers is shown in Fig. 2.7(a). A maximum average power of 310 W was obtained with the output coupling of 28%. A hard-edge positive unstable resonator was constructed to improve the beam quality [13, 19]. The resonator magnification was designed to be 1.4. A concave mirror with 360-nm radius-of-curve of 360 mm and a cut-away convex mirror with 240-nm radius-of-curve were used as the rear mirror and the output coupler, respectively. The output power decreases to 280 W with a slope efficiency of 38%, but the beam quality factor was improved to M^2 =1.5 in the width direction. Roll-over due to the thermal effects was not observed in the experiments. As presented in Fig. 2.7(b), the output power decreases linearly with the stepping down the repetition rate in the same LD current. This means the loss caused by the thermal distortion is not significant at high repetition rate operation.

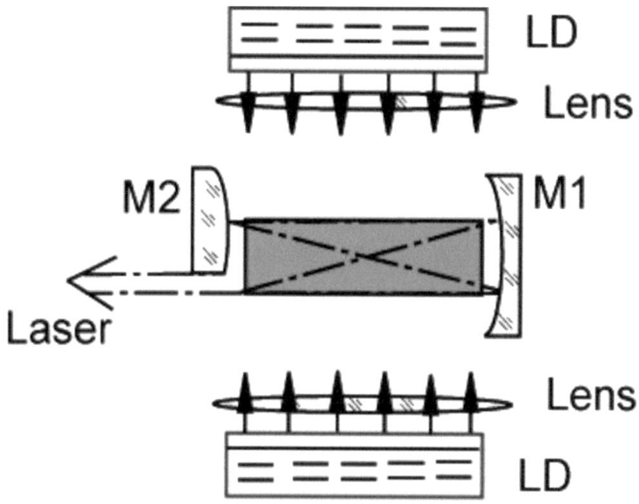

Fig. 2.6. Installation of non-symmetrical Nd:YAG planar waveguide laser.

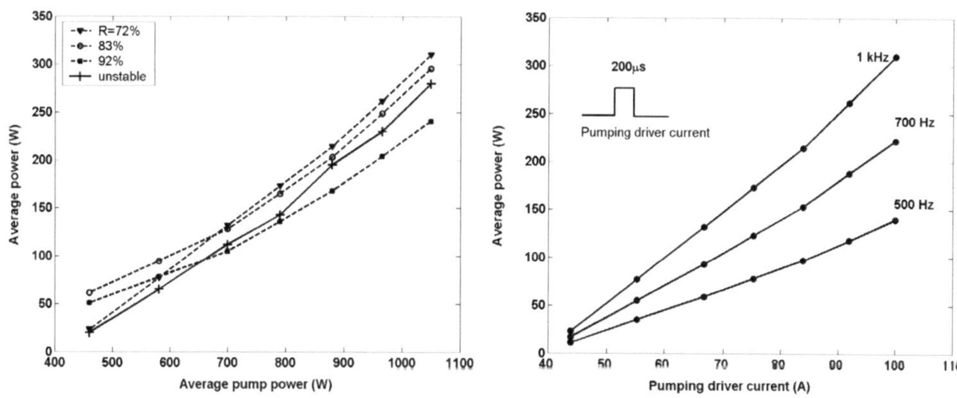

(a) (b)

Fig. 2.7. (a) Experimental output power versus the pump power for plane-parallel and unstable cavities. (b) Change of output power with the repetition rate of pumping.

Using non-symmetrical claddings in planar waveguide lasers can reduce the thermal distortion in high-power regime. Uniform and efficient pumping can be obtained by oblique injection to the edge of planar waveguide. Calculations based on the thermal equations show that, even for the very thin coating, the thermal conductivity and thermal transfer coefficient are important to the heat removal. Applied Al_2O_3 coating to the non-symmetrical planar waveguide, the maximum temperature reduces to only a half of that in the planar waveguide with SiO_2 coating. In the last, we have experimentally demonstrated a high-power Nd:YAG planar waveguide laser with YAG and Al_2O_3 claddings. An average power of 280 W with a slope efficiency of 38% has been obtained in a positive unstable resonator. The output beam quality factor is about M^2=1.5.

2.2 Passively Q-switched planar waveguide lasers

In some industrial applications, *e.g.*, drilling, cutting, and material processing, pulsed lasers with high peak power and high repetition rate are desired. The technique to achieve huge pulsed laser output is known as Q-switching. Depending on whether the quality Q is altered by external drivers, the laser can be referred as active or passive Q-switching. In comparison with active Q-switching, passively Q-switched lasers offer advantages of low-cost, reliability, and simplicity of operation and maintenance [29, 30]. Passive Q-switching is extensively applied in compact DPSSL. In this section, we describe a high-power passively Q-switched Nd:YAG thin slab laser. A Cr^{4+}:YAG microchip is adopted as saturable absorber mirror in this thin slab laser. The design of thin slab lasers with diffraction limited beam quality is discussed. Average output power of 70 W with a slope efficiency of 36% is obtained.

The experimental setup was depicted in Fig. 2.8, where the Nd:YAG crystal slab is 1×10×60 mm in size and Nd^{3+} ion doping concentration is 1.0 at.%. The slab was clamped by micro-channel heat sinks from two large surfaces (i.e., 10×60 mm surfaces). A 100-μm thick indium foil was filled between the crystal slab and the heat sink in order to reduce the thermal and mounting stress. The temperature of cooling water was set to 20°C. The cooling water flowed parallel to the laser propagation direction in the micro-channel heat sink, reducing the temperature gradient through the cross section of slab. To minimize the mounting stress during assembly, mount of the slab on the heat sink was monitored by an interferometer. Aberrant fringes should be eliminated by adjusting the locking screws. The dimension of a finished laser head was about 60×74×150 mm.

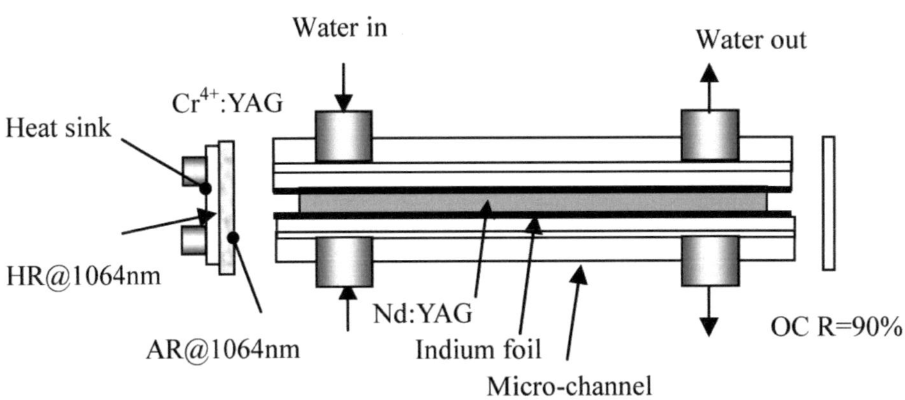

Fig. 2.8. Experimental setup of passively Q-switched thin Nd:YAG slab lasers.

The laser is passively Q-switched by a Cr^{4+}:YAG micro-chip with size of 0.5×5×15 mm. One surface of the Cr^{4+}:YAG chip is HR (high reflectivity) coated while another one is AR (anti reflectivity) coated at 1.06 μm. Thus, the Cr^{4+}:YAG micro-chip acts as a saturable absorber mirror in the slab laser. The initial transmission of the Cr^{4+}:YAG saturable absorber mirror is 90%. Since the slab laser is designed working in high power regime, the Cr^{4+}:YAG chip is compulsively cooled by a micro-channel heat sink attached to its back surface.

In order to acquire single mode oscillation in compact resonator, we control the pump beam diameter to match the fundamental laser mode inside the Nd:YAG slab. In such a way, higher order modes will be filtered out due to lower gain outside the pump range. For the plane-parallel resonator adopted in our experiment, the fundamental mode beam radius calculated from *ABCD* matrix method is shown in Fig. 2.9 as a function of the thermal refractive power. The laser works in the range shaded in Fig. 2.9. The laser beam radius inside the slab is about 0.2 mm, so we designed the pump beam radius to be 0.25 mm and chose the Nd:YAG slab of 1 mm in thickness.

Two horizontal laser diode (LD) arrays, of which each consists of three LD bars, closely pump the slab from both edges. Micro cylindrical lenses of 600 μm in diameter are employed to collimate the pump beam from the LD arrays. The emitted pump beam was about 0.5 mm ($1/e^2$) in diameter with the divergence angle of 2°. The pump beam from two edges was adjusted carefully to be in a same plane.

An important feature of thin slab lasers is that the thermal fracture limit is greatly improved. The distributions of temperature and stress in the slab were analyzed numerically to ensure the laser operating safely. The heat removal of the micro-channel heat sink was compared with that of the non-micro-channel heat sink. The micro-channel heat sink was manufactured with channel's diameter of 0.5 mm, length of 55 mm and pitch of 1.5 mm, respectively. The non-micro-channel heat sink was made by drilling a single hole of 5-mm in diameter in a copper block. In Fig. 2.10, we compared the focal length of the thermal lens when different heat sinks were used. Since heat removal with the micro-channel heat sink was more efficiently, the thermal focal length was longer. Furthermore, when using the non-micro-channel heat sink in the slab laser, the laser output power rolled back after the pump power reached 110 W, revealing serious thermal distortion in the crystal slab.

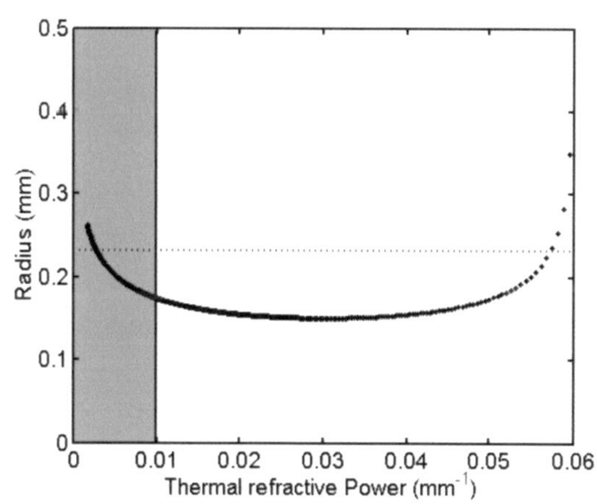

Fig. 2.9. The beam radius of fundamental mode in the center of slab varies with the thermal refractive power. The dotted line indicates the radius of the pump beam. The laser works in the shaded region.

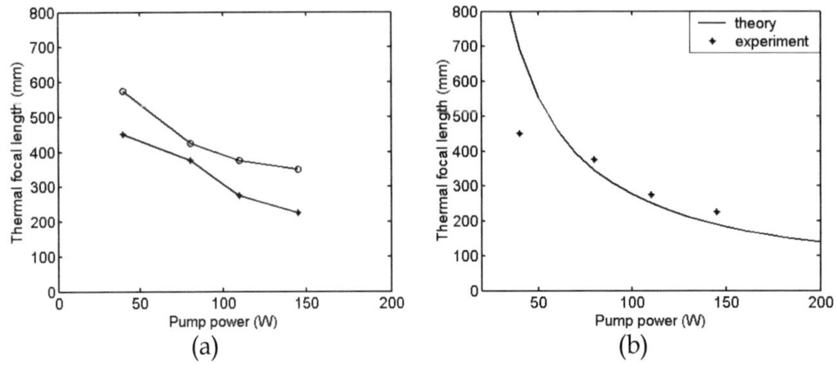

Fig. 2.10. (a) Thermal focal length with different heat sinks. *: non-micro-channel heat sink; o: micro-channel heat sink; (b) Measured and calculated thermal focal length.

After the temperature distribution in the slab being obtained, we can calculate the thermal focal length by assuming a grade-index lens of the pumped slab. The calculated and measured thermal focal lengths in the thickness direction using the micro-channel heat sink are shown in Fig. 2.10. The thermal focal length in the thickness direction was about 140 mm with 200-W pump power. Substituting the thermal focal length into *ABCD* matrix and as shown in Fig. 2.9, we found the requirement for single-mode oscillation was satisfied.

To achieve higher repetition rate and higher output power, it is needed high initial transition of the saturable absorber. The rate equations describing the passively Q-switched laser operation are [27]

$$\frac{d\phi}{dt} = \frac{\phi}{t_r}\left[2\sigma Nl - 2\sigma_s l_s - 2\sigma_e (N_{s0} - N_g) - \ln\frac{1}{R} - \alpha_L\right] \tag{2.6a}$$

$$\frac{dN}{dt} = W_p - \gamma\sigma c N\phi - \frac{N}{\tau} \tag{2.6b}$$

$$\frac{dN_g}{dt} = -\frac{A}{A_g}\sigma_g c N\phi + \frac{N_{s0} - N_g}{\tau_s} \tag{2.6c}$$

The parameters used for above equations are listed in Table 2.2. The condition to produce giant pulse is [31]

$$\frac{d^2\varphi}{dn^2} > 0 \tag{2.7}$$

Combining Eq. (2.6) with (2.7), we get the maximum initial transition of the saturable absorber as

$$T_{0\max} = \exp[-\frac{L + \ln\frac{1}{R}}{2(\frac{A}{A_s}\frac{\sigma_g}{\gamma\sigma}(1 - \frac{\sigma_e}{\sigma_g}) - 1)}] \tag{2.8}$$

The maximum transition was calculated as $T_{0\max}$=96%. Considering the cost of fabrication, we chose 90% of initial transition of the saturable absorber. The doping concentration of Cr^{4+}

ions was then calculated from N_{s0}=-ln$T/\sigma_g l_s$. Fig. 2.11(a) shows the evolution of the intracavity photon density and the population inversion density in the gain medium. The giant pulse occurs when the saturable absorber is saturated to transparent and the photon density reaches the maximum. From the calculation, we founded the repetition rate higher than 10 kHz. The measured pulse sequences at 220-W pump power are shown in Fig. 2.11(b). The pulse duration is about 10 ns, resulting in the peak power of 0.7 MW. The peak power fluctuation from pulse to pulse is about 5%. The agreement between the calculation and experiment is very well.

Symbol	Parameter	Value
ϕ	intracavity photon density	
N	population inversion density of gain medium	
N_g	ground-state population density of saturable absorber	
N_{s0}	Total population density of saturable absorber	7.0×10^{17} cm^{-3}
γ	inversion reduction factor	1 for Nd:YAG
A	Beam cross-section area in gain medium	
A_s	Beam cross-section area in saturable absorber	
σ	stimulated emission cross section	2.8×10^{-19} cm^2
σ_g	ground state cross section of saturable absorber	4.3×10^{-18} cm^2
σ_e	excite state cross section of saturable absorber	8.2×10^{-19} cm^2
L	length of laser crystal	60 mm
l_s	thickness of saturable absorber	0.5 mm
τ	lifetime of upper laser level of gain medium	230 μs
τ_s	lifetime of excite state of saturable absorber	3.4 μs
t_r	cavity round-trip time	$2nL/c$
R	reflectivity of output coupler	
α_L	intracavity round-trip dissipative optical loss	0.04
W_p	volume pump rate into upper laser level	4.4×10^{21}s^{-1}cm^{-3}

Table 2.2. Parameters used in Q-switching equations.

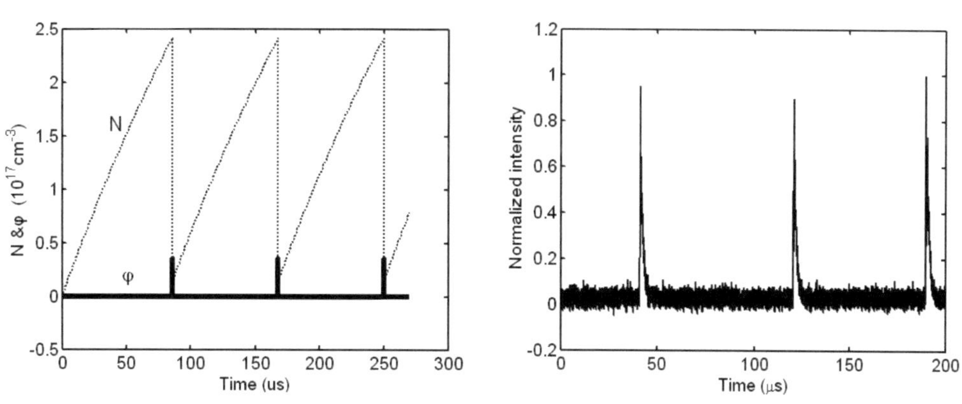

Fig. 2.11. (a) Temporal evolution of intracavity photon density and population inversion, (b) Measured passively Q-switched laser pulse sequence.

Various output couplers were used in the experiments. Output power versus pump power with output coupling of 5%, 9%, 17% are compared in Fig. 2.12. The threshold of pump power is around 10 W. With the optimized output coupling of 9%, the maximum average power of 70 W has been achieved with a slope efficiency of 36% when the pump power is 220 W. Output laser power grows linearly with increasing the pump power. The thin slab laser would be scalable to more than 300 W with 1-kW pump power be used. The beam propagation factor M_y^2 in the thickness direction is 1.4, indicating near single-mode oscillation in the thickness direction. In the width direction, the beam profile is multi-mode due to the weakness of the plane-parallel resonator. If unstable resonator or graded reflectivity mirrors are adopted, good output beam quality in the width direction can be acquired.

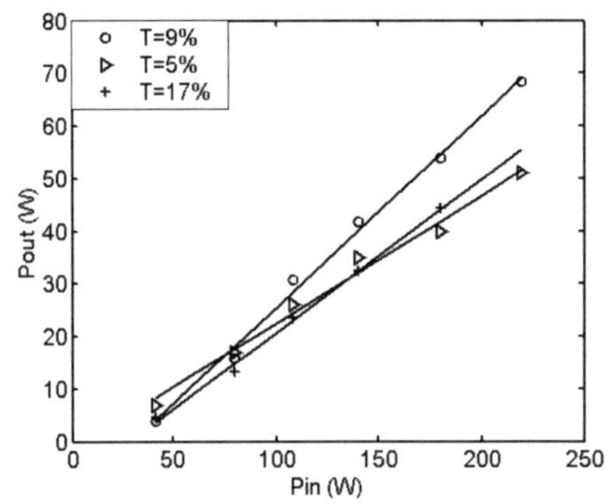

Fig. 2.12. Output laser power versus pumping power.

Thin Nd:YAG slab laser passively Q-switched by Cr^{4+}:YAG microchip is a cost effective approach to high-peak-power, high-repetition-rate solid-state lasers. Output average power of near hundred watts with a slope efficiency of 36% has been acquired with the repetition rate higher than 10 kHz and the pulse width around 10 ns. Near diffraction limited beam quality in the thickness direction has been obtained by precise control of the pump beam width inside the slab. An obvious advantage of this kind of lasers is their relatively compact design with the footprint less than 100 mm^3 and simple replacement of every part. It will find wide applications in industrial and medical fields.

2.3 Acousto-optical Q-switched planar waveguide lasers

In a planar waveguide laser, the laser beam is represented by a set of guided modes in the waveguide, while it propagates in the free space outside. Transform between guided modes and free-space propagation at the waveguide entrance leads to coupling losses and energy exchange among the guided modes. These effects cause competition of the guided modes in the Q-switched planar waveguide lasers. The mode competition enlarges the pulse-to-pulse instability, and distorts the transverse beam patterns.

The crystal waveguide used in the experiments is 11 mm wide and 60 mm long with a 0.2-mm thick Nd:YAG core sandwiched by 0.4-mm thick undoped YAG claddings. The crystal waveguide can support 26 TM modes. Pump light from 10 LD bars with 450-W CW maximum power is delivered into the pump chamber through slotted windows, providing uniform pumping to the whole waveguide. The folded hybrid resonator is 215 mm long, comprised by a concave mirror (focal length *f* =207 mm), a cutaway hard edge output mirror (*f* =138 mm), and a Brewster angle polarizer (see Fig. 2.13). The magnification of the resonator is 1.5, corresponding to the output loss of 0.33. The resonator equals effectively a case-I/II waveguide resonator in the transverse (guided) direction, and a negative branch confocal unstable resonator in the lateral (unguided) direction. This configuration avoids large inserting losses of the polarizer, typically 5% per pass [3]. The acousto-optical (AO) modulator is 46 mm long and the ultrasonic wave travels in the transverse direction. Mode control in the transverse direction has been achieved by an intracavity aperture just before the output mirror as well as space-waveguide coupling effects from the rear mirror [2].

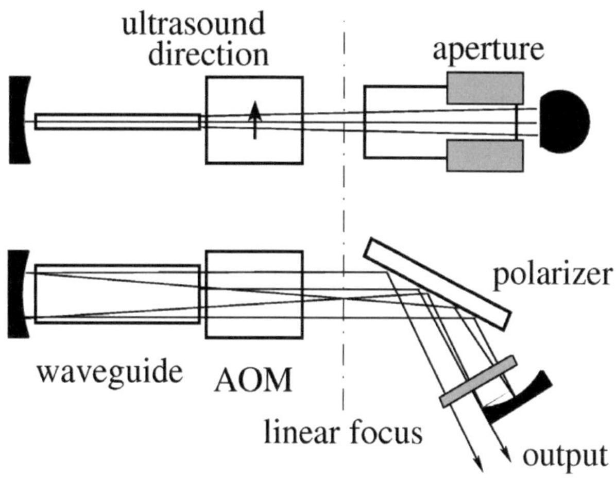

Fig. 2.13. Schematic diagram of the experimental setup.

Planar waveguides can be thought as one dimensional waveguides in the transverse direction and free space propagation in the lateral direction. We assume *x*- and *y*-axis are the transverse and lateral directions, respectively. Then, the laser travels along *z*-axis. The electric field of laser can be expressed as a superposition of the guided modes,

$$E(x,z)=\sum_m E_m(x,z)\exp\{j(\omega t-\beta_m z+\theta_m\} \tag{2.9}$$

where ω is the angular frequency of the laser electric field; θ_m and β_m are the phase and the propagation constant of the *m*th order mode, respectively. For the weak-guided waveguide, the propagation constant for the *m*th order mode can be found approximately as [27]

$$\beta_m=\frac{\omega n}{c}+\frac{m^2\pi\lambda}{4n^2a^2} \tag{2.10}$$

where n is the refractive index of the core; λ is the laser wavelength in the vacuum; and $a=d+\Delta$ is the effective thickness of the waveguide. Using Δ=5 m for our waveguide, the deviation of (2.10) is less than 0.1% compared with the standard eigen equation. If the origin of *x*-axis is chosen at the edge of the waveguide, the amplitude of the normalized *m*th order eigen-mode $E_m(x,z)$ can be written as

$$E(x,z) = A_m(z)\sqrt{\frac{2}{a}}\sin(k_{xm}x) \tag{2.11}$$

where $A_m(z)$ is the amplitude at the origin of x-axis (x=0). The x-component of wave vector k_{xm}, the free-space wave vector k, and the propagation constant β_m satisfy $k_{xm} = \sqrt{k^2 - \beta_m^2}$.

Generally, the basic rate equations for Q-switched lasers are described with average photon density. The photon density ϕ is calculated by integrating the electric field over whole space of the waveguide,

$$\phi = \frac{n}{Z_0 Slh\nu}\int_z\int_x E(x,z)E^*(x,z)dxdz \tag{2.12}$$

where Z_0 is the vacuum intrinsic impedance; l and S are the length and cross-sectional area of the waveguide, respectively. $h\nu$ is the photon energy of laser field. Applying the orthogonality of eigen-modes to Eq. (2.12), the total photon density is found to be the sum of photon density of all eigen-modes,

$$\phi = \sum_m \phi_m = \sum_m \frac{n}{Z_0 Sh\nu}\int_z A_m(z)A_m^*(z)dz \tag{2.13}$$

Substituting ϕ from Eq. (2.13) into the standard Q-switched rate equations, we obtain multi-mode rate equations,

$$\frac{d\phi_m}{dt} = \frac{\phi_m}{t_r}(2\sigma Nl - \varepsilon_m) + \phi_{spon}, \quad m = 1, 2, 3 \ldots \tag{2.14a}$$

$$\frac{dN}{dt} = -\gamma\sigma cN\sum_m \phi_m \tag{2.14b}$$

where t_r is the cavity round-trip time; σ the stimulated emission cross-section; N the average population inversion; γ the reduced gain factor [32]; c the speed of light, and ε_m the cavity losses. The contribution of spontaneous emission is taken into account by ϕ_{spon}. Notice that although above equations are derived for TE polarization, they are also valid for TM polarization if the electric field **E** is replaced by the magnetic field **M**. Actually, because our laser worked with TM polarization in the experiments, numerical simulation throughout this chapter has been performed for TM modes.

In the multi-mode rate equations, the cavity losses ε_m are mode dependence. The cavity losses include four parts. They are the output losses ln(1/R), space-waveguide coupling losses from the rear mirror L_m, losses from the AO modulator $k_m(t)$, and other intrinsic losses O,

$$\varepsilon_m = \ln(1/R) + O + L_m + \kappa_m(t) \tag{2.15}$$

The output losses ln(1/R) and the intrinsic losses O are almost the same for all modes, but the losses L_m are special for given eigen-modes. These three types of losses are constants during Q-switching interval. The losses $k_m(t)$ determined by the AO modulator varies with time, playing a important role in the mode competition. The losses $k_m(t)$ can be calculated by wave propagation method. The laser field propagating in the free space is described by the Huygen's integral, [1]

$$E_m(x,z_1) = \int K(x,z_1,z_0)E_m(x,z_0)dx \tag{2.16}$$

where $K(x, z_1, z_0)$ is the propagation integral kernel [1]. While the laser beam arrives at the AO modulator, the laser beam will be modulated. The radio-frequency (RF) driver power decays exponentially after the AO modulator being triggered. The modulation of the electric field in the AO modulator is given by

$$\tilde{E}(t,x,z_1) = \eta_{AO} \exp\{-(t - x/v)/\tau_{ao}\}E(x,z_1) \tag{2.17}$$

where η_{AO} is the modulation amplitude of the AO modulator, v is the propagation velocity of ultrasonic wave in silica, and τ_{ao} is the decay constant of the RF driver. The time t is counted from the AO modulator being opened. In our experiments, τ_{ao} was measured to be 100 ns, and the ultrasonic wave spends about 80 ns passing through the laser beam diameter. Thus, the AO modulator can be opened transparently in about 120 ns.

When the laser beam is reflected back to the waveguide, it couples to eigen-modes of the waveguide. For the mth order mode, the coupling coefficient is given by the overlap integral,

$$C_m(t) = \sum_i \sqrt{2/a} \int E_i'(t,x)\exp(j\theta_i')\sin(k_{xm}x)dx \tag{2.18}$$

where $E_i'(t,x)\exp(j\theta_i')$ is the returned electric field of ith order mode. The contributions to C_m from different eigen-modes are added coherently. The exact value of C_m depends on the relative phase angle among eigen-modes. Finally, the losses $k_m(t)$ is written as

$$k_m(t) = \ln[1 - C_m(t)] \tag{2.19}$$

If the laser beam propagates in the resonator without disturbance, the contributions to C_m from the ith modes ($i \neq m$) are very small. With an intracavity aperture inserted into the resonator, the electric field is distorted and the contributions from adjacent modes ($i = m \pm 1$) will increase due to the coupling effects among eigen-modes.

At first, we calculate coupling coefficient without the AO modulator in the cavity. If we assume no coupling effects among eigen-modes, the coupling coefficient drops down monotonously with narrowing the intracavity aperture. When the influences from adjacent modes are considered, several ripples, corresponding to constructive interference between eigen-modes, are found in the curves with narrowing the aperture. We compared the coupling coefficient in these two cases in Fig. 2.14. In the figures we assume all modes with the same initial phase and amplitude. As can be seen, the influences of the coupling among eigen-modes are obvious. Notice that TM1 mode has the largest coupling coefficient all the time without the influence of the AO modulator.

Now let us consider the influence of the AO modulator. Because the ultrasonic wave travels from bottom to top in the AO modulator, just after the AO modulator is opened, the mode whose electric field has a side ripple near the bottom of the AO modulator, has the largest transmission in the first instance. While the ultrasonic wave is totally opened, the AO modulator becomes transparent, and then the electric field passes the AO modulator without distortion. The evolution of the coupling coefficient is described in Fig. 2.15. In our experiment, when the intracavity aperture is set to 2-mm wide, a side ripple of TM2 mode locates near the bottom of the AO modulator. Thus, the coupling coefficient of TM2 mode is the largest in the first 120 ns. After that time, TM1 mode obtains the largest coupling coefficient. This time-dependence losses lead to mode competition in the Q-switched planar waveguide lasers. In contrast, the coupling loss is time independent in the CW lasers, and mode competition has not been observed [19].

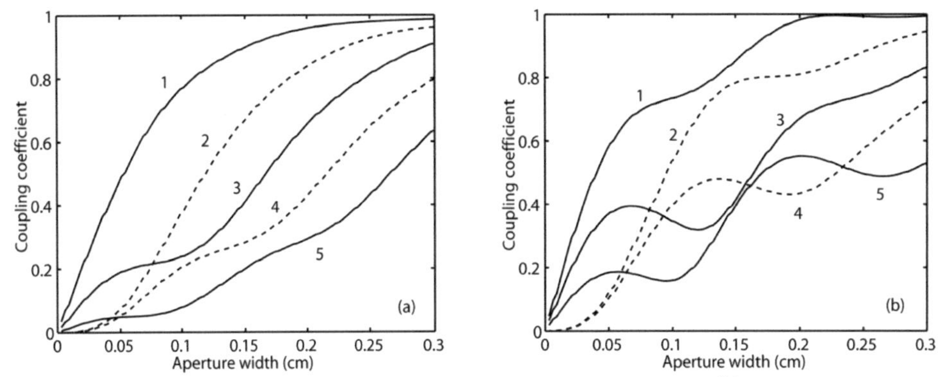

(a) (b)

Fig. 2.14. Coupling coefficients of TM1-TM5 modes for varying aperture width without (a) and with (b) mode coupling effects.

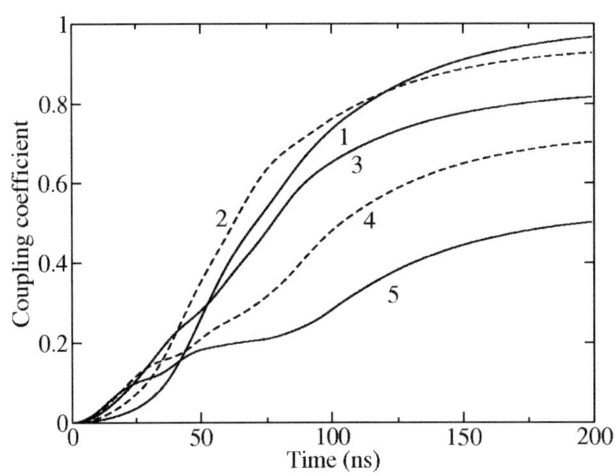

Fig. 2.15. Evolution of coupling coefficients of TM1-TM5 modes after the AO modulator being switched on. The intracavity aperture is 2-mm wide.

In the switch-off interval, population inversion accumulates rapidly under heavy pumping. Parasitic oscillation occurs when its threshold is overcome by the accumulated gain. Then in turn, the parasitic oscillation clamps the population inversion to the level corresponding to the threshold. The parasitic oscillation can be enhanced by the total internal reflection at the waveguide surface along zig-zag path or in-plane path [19]. The parasitic oscillations along different paths deplete particular parts of the population inversion. Therefore, the average population inversion is given by

$$N = \frac{1}{U}\sum_{i=1}^{U} N_i \tag{2.20}$$

The parasitic oscillation propagating along the ith path is governed by the rate equation,

$$\frac{d\Phi_i}{dt} = \Phi_i[c\sigma N_i - \varepsilon_i] + \Phi_{spon} \tag{2.21a}$$

$$\frac{dN_i}{dt} = R_p\eta_p - \frac{N_i}{\tau_2} - c\sigma N_i\Phi_i \tag{2.21b}$$

where ϕ_i, N_i, and ε_i are the photon density, the population inversion and the threshold, respectively. τ_2 is the excited-state lifetime. Pump efficiency is $\eta_p = 0.7$ for our experiments. Pump rate R_p is determined by the pump power.

By measuring the small signal gain, we found the parasitic oscillation occurred when the pump power was around 300 W. Hence, we assume the threshold of the parasitic oscillation corresponding to the pump power of 300 W. The calculated population inversion and photon density of parasitic oscillation are plotted in Fig. 2.16. Without parasitic oscillation [1], the population inversion can be expressed as $R_p\eta_p t[1-\exp(-t/\tau_2)]$. Comparing this solution with Fig. 2.16, we find the inversion can be several times higher than the parasitic threshold before it is really depleted by the parasitic oscillation. This means that quasi CW pumping sources with short duty-cycles can be beneficial to push the inversion to higher levels. With slow repetition rate, intense pumping is useless because the parasitic oscillation grows so quickly that the inversion has been clamped to the parasitic threshold at the end of pumping interval. For the maximum pump power of 450 W in our experiments, the inversion can be accumulated without severe degradation up to 1-kHz repetition rate.

Once the population inversion at the end of pumping interval having been obtained, we use it as the initial value n_i to the Q-switching equations Eq. (2.14). Solving Eq. (2.14) numerically, we can obtain remaining population inversion n_f at the end of the Q-switched pulse. The remaining inversion n_f is then inserted back into pump equations Eq. (2.20) and (2.21). These procedures have been iterated until the population inversion reaches steady state.

As can be seen from Fig. 2.15, the losses for TM2 mode is less than that of TM1 mode in the first 120 ns. After 120 ns, TM1 mode becomes the one with the lowest loss. With the pump power below 450 W, the build-up time of laser pulse is more than 120 ns, so that single TM1 mode exists during whole pulse duration. When the pump power is increased to 450 W, the build-up of laser pulse is around 120 ns. TM2 mode can oscillate before 120 ns, but after 120 ns, TM1 mode grows so quickly that TM2 mode is clamped to a low level [see Fig. 2.17(a)]. When the pump power is greater than 450 W, TM2 mode can grow significantly in the first 120 ns. After that time, TM1 mode starts growing. Because the inversion has been saturated down by TM2 mode, finally TM1 mode is lower than TM2 mode [Fig. 2.17(b)].

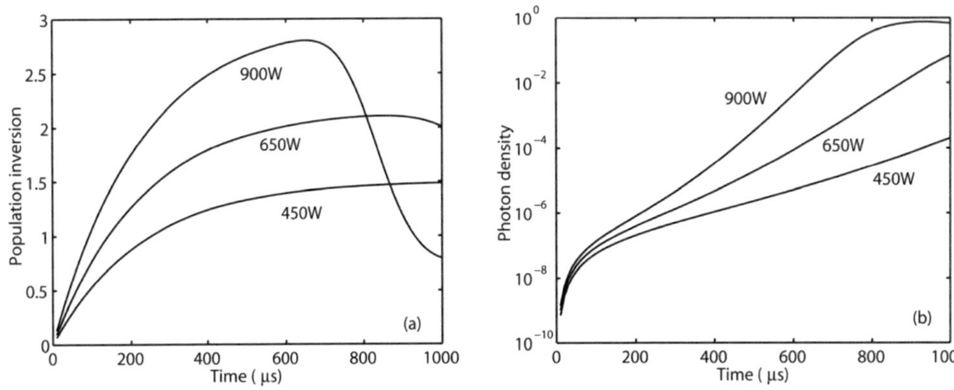

Fig. 2.16. (a) Normalized population inversion, and (b) photon density of parasitic oscillation for various pump powers.

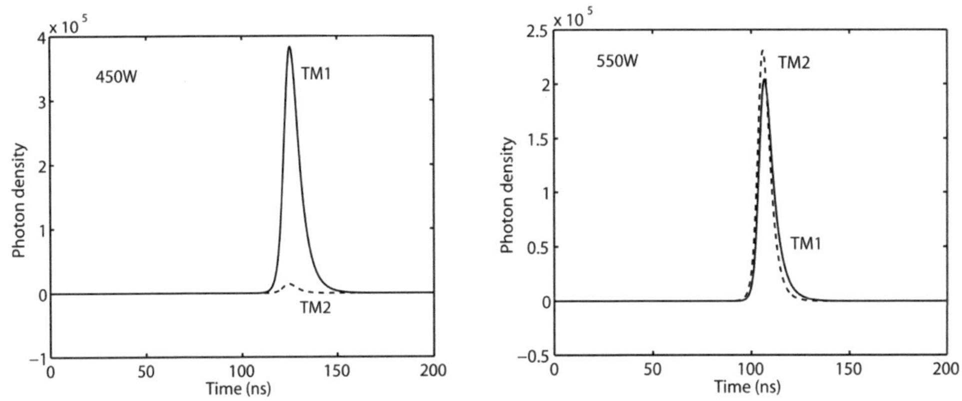

Fig. 2.17. Temporal profiles of TM1 and TM2 modes with pump power of 450 W and 550 W, respectively.

We measured mode structure by scanning the near-field laser beam with a 200-μm-wide slot. The transmission from the slot was recorded with an InGaAs photondetector and an oscilloscope. Then, the mode structure was found by best fitting of measured data to eigen-modes of the waveguide. The results were compared with theoretical simulation in Fig. 2.18. As expected, the percentage of TM2 mode decreases with time. However, in comparison with the simulation, the percentage of TM2 mode is about 8 and 2 times larger with 300 and 450 W pump power, respectively. This may be caused by the thermal distortion in the waveguide, so that the mode controlled by the intracavity aperture is degraded.

Since laser pulses develop from spontaneous emission, pulse instability is a characteristic feature of Q-switched lasers. Because the coupling losses $k_m(t)$ are sensitive to the related phase angle between eigen-modes, the pulse instability in the planar waveguide laser is larger than that in the ones whose output are phase insensitive, e.g., conventional rod lasers. Assuming constant amplitude and random phase of spontaneous emission in the simulation, about 7% fluctuation in the pulse's peak was found. Fig. 2.19 shows a typical pulse train recorded in the experiments for the 450 W pump power and 1 kHz repetition rate.

The fluctuation of the pulse's peak was measured to be about 15%. If we take into account the influence of thermal turbulence and instability of the AO modulator, the measured fluctuation agrees with the simulation. Notice that the mode competition can be suppressed by using the narrower intracavity aperture as doing in CW lasers [19]. In experiments, the pulse fluctuation was reduced to 5% when 1.2-mm wide aperture was used.

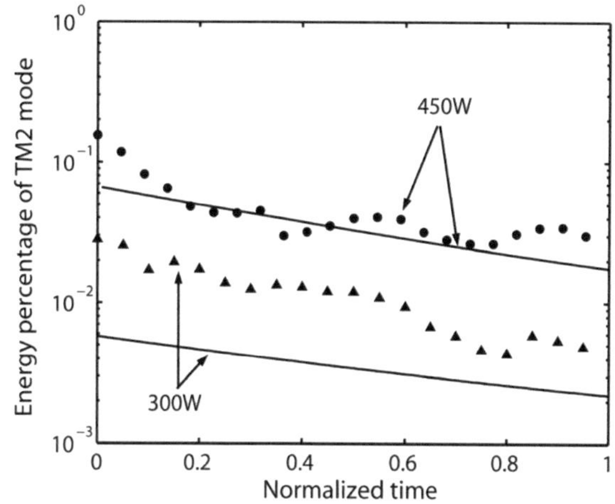

Fig. 2.18. Energy percentage of TM2 mode as a function of normalized time for various pump powers.

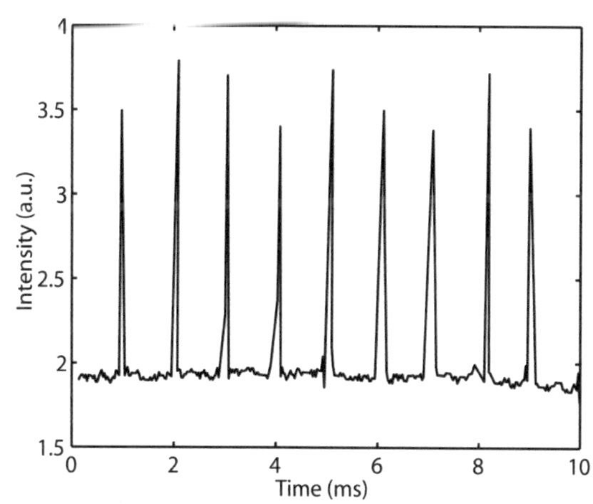

Fig. 2.19. Typical pulse train at 1-kHz repetition rate under 450-W pump power.

As we mentioned previously, the intensity ratio of TM2 mode to TM1 mode depends on the input pump power and varies with time, leading to temporal structures in the output beam patterns. We measured the near-field beam patterns and compared them with theoretical simulation in Fig. 2.20. The near-field beam patterns were recorded at 20 cm far from the output coupler in the pulse rising edge, peak and falling edge, respectively. Under 300 W pump power, because TM1 mode was dominant during whole pulse duration, the output beam pattern was almost invariable. With the pump power increasing to 450 W, TM2 mode was significant in the rising edge, and decreased with time. As can be seen, side ripples were higher in the rising edge than that in the falling edge. The theoretical calculation was obtained by coherent sum of the field of each eigen-modes, because we find that the eigen-modes tend to be phase-locked in the buildup of laser pulses. Further, asymmetry of side ripples is easily explained by the coherent sum. The output beam patterns of planar waveguide lasers vary in a different way from that of rod lasers [33].

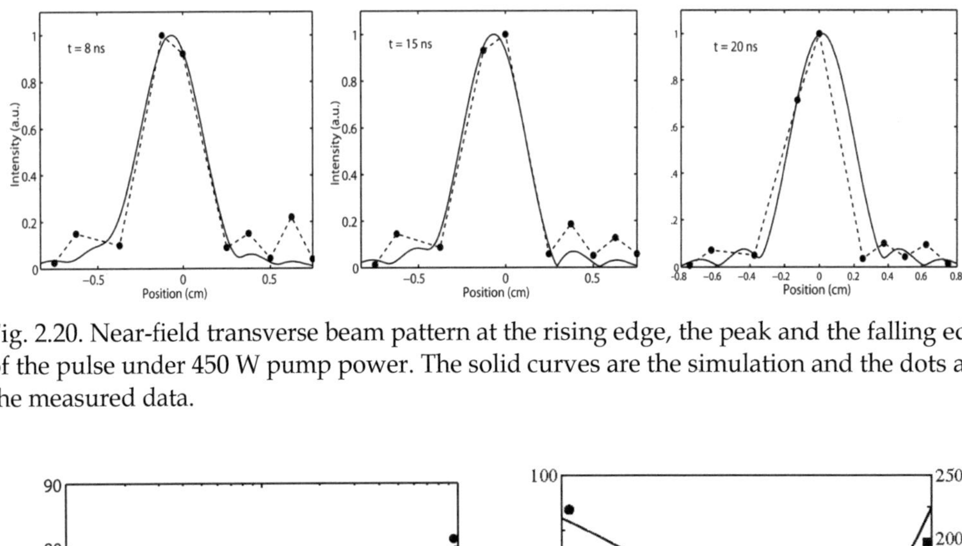

Fig. 2.20. Near-field transverse beam pattern at the rising edge, the peak and the falling edge of the pulse under 450 W pump power. The solid curves are the simulation and the dots are the measured data.

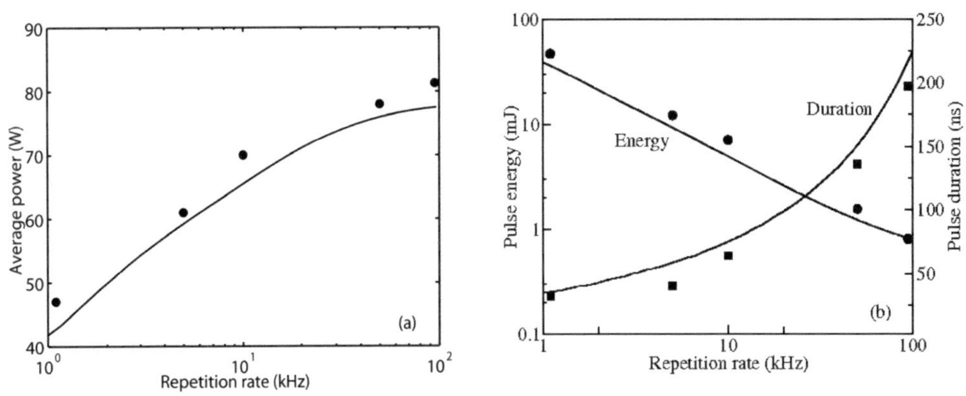

Fig. 2.21. (a) Average power, and (b) pulse energy and duration as a function of repetition rate with 450 W pump power.

The pulse energy was calculated by integrating photon density over whole pulse duration, and the average power was obtained by multiplying the pulse energy with the repetition rate. Bearing pulse instability in mind, we averaged the integration over 20 pulses, where random phase and amplitude was applied to the spontaneous emission. Accordingly, the data measured in the experiments were averaged over 32 sequential laser pulses. Without intracavity aperture, up to 109 W of average power can be obtained with $M_x^2 \times M_y^2 = 6 \times 1.4$ at 100 kHz repetition rate. Using 2-mm-wide intracavity aperture, a maximum average power of 83 W was generated from the laser with $M_x^2 \times M_y^2 = 1.2 \times 1.4$ at 100 kHz repetition rate. The typical pulse duration and pulse energy at 1 kHz repetition rate were 32 ns and 47 mJ, respectively. Correspondingly, the multi-mode rate equations gives 29 ns pulse duration and 43 mJ pulse energy, compared with 22 ns and 37 mJ using standard single-mode rate equations. The multi-mode model gives a better prediction than that using single-mode assumption.

3. References

[1] A.E. Siegman, *Lasers* (University Science, Maple-Vail, CA, 1986).

[2] H.J. Baker, A.A. Chesworth, D.P. Millas, and D.R. Hall, "A Planar waveguide Nd:YAG laser with a hybrid waveguide-unstable resonator," *Opt. Commun.* 191, 125 (2001).

[3] J. Xu, J.R. Lee, H.J. Baker, and D.R. Hall, "A new Q-switched hybrid resonator configuration for the 100W Nd:YAG planar waveguide laser," presented at CLEO-EURO 2003, Paper CA-2-3-MON, Germany, 2003.

[4] A.E. Siegman, "New development in laser resonators," *Proc. of SPIE* 1224, 2-14 (1990).

[5] E.A. Sziklas, A.E. Siegman, "Mode calculations in unstable resonators with flowing saturable gain. 2: Fast Fourier transform method," *Appl Opt.*, 14, 1874 (1975).

[6] B. Wasilewski, H.J. Baker, and D.R. Hall, "Intracavity beam behavior in hybrid resonator planar-waveguide CO2 lasers," *Appl Opt.*, 39, 6174 (2000).

[7] E.A. Sziklas and A.E. Siegman, "Diffraction calculations using fast fourier transform methods," *Proc of IEEE* 62, 410 (1974).

[8] D.B. Rensch and A.N. Chester, "Iterative diffraction calculations of transverse mode distributions in confocal unstable laser resonators," *Appl Opt.* 27, 997 (1973).

[9] A.N. Chester, "Three-dimensional diffraction calculation of laser resonator modes," *Appl Opt.* 12, 2353 (1973).

[10] T.S. Rutherford, W.M. Tulloch, E.K. Gustafson, and R.L. Byer, "Edge-pumped quasi-three level slab lasers: design and power scaling," *IEEE J. Quantum Electron.* 36, 205 (2000).

[11] Z. Wang, X. Ye, T. Fang, J. Xu, "Output beam quality of edge pumped planar waveguide lasers with confocal unstable resonators," Optical Review, 12, 391 (2005)

[12] R. Ulrich and G. Ankele, "Self-imaging in homogeneous planar optical waveguides," *Appl. Phys. Lett.* 27, 337 (1975).

[13] J. Xu, "Quasi self-image planar waveguide lasers with high-power single-mode output," *Opt. Commun.* 259, 251 (2006).

[14] H.J. Baker, J.R. Lee, and D.R. Hall, "Self-imaging and high-beam-quality operation in multi-mode planar waveguide optical amplifiers," *Optics Express* 10, 297(2002).

[15] J.Banerji, A.R. Davies, and R.M. Jenkins, "Laser resonators with self-imaging waveguides," *J. Opt. Osc. Am.* B 14, 2378 (1997).

[16] I.T. McHinnie, J.E. Koroshetz, W.S. Pelouch, D.D. Smith, J.R. Unternahrer, and S.W. Henderson, "Self-imaging waveguide Nd:YAG laser with 58% slope efficiency," presented at CLEO/QELS 2002, USA, 262(2002).

[17] W. Koechner, *Solid-state laser engineering* (Spring-Verlag, Berlin, 1999).
[18] J.I. Mackenzie, and D.P. Shepherd, "End-pumped, passively Q-switched Yb:YAG double clad waveguide laser," *Opt. Lett.* 27, 2161 (2002).
[19] J.R. Lee, H.J. Baker, G.J. Friel, G.J. Hilton, and D.R. Hall, "High-average-power Nd:YAG planar waveguide laser that is face pumped by 10 laser diode bars," *Opt. Lett.* 27, 524 (2002).
[20] J.M. Eggleston, T.J. Kane, K. Kuhn, J. Unternahrer, and R.L. Byer, "The slab geometry laser - part I: Theory," *IEEE J. Quantum Electron.* QE-20, 289 (1984).
[21] D. C. Brown, "Nonlinear thermal and stress effects and scaling behavior of YAG slab amplifiers," *IEEE J. Quantum Electron.* 34, 2393 (1998).
[22] D.P. Shepherd, S.J. Hettrick, C. Li, J.I. Mackenzie, R.J. Beach, S.C. Mitchell and H.E. Meissner, "High-power planar dielectirc waveguide lasers," *J. phys. D: Appl. Phys.* 34, 2420 (2001).
[23] A. Faulstich, H.J. Baker and D.R. Hall, "Face pumping of thin, solid-state slab lasers with laser diodes," *Opt. Lett.* 21, 594 (1996).
[24] C. Becker, T. Oesselke, J. Pandavenes, R. Ricken, K. Rochhausen, G. Schreiber, W. Sohler, H. Suche, R. Wessel, S. Balsamo, I. Montrosset and D. Sciancalepore, "Advanced Ti:Er:LiNbO3 Waveguide Lasers," *IEEE J. Selected Topics in Quantum Electron.* 6, 101 (2000).
[25] J. Xu, and Q. Lou, "Gain-guide effects in transient Raman amplifiers," *J. Opt. Soc. Am.* B 16, 961 (1999).
[26] Ling Xiao, Xiaojin Cheng, and Jianqiu Xu, "High-power Nd:YAG planar waveguide laser with YAG and Al_2O_3 claddings", Optics Comm. 281, 3781(2008)J. Xu, H.J. Baker, D.R. Hall, "Mode competition in acousto-optically Q-switched planar waveguide lasers," *Optics and Laser Technology*, 39, 814 (2007)
[27] J. Xu, H.J. Baker, D.R. Hall, "Mode competition in acousto-optically Q-switched planar waveguide lasers," Optics and Laser Technology, 39, 814 (2007)
[28] N. P. Barnes, "Solid-state lasers from an efficiency perspective," *IEEE J. Selected Topics in Quantum Electron.* 13, 435 (2007).
[29] N. Hodgson, V.V. Ter-Mikirtychev, H.J. Hoffman, W. Jordan, "Diode-pumped, 220W ultra-thin slab Nd:YAG laser with near-diffraction limited beam quality," in Advanced Solid-State Lasers, OSA TOPS, 68, 552 (2002).
[30] J.J. Degnan, "Optimization of passively Q-switched lasers," *IEEE J. Quantum Electron.* 31, 1890 (1995).
[31] J. Dong, J. Lu and K. Ueda, "Experiments and numerical simulation of a diode-laser-pumped Cr, Nd:YAG self-Q-switched laser," *J. Opt. Soc. Am.* B 21, 2130 (2004).
[32] J. J. Degnan, "'Theory of the optimally coupled Q-switched Laser," *IEEE J. Quantum Electron.* 25, 214 (1989).
[33] A. Caprara and G. C. Reali, "Time varing M^2 in Q-switched lasers," *Opt. and Quantum Electron.* 24, S1001 (1992).

Part 2

Rare-Earth Doped Lasers

3

The Recent Development of Rare Earth-Doped Borate Laser Crystals

Chaoyang Tu* and Yan Wang
Key Laboratory of Optoelectronic Materials Chemistry and Physics of CAS, Fujian Institute of Research on the Structure of Matter, Chinese Academy of Sciences, P. R. China

1. Introduction

As a laser host, borates possess favourable chemical and physical characteristic and higher damnification threshold. $Ln_2Ca_3B_4O_{12}$ (Ln = La, Gd, or Y) double borate family crystals,$Ca_3(BO_3)_2$ (CBO) and LaB_3O_6 crystals are the potential laser host materials. They have the suitable hardness and good chemical stability and are moisture free. Furthermore, they melt congruently and can be grown by Czochralski method [1-5], so the high optical quality crystal with large dimension can be easily grown.

The $Ln_2Ca_3B_4O_{12}$ (Ln = La, Gd, or Y) double borate family was first grown by the Czochralsky method with the Nd^{3+} doping[3,5]. The orthorhombic crystallographic structure of Yb^{3+}-doped $Y_2Ca_3B_4O_{12}$ (CYB) was determined in Ref. [6], it is made of three sets of M-oxygen distorted polyhedrons and three sets of BO_3 planar triangles. The Yb^{3+}, Y^{3+} and Ca^{2+} cations occupy statistically the three M sets. This structure disorder contributes to the line broadening of spectra of rare earth doping ions, such as Yb^{3+}, Nd^{3+}, and can lead to a tunable laser.

In the rare earth-doped CBO crystal, the rare earth ions substitute for the divalent cation (Ca^{2+}) and charge compensation is required. Because of the charge compensation effects, the rare earth-doped CBO crystals show partly distorted structure, and the inhomogeneous broadening of the emission similar to amorphous materials can be also expected.[7]

Furthermore, rare earth-doped LaB_3O_6 crystal can serve as a microchip laser crystal without any processing because of the cleavage of LaB_3O_6 crystal[8~10].

The study on the rare earth-doped $Ca_3(BO_3)_2$, LaB_3O_6 and $Ca_3Re_2(BO_3)_4$ laser crystals will be covered in this chapter. The growth, thermal, optical and spectrum characteristics of these crystals are presented. Their laser characteristics are also covered.

2. Rare earth-doped $Ca_3(BO_3)_2$ crystals

$Ca_3(BO_3)_2$ (CBO), which belongs to the trigonal system with the space group R_{-3C}, and the cell parameters are as follows: a=b=8.6377(8) Å, c=11.849(2) Å, v=765.61 Å^3, z=6, and Dc=3.096 g/cm^3 is a good laser host material. It has the suitable hardness and good chemical

*Corresponding Author

stability and is moisture free. The Ca^{2+} ion in CBO is surrounded by eight nearest oxygen to form a distorted polyhedron, in which the only symmetry is a two-fold axis passing through the Ca ion. Thereby the calcium ions belong to the C_2 point symmetry. As in this borate the trivalent rare earth ion is introduced on the divalent cation site, a mechanism for charge compensation should be considered.

2.1 The crystal growth

Pure, Er^{3+}-, Dy^{3+}-, and Nd^{3+}-doped CBO crystals were grown by the Czochralski method along the [0 1 0] orientation (by using the *b*-axis seeds). The raw materials were analytical grade, $CaCO_3$, H_3BO_3, Na_2CO_3, and some spectral grade, Dy_2O_3, and Nd_2O_3. In fact, in the case of rare earth-doped CBO crystal, the co-doping Na_2CO_3 as charge compensator was introduced to obtain crystal with a large concentration of the rare earth ions. The melt composition ratio of $CaCO_3$ and H_3BO_3 is 3:2 for the pure CBO and rare earth (RE)-doped CBO crystals according to the following reactions:

$$(3\text{-}6x)\ CaCO_3 + 2H_3BO_3 + 3x/2\ Na_2CO_3 + 3x/2\ Ln_2O_3 = (Ca_{1-2x}Na_xLn_x)_3(BO_3)_2$$

$$+\ 3H_2O\uparrow + (3\text{-}9x/2)\ CO_2\uparrow$$

The polished sample of the pure CBO crystal is shown in Fig.2.1. Its optical homogeneity was determined using Zygo optical interferometer (shown in Fig.2.2). Its homogeneity is 2.57×10-5, and its thickness is 5.00 mm.

2.2 The thermal characteristic

Measurements of thermal expansion have greatly increased our knowledge of material properties such as lattice dynamics, electronic and magnetic interactions, thermal defects, and phase transitions [12]. As a significant part of the power pump is converted into heat inside the material during laser operation, it is important to know its linear thermal expansion coefficients to predict how the material behaves when the temperature increases [13]. The figure of linear expansions versus temperature was shown in Fig.2.3. The linear thermal expansion coefficient is defined as:

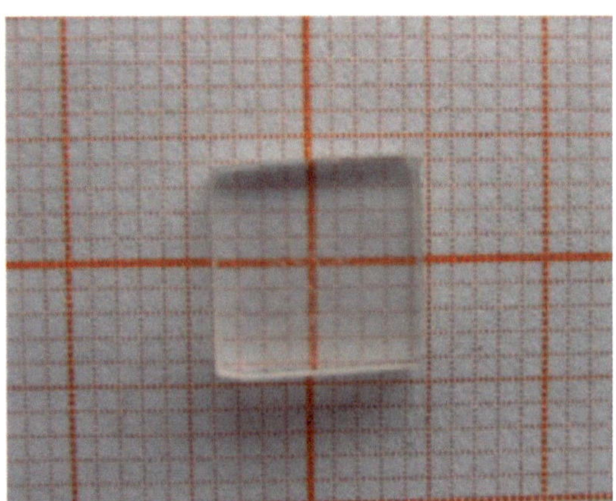

Fig. 2.1. The polished sample of the pure CBO crystal

Fig. 2.2. Interference fringe of CBO crystal.

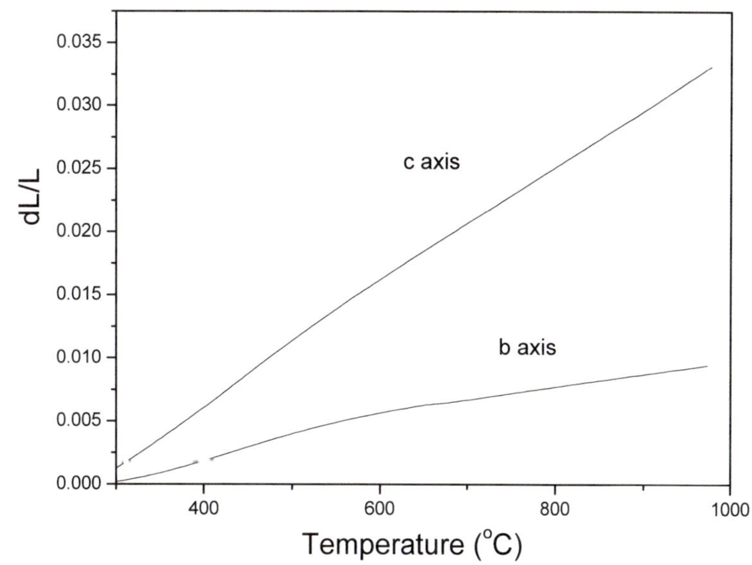

Fig. 2.3. Thermal expansion of the CBO crystal

$$\alpha = \frac{1}{T}\frac{\Delta L}{\Delta T} \quad (2.1)$$

Where L is the initial length of the sample at room temperature and ΔL is the change in length when the temperature changes ΔT. We can calculate the thermal expansion coefficient from the slope of the linear fitting of the linear relationship between $\Delta T/T$ and the temperature. In this case, the linear thermal expansion coefficients for different crystallographic directions c-, and b-axes are 4.69×10^{-5} K^{-1}, 1.37×10^{-5} K^{-1}, respectively[14]. The thermal expansion coefficients of the *a* and *b* axes are comparable. Thermal expansion coefficient along the *c*-axis is about two times larger than those of *b* and *a* axes. Although the thermal expansion property of YVO_4 crystal has little different from that of CBO crystal, the CBO crystal has no cleavage plane.

$$\alpha_{ij(CBO)} = \begin{vmatrix} 1.37 & 0 & 0 \\ 0 & 1.37 & 0 \\ 0 & 0 & 4.69 \end{vmatrix} \times 10^{-5}\ {}^{o}C \qquad (2.2)$$

It is well known that the higher the consistency of atom in the crystal structure, the larger the heat expansion coefficient, and vice versa. Obviously, it was demonstrated from the Fig.2.4 that the consistency of atom along c axis is higher than that along b axis, which is comparable to a axis. Therefore, the heat expansion coefficient along c axis is much larger than those along b axis and a axis.

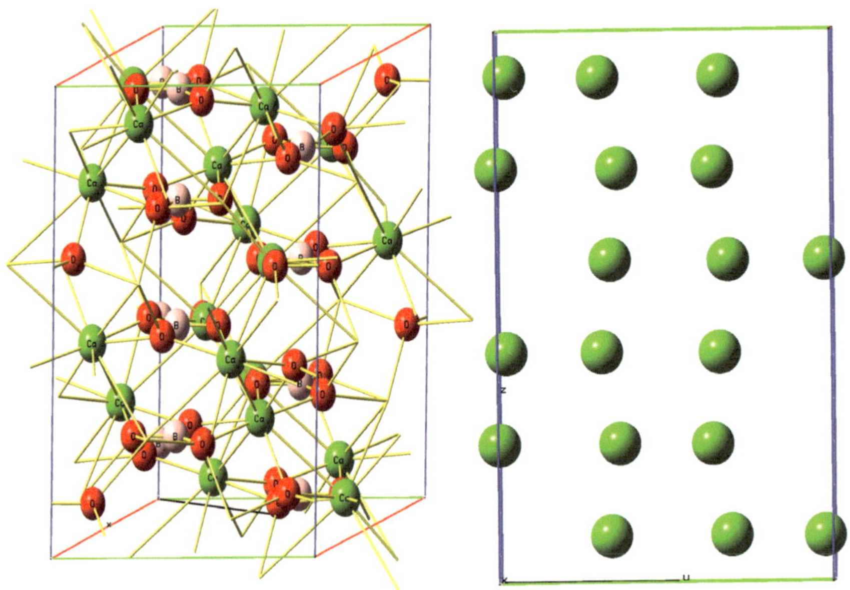

(a) (b)

Fig. 2.4. (a) The structure of CBO crystal; (b) The cb section of the structure of CBO crystal

2.3 The spectrum characteristics

2.3.1 The spectrum characteristics of Nd^{3+}:$Ca_3(BO_3)_2$ crystal

Fig.2.5 shows the transmission spectrum of CBO crystal. It has high transmittance in the 190-3300 nm optical ranges.[14]

Fig.2.6 shows the absorption spectrum measured at room temperature in the 300~950 nm ranges[15]. There are three main strong absorption peaks in the spectrum centered at 588, 751 and 808 nm, respectively, corresponding to the transitions from the $^4I_{9/2}$ ground state. The

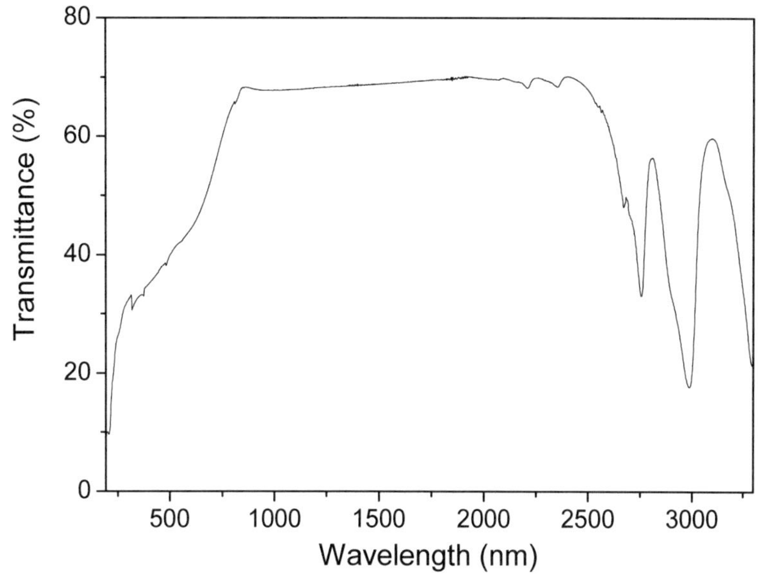

Fig. 2.5. The transmission spectrum of CBO crystal

introduction of Na^{+} as charge compensator results in the disorder in the local crystal fields acting on the optically active ions.[16] Therefore, the absorption and emission peaks can become broadening.[17] The full width at half maximum (FWHM) of the absorption peak at 808 nm is about 19 nm, which is larger than that of Nd^{3+}:$GdVO_4$ (FWHM is 4 nm)[18]. The absorption coefficient and absorption cross section of Nd^{3+}:CBO crystal at 808 nm are 0.93 cm^{-1} and 2.12×10^{-20} cm^2, respectively, which are compared with those of Nd^{3+}:$GdVO_4$ (2.0×10^{-20} cm^2 for π spectrum) [20] and Nd^{3+}:YVO_4(2.0×10^{-20} cm^2 for π spectrum) [19]. Therefore, Nd^{3+}:CBO crystal is suitable for GaAlAs laser diode pumping.

Fig.2.7 presents the emission spectrum of Nd^{3+}:CBO crystal, in which there are three main emission peaks in the spectrum centered at 1346, 1060 and 883 nm, respectively, corresponding to the transitions of $^4F_{3/2}\rightarrow^4I_J$ (J=13/2, 11/2, 9/2). Table 2-1 lists the line strengths and optical parameters of Nd^{3+} in CBO crystal. Table 2-2 lists the values of intensity parameters of Nd^{3+} in CBO crystal and those of some other well-known Nd-doped laser crystals. Table 2-3 presents the luminescence parameters of Nd^{3+} in CBO crystal for the transitions $^4F_{3/2}\rightarrow^4I_J$. The stimulated emission cross-section at 1060 nm is about 8.04×10^{-20} cm^2, which is smaller than those of $NdAl_3(BO_3)_4$ (1.43×10^{-19} cm^2) [26] and Nd^{3+}:$La_2(WO_4)_3$ (11.2×10^{-20} cm^2) crystals [27], but larger than that of Nd^{3+}:LaB_3O_6 (3.46×10^{-20} cm^2) crystal [21]. Fig.2.8 presents the luminescence decay curve excited at 808 nm at room temperature corresponding to the emission line $^4F_{3/2}\rightarrow^4I_{11/2}$ at 1060 nm. The fluorescence lifetime of $^4F_{3/2}$ is 43.8μs, and the radiative lifetime is 135.03μs. So the luminescent quantum efficiency of the $^4F_{3/2}$ level is about 32.4%.

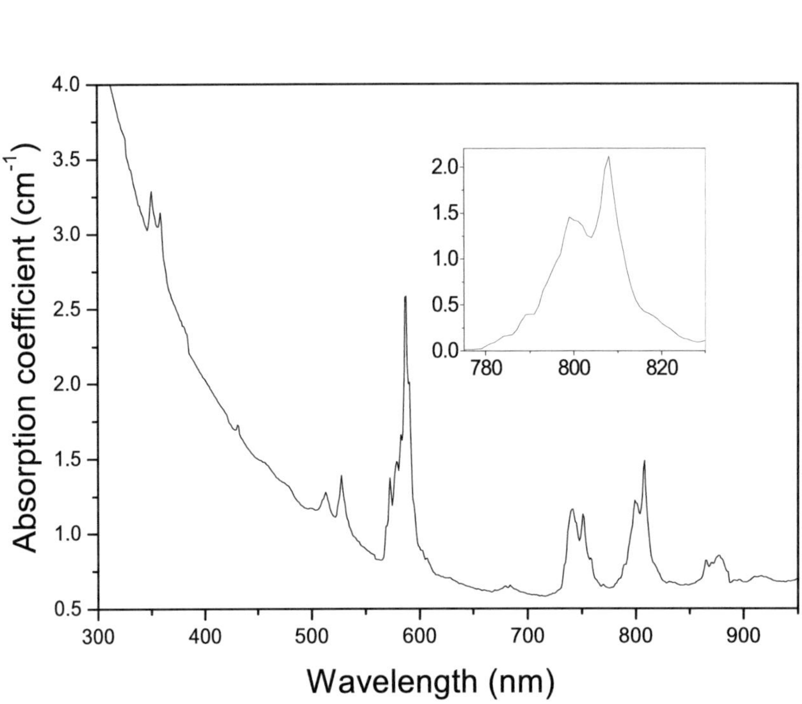

Fig. 2.6. The absorption spectrum of Nd^{3+}:CBO crystal measured at room temperature

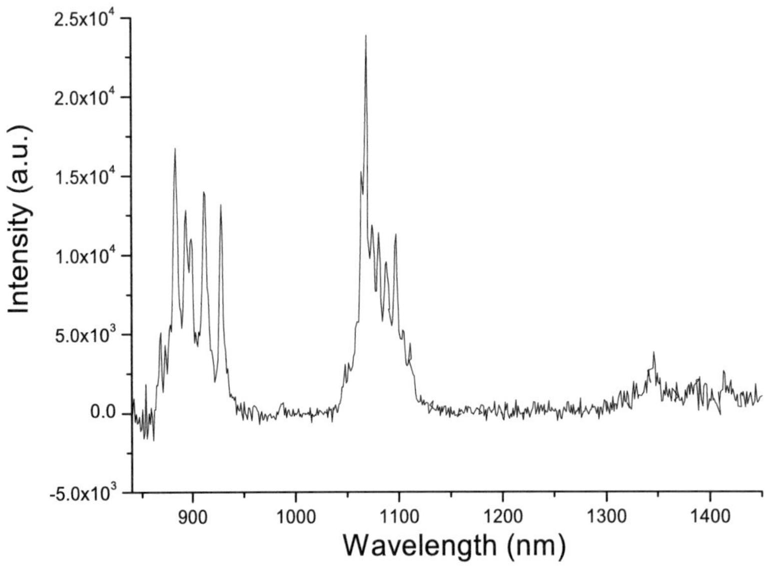

Fig. 2.7. The emission spectrum of Nd^{3+}:CBO crystal

Transition final state	Central wavelength $\bar{\lambda}(nm)$	$S_{mea}(J \rightarrow J')$ (10^{-20} cm^2)	$S_{cal}(J \rightarrow J')$ (10^{-20} cm^2)	$\sigma_{abs}(\lambda)$ (10^{-20} cm^2)
$^4I_{9/2} \rightarrow ^4D_{1/2}$	359	0.837	0.772	5.342
$^4G_{9/2}$	528	0.920	0.949	2.020
$^4G_{5/2}$	588	5.529	5.529	4.419
$^4S_{3/2}$	751	4.564	4.620	1.373
$^2H_{9/2}$	808	2.475	2.377	2.303

Table 2.1. Line strengths and optical parameters of Nd^{3+} in CBO crystal

Crystal	Ω_2 (10^{-20} cm^2)	Ω_4 (10^{-20} cm^2)	Ω_6 (10^{-20} cm^2)	$\beta_{J'}$ $^4F_{3/2} \rightarrow ^4I_{11/2}$	Ref.
CBO	4.63	2.40	10.4	0.476	15
NAB	6.07	9.14	14.58	0.518	21
$LaSc_3(BO_3)_4$	5.349	4.124	3.852	0.470	22
LaB_3O_6	0.54	2.31	4.51	0.538	23
YVO_4	4.667	2.641	4.047	0.509	24
$GdVO_4$	12.629	4.828	8.425	0.519	25
$Ca_3Sc_2Ge_3O_{12}$	0.99	4.24	7.14	0.524	26

Table 2.2. The intensity parameters of Nd^{3+} in CBO crystal and those of some other well-known Nd-doped laser crystals

Radiation transition	Radiation wavelength (nm)	$A(J \rightarrow J')$ (s^{-1})	$\beta_{J'}$
$^4F_{3/2} \rightarrow ^4I_{9/2}$	883	3157	0.426
$^4I_{11/2}$	1060	3529	0.476
$^4I_{13/2}$	1346	720	0.097
$^4I_{15/2}$	1852	37.8	0.001

Table 2.3. The luminescence parameters of Nd^{3+} in CBO crystal for the transitions $^4F_{3/2} \rightarrow ^4I_J$.

2.3.2 The spectrum characteristics of Er^{3+}:$Ca_3(BO_3)_2$ crystal

Figure.2.9 shows the room temperature (RT) polarized absorption spectra in the **200–1600 nm** spectra region of **Er3+** in the CBO crystal[28]. It consists of a number of groups of lines corresponding to transitions between the ground state $^4I_{15/2}$ and higher energy states inside the **4f** ^{11} electronic configuration of the **Er**^{3}+ ion. Due to the high **Er**$^{3+}$ concentration of the CBO sample, the spectra are well defined. In this sample, eleven absorption bands clearly are located at 1517, 978, 793, 652, 523, 486, 450, 403, 380, 365, and **257 nm**, which correspond to the transitions from $^4I_{15/2}$ to $^4I_{13/2}$, $^4I_{11/2}$, $^4I_{9/2}$, $^4F_{9/2}$, $^2H_{11/2}$, $^4F_{7/2}$, $^4F_{5/2}$, $^2H_{9/2}$, $^4G_{11/2}$, $^4G_{9/2}$ and $^4D_{5/2}$, respectively.

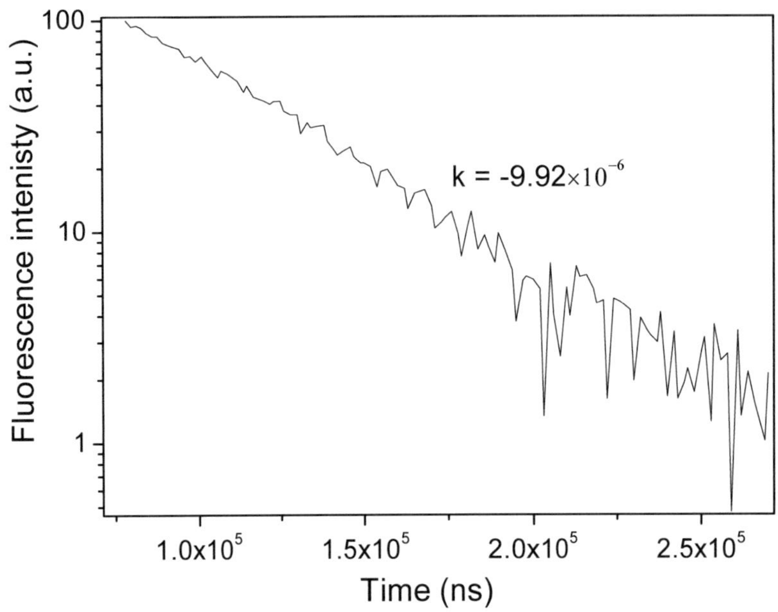

Fig. 2.8. The luminescence decay curve excited at 808 nm at room temperature corresponding to the emission line $^4F_{3/2}\rightarrow{}^4I_{11/2}$ at 1060 nm

Fig.2.10 displays the RT polarized emission spectra (in the **1400–1700 nm** spectral range). The emission spectra obtained under $^4I_{13/2}$ multiplet excitation at **980 nm**. And the fluorescence decay curve of the $^4I_{13/2}\rightarrow{}^4I_{15/2}$ transition is also shown in Fig.2.11. From the curve, the fluorescence lifetime was **2.54ms**.

Fig.2.12 shows the cross sections of the Er^{3+} ions in the CBO crystal. (a) The absorption cross-section, (b) the emission cross-section derived by the reciprocity method (RM), (c) the emission cross-section derived by the Füchtbauer-Ladenburg (FL) formula, and (d) the emission cross-section derived by the modified method. The maximum values of the emission cross section centered at about 1530 nm are $9.67\times10^{-21}cm^2$ for the π spectrum and 7.43×10^{-21} cm^2 for the σ spectrum, which can be compared with those reported for other Er^{3+} doped laser crystals [9.3×10^{-21} cm^2 for Er^{3+}:$LaGaO_3$,[29] 4.5×10^{-21} cm^2 for Er^{3+}:YAG (yttrium aluminum garnet),[30] and 3.1×10^{-21} cm^2 for Er^{3+}:$YAlO_3$ [30]]. The wavelength dependence of the gain cross-section for several values of population inversion P (P = 0,0.1,0.2,…,1) is shown in Fig.2.13. A wide tunable wavelength range from 1530 to 1650 nm is expected when the population inversion P is larger than 0.5, which is encountered in a free-running laser operation. Table 2-4 displays the experimental, theoretical oscillator strengths for Er^{3+} ions in CBO crystal. Table 2-5 presents the comparison of the Judd-Ofelt parameters of Er^{3+}:CBO and other Er^{3+} doped crystals. Generally, the Ω_2 parameter is sensitive to the symmetry of the rare earth site and is strongly affected by covalency between rare earth ions

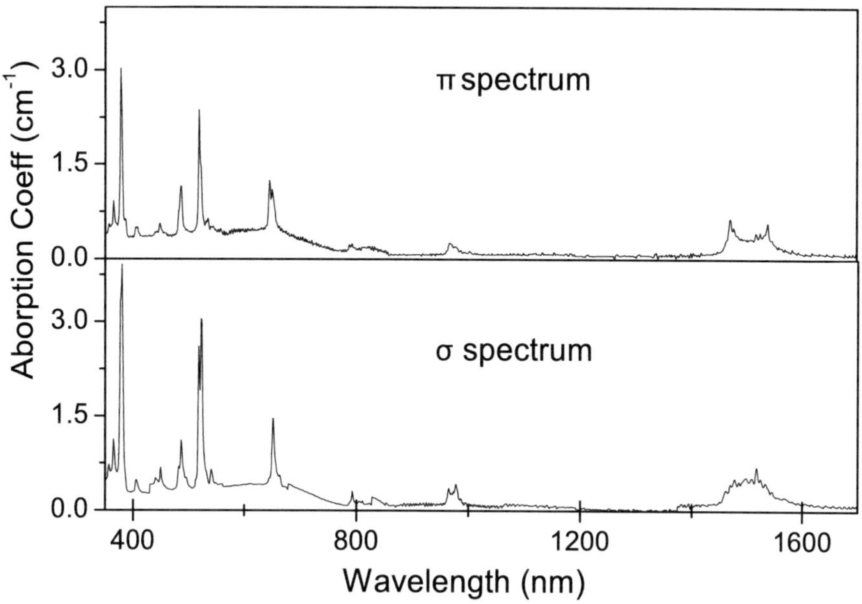

Fig. 2.9. Room temperature (RT) polarized absorption spectra of **Er**$^{3+}$ in the CBO crystal

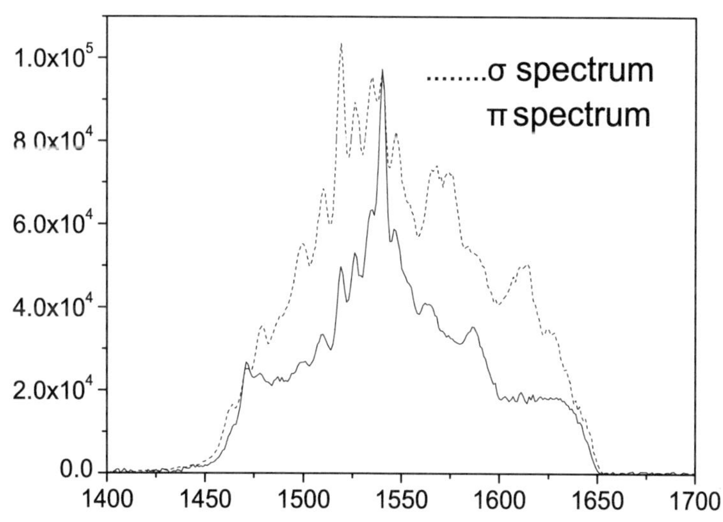

Fig. 2.10. Room temperature polarized emission spectra of the Er^{3+}:CBO single crystal.

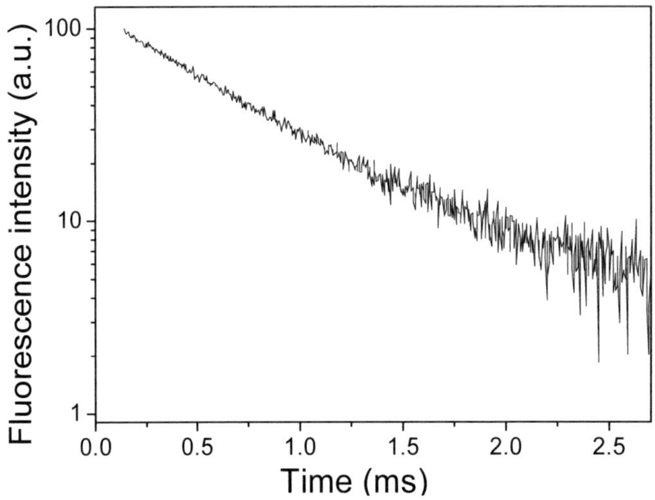

Fig. 2.11. Fluorescence decay curve of the ${}^4I_{13/2}\rightarrow{}^4I_{15/2}$ transition

and ligand anions. In the present work, the parameter Ω_2 is high, but it is lower than those for YVO_4 and $NaBi(WO_4)_2$. This possibly indicated that the Er^{3+} doped CBO crystals are more covalent in character. Table 2-6 displays the radiative transition probabilities $A_{JJ'}$, fluorescence branching ratios $\beta_{JJ'}$, and radiative decay time τ_r for Er^{3+} ions in CBO crystal. From the measured and calculated radiative lifetimes, the luminescent quantum efficiency $\eta=\tau_f/\tau_r$ for the ${}^4I_{13/2}\rightarrow{}^4I_{15/2}$ transition of the Er^{3+} :CBO crystal is found to be approximately 84.4%.

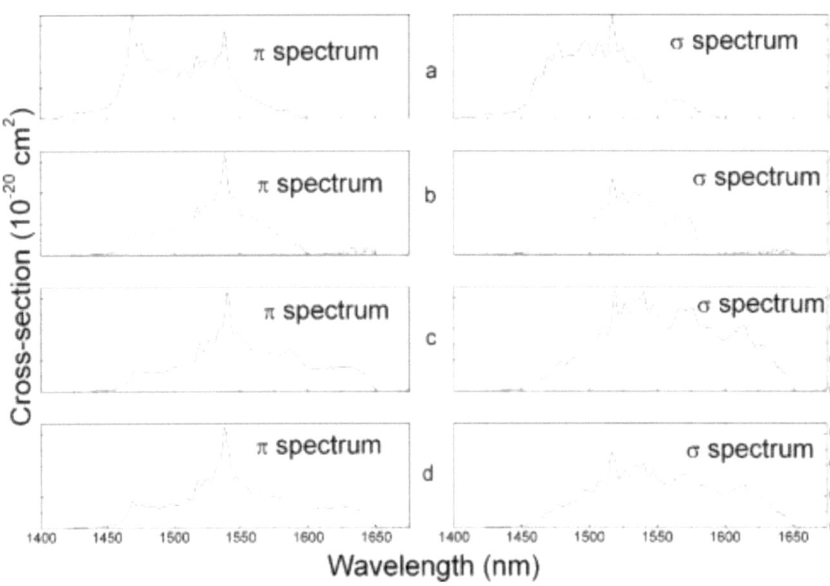

Fig. 2.12. The cross sections of the **Er^{3+}** ions in the **CBO** crystal. (a) The absorption cross-section, (b) the emission cross-section derived by the RM, (c) the emission cross-section derived by the FL, and (d) the emission cross-section derived by the modified method.

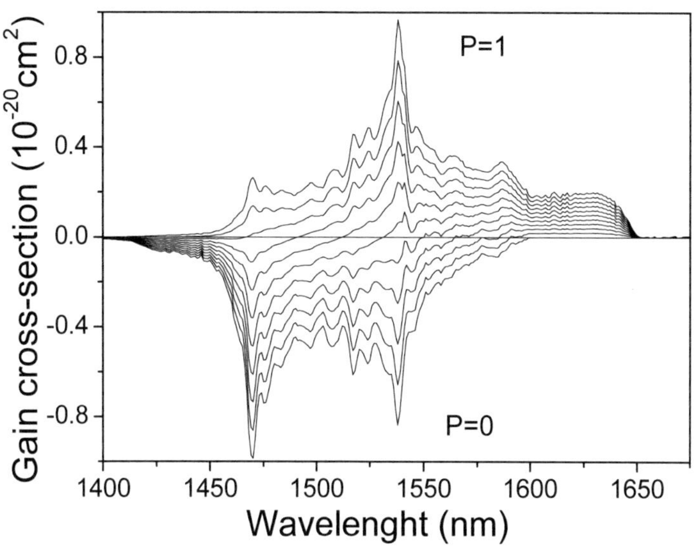

(a)

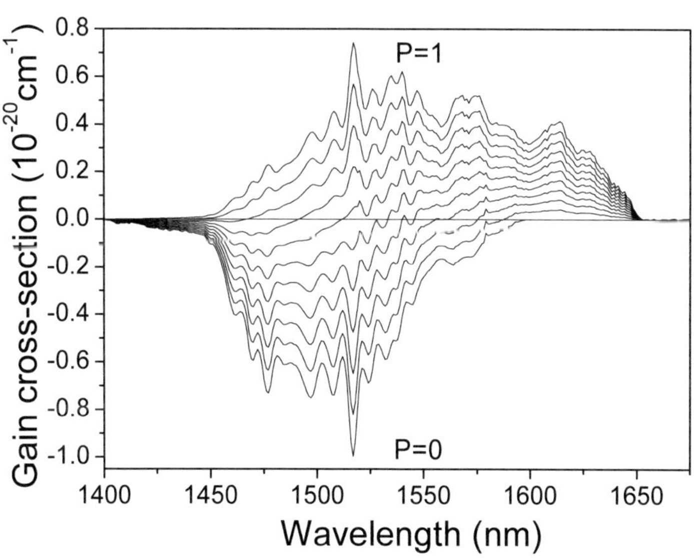

(b)

Fig. 2.13. The wavelength dependence of the gain cross-section for several values of population inversion P (P = 0,0.1,0.2,…,1) (a) for π spectrum (b) for σ spectrum

Transition (from $^4I_{15/2}$)	λ(nm)	the π -polarized f ×10[6]		the σ -polarized f ×10[6]	
		$f_{\exp}$	$f_{ed,th}$	$f_{\exp}$	$f_{ed,th}$
$^4I_{13/2}$	1517	2.73	2.46	3.33	2.98
$^4I_{11/2}$	978	0.72	0.97	1.39	1.26
$^4I_{9/2}$	793	0.56	0.70	0.75	0.82
$^4F_{9/2}$	652	3.82	4.32	5.45	5.74
$^2H_{11/2}$	523	8.73	8.95	12.92	13.07
$^4F_{7/2}$	486	4.17	4.38	5.74	5.62
$^4F_{5/2}$	450	0.91	1.09	1.35	1.33
$^2H_{9/2}$	403	1.48	1.49	1.48	1.57
$^4G_{11/2}$	380	16.36	16.12	26.50	26.12
$^4G_{9/2}$	365	3.85	4.11	3.94	4.08
$^4D_{5/2}$	257	20.37	19.93	26.71	26.45

Table 2.4. Experimental, theoretical oscillator strengths for Er^{3+} ions in CBO Crystal

Hosts	Ω_2 (10^{-20} cm^2)	Ω_2 (10^{-20} cm^2)	Ω_2 (10^{-20} cm^2)	Ref.
$Ca3(BO_3)_2$ Ω_t^{eff}	5.35	4.95	2.29	28
YAG	0.66	0.81	0.71	31
YVO_4	13.45	2.23	1.67	32
$LiYF_4$	1.92	0.26	1.96	33
$NaBi(WO_4)_2$	5.50	1.00	0.71	34
Y_2O_3	4.59	1.21	0.48	35

Table 2.5. Comparison of Judd-Ofelt parameter Ω_2 of Er^{3+}: CBO and other Er^{3+} doped crystals

Transition	$\bar{\lambda}$ (nm)	A_{ed}	A_{md}	$\beta_{JJ'}$ (%)	τ_r (ms)
$^4I_{13/2}\rightarrow{}^4I_{15/2}$	1540	292.11	41.51	100	3.01
$^4I_{11/2}\rightarrow{}^4I_{13/2}$	2822	43.50	8.61	14.6	2.79
$^4I_{15/2}$	982	305.91		85.4	
$^4I_{9/2}\rightarrow{}^4I_{11/2}$	4490	2.37	1.35	0.5	1.46
$^4I_{13/2}$	1704	115.09		16.8	
$^4I_{15/2}$	809	556.95		82.7	
$^4F_{9/2}\rightarrow{}^4I_{9/2}$	3466	6.25	2.96	0.2	0.20
$^4I_{11/2}$	1956	151.25	6.76	3.2	
$^4I_{13/2}$	1142	221.42		4.5	
$^4I_{15/2}$	656	4529		92.1	
$^4S_{3/2}\rightarrow{}^4I_{9/2}$	1666	176.60		4.3	0.24
$^4I_{11/2}$	1215	89.39		2.2	
$^4I_{13/2}$	842	1132		27.4	
$^4I_{15/2}$	545	2736		66.2	

Table 2.6. Radiative transition probabilities $A_{JJ'}$, fluorescence branching ratios $\beta_{JJ'}$, and radiative decay time τ_r for Er^{3+} ions in CBO crystal.

2.3.3 The spectrum characteristics of Dy^{3+}:$Ca_3(BO_3)_2$ crystal

Fig.2.14 shows the room temperature absorption spectrum of Dy^{3+}:$Ca_3(BO_3)_2$ crystal, which consists of nine groups of bands.[36] They are associated with the observed transitions from the $^6H_{15/2}$ ground state. The wavelengths corresponding to the transitions are listed in Table 2-7, which also displays the integrated absorbance, the measured and calculated line strengths of Dy^{3+}: CBO crystal. The room temperature emission spectrum is presented in Fig.2.15, in which there are several bands centered at 479, 575, 663, and 750 nm corresponding to $^4F_{9/2}$→$^6H_{15/2}$, $^6H_{13/2}$, $^6H_{11/2}$, and $^6H_{9/2}$+$^6F_{11/2}$ transitions, respectively. Fig.2.16 displays the fluorescence decay curves of $^4F_{9/2}$→$^6H_{13/2}$ and $^4F_{9/2}$→$^6H_{15/2}$ transitions of Dy^{3+}: CBO crystal excited at 397 nm, from which the fluorescence lifetime of the $^4F_{9/2}$ of Dy^{3+} in CBO is calculated to be about 1.275 ms. Table 2-8 presents the intensity parameters of Dy^{3+}: CBO crystal and the comparison between the intensity parameters of Dy^{3+} doped in some other laser crystals and in CBO. The spectroscopic quality factor $X=\Omega_4/\Omega_6$ for the Dy^{3+} in CBO is determined to be 2.982. This value for X suggests that Dy^{3+}: CBO is a promising material for efficient laser action when compared with Dy^{3+}: YVO_4 for which X=2.132. Table2-9 displays the experimental and calculated oscillator strengths for absorption $^6H_{15/2}$ ground state of Dy^{3+} ion in CBO crystal. Table 2-10 demonstrates the calculated radiative transition rate, the branching ratios, and the radiative lifetime for the emission from the $^4F_{9/2}$ level of Dy^{3+}: CBO. The fluorescence branching ratio is a critical laser parameter because it also characterizes the possibility of attaining stimulated emission from a specific transition. The $^4F_{9/2}$→$^6H_{13/2}$ transition has large value of the branching ratios, thereby suggesting that this transition may result in the strongest laser action. This work gives a consistent optical characterization of Dy^{3+}: CBO crystal, which may realize a yellow solid state laser device.

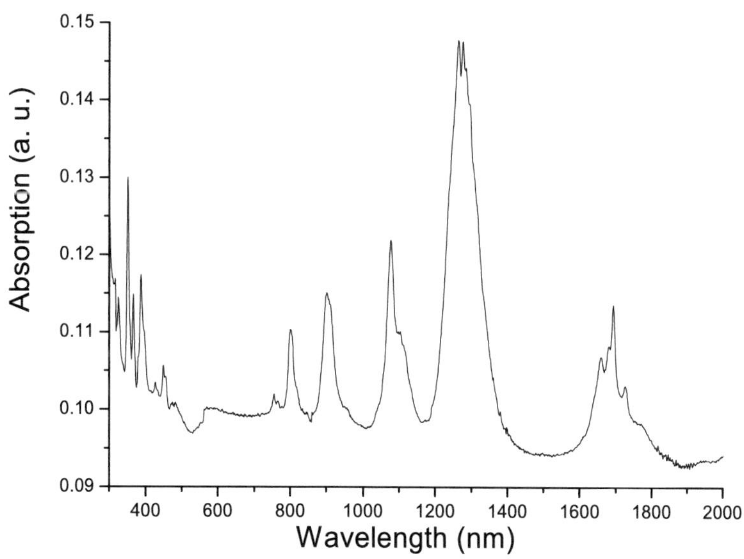

Fig. 2.14. The room temperature absorption spectrum of CBO: Dy^{3+} (sample thickness=0.306 cm).

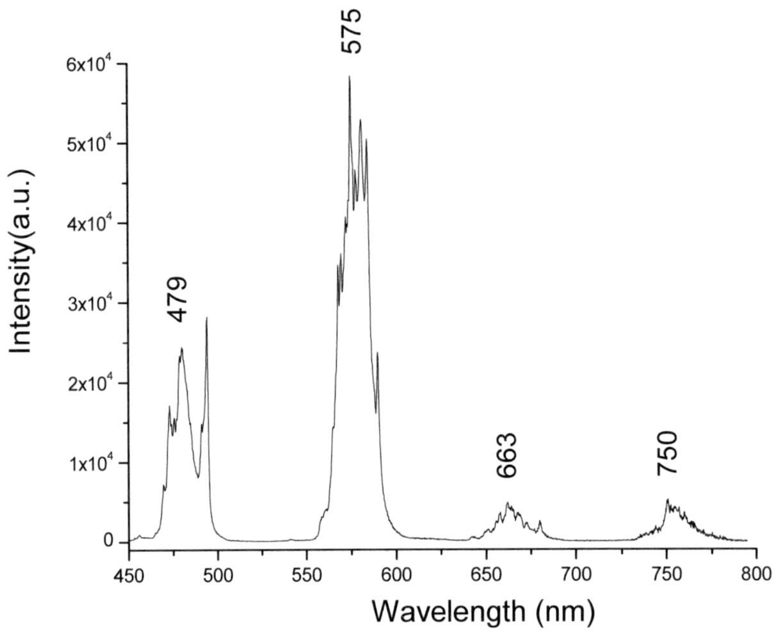

Fig. 2.15. The room temperature emission spectrum of Dy^{3+}: CBO (excited by 397 nm).

Excited state	Wavelength (nm)	Γ (nm/cm)	S_{mea} ($10^{-20}cm^2$)	S_{cal} ($10^{-20}cm^2$)
$^6H_{11/2}$	1692	2.755	0.384	0.949
$^6F_{11/2}$, $^6H_{9/2}$	1262	35.411	6.619	6.564
$^6H_{7/2}$, $^6F_{9/2}$	1076	1.823	0.399	0.026
$^6F_{7/2}$	901	3.841	1.001	0.698
$^6F_{5/2}$	800	1.244	0.364	0.218
$^4G_{11/2}$	449	0.327	0.167	0.029
$^4M_{21/2}$, $^4K_{17/2}$	386	1.511	0.891	0.070
$^4I_{11/2}$	366	0.440	0.272	0.512
$^6P_{7/2}$	351	1.291	0.830	0.978

Table 2.7. The integrated absorbance, the measured and calculated line strengths of Dy^{3+}: CBO crystal

Crystals	Ω_2	Ω_4	Ω_6	Reference
CBO	5.216	1.858	0.623	This work
$Y_3Sc_2Ga_3O_{12}$	0.134	0.7261	0.61	36
$LiYF_4$	2.01	1.34	2.39	11
YVO_4	6.59	3.71	1.74	37
$YAl_3(BO_3)_4$	10.04	2.04	2.31	38
$KY(WO_4)_2$	23.24	3.329	2.359	1

Table 2.8. Comparison between the intensity parameters of Dy^{3+} doped in some laser crystals and in CBO (Ω_t are in units of 10^{-20} cm^2)

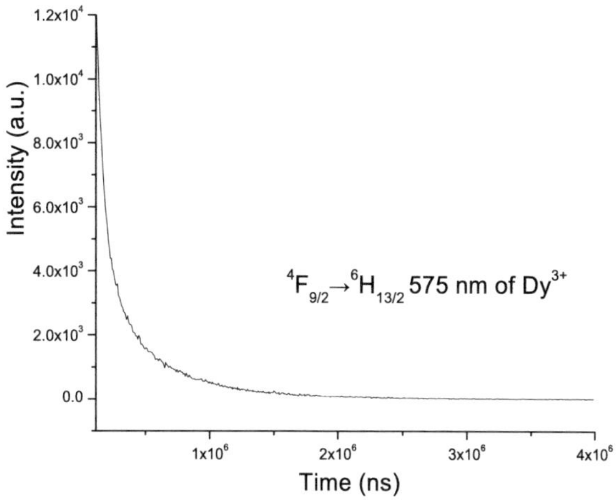

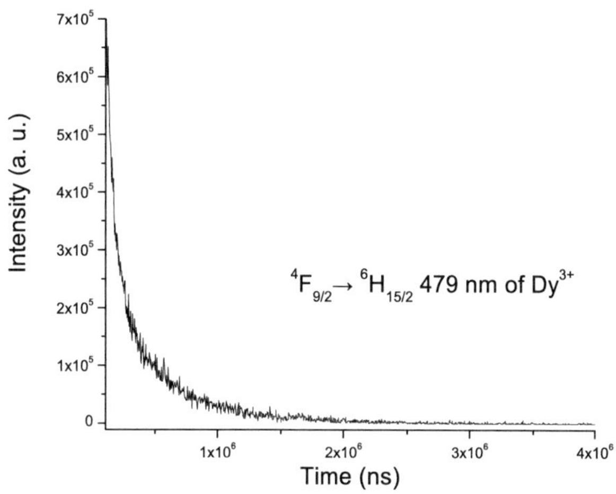

Fig. 2.16. The fluorescence decay curves of $^4F_{9/2}\rightarrow{}^6H_{13/2}$ and $^4F_{9/2}\rightarrow{}^6H_{15/2}$ transitions of Dy^{3+}: CBO crystal (by 397 nm excited)

Upper state	f_{exp}	f_{cal}	Δf
$^6H_{11/2}$	0.243	0.601	0.358
$^6F_{11/2}$, $^6H_{9/2}$	5.633	5.586	−0.047
$^6H_{7/2}$, $^6F_{9/2}$	0.398	0.026	−0.372
$^6F_{7/2}$	1.096	0.834	−0.262
$^6F_{5/2}$	0.491	0.290	−0.201
$^4G_{11/2}$	0.410	0.072	−0.338
$^4M_{21/2}$, $^4K_{17/2}$	2.563	0.202	-2.361
$^4I_{11/2}$	0.830	0.016	−0.814
$^6P_{7/2}$	2.648	3.120	0.472

Table 2.9. The experimental and calculated oscillator strength f (×10^6) for absorption $^6H_{15/2}$ ground state of Dy^{3+} ion in CBO crystal. Also the Δf (×10^6) between the calculated and experimental oscillator (note: rms f=0.883×10^{-6})

Start levels	Wavelength (nm)	A (s^{-1})	β (A)	τ (μs)
$^6H_{15/2}$	479	90.209	0.115	1275
$^6H_{13/2}$	575	612.77	0.781	
$^6H_{11/2}$	663	66.561	0.085	
$^6H_{9/2}+^6F_{11/2}$	750	15.066	0.019	

Table 2.10. The calculated radiative transition rate, the branching ratios, and the radiative lifetime for the emission from the $^4F_{9/2}$ level of Dy^{3+}: CBO

3. Rare earth-doped $Ca_3Re_2(BO_3)_4$ [Re=Y,Gd] crystals

3.1 The crystal structure

An ORTEP drawing of the structure fragment of crystal Yb:$Ca_3Y_2(BO_3)_4$ is shown in Fig.3.1a. Fig.3.1b shows the packing diagram of cell units of Yb:$Ca_3Y_2(BO_3)_4$ crystal.[6] In Yb^{3+}:$Ca_3Y_2(BO_3)_4$ crystal structure, cations occupy three independent sites statistically, which is similar to $Ca_3La_2(BO_3)_4$[39] and $Ba_3La_2(BO_3)_4$[40]. The basic structure of Yb:$Ca_3Y_2(BO_3)_4$ is composed of three sets of M-oxygen distorted polyhedrons, and three sets of BO_3 planar triangles. Ca^{2+} and Y^{3+} ions occupy three independent sites statistically. M1、M2 and M3 were suggested to stand for these three independent sites respectively. They are coordinated by eight oxygen ions to form the distorted polyhedron. From the value of the electronic density of each independent site, and the ion charges of Ca^{2+} and Y^{3+} ions as well as the ratio of their atomic number in the formula, we can calculate their ratio in each site. Concretely, the method for calculating the ratio of Yb/Ca is as follows: we suggest the average atomic number of M_n atom:

$$\overline{Z}n = Z_Y X_n + Z_{Ca} Y_n \tag{3.1}$$

Here the atomic number of Y and Ca are: $Z_Y = 39, Z_{Ca} = 20$, and X_n is the occupancies of Y^{3+} ion in M_n, Y_n is the occupancies of Ca^{2+} ion M_n,

$$X_n + Y_n = 1 \tag{3.2}$$

As we know:

$$\frac{\overline{Z_n}}{\overline{Z_{n+1}}} = \frac{\rho_n}{\rho_{n+1}} \tag{3.3}$$

in which ρ is the electronic density of M_n. According the formula $Ca_3Y_2(BO_3)_4$, we can get:

$$X_1 + X_2 + X_3 = 2 \tag{3.4}$$

$$Y_1 + Y_2 + Y_3 = 3 \tag{3.5}$$

Combing all the above equations, we can calculate the ratio of Y/Ca in the three sites M1, M2 and M3,the results are as follows: M1=0.61Y+0.39Ca, M2=0.445Y+0.555Ca and M3=0.25Y+0.75Ca.Yb^{3+} ions substitute Y^{3+} ions entering these three lattices. The fact that the statistical distribution of Ca^{2+}, $Y^{3+}$and Yb^{3+} ions might lead to the increase of width of spectra of this crystal. As a matter of fact, this was confirmed by the next part of this report. Table 3-1 presents the atomic coordinates and thermal parameters.

	x	y	z	Wyckoff	U(eq)
M(1)	1750(3)	2500	9687(2)	4c	20(1)
M(2)	-193(2)	4164(1)	6788(2)	8d	30(1)
M(3)	-1999(3)	3738(1)	11589(2)	8d	32(1)
O(1)	4458(11)	3275(5)	10709(10)	8d	40(20)
O(2)	1200(20)	2500	12380(17)	4c	40(30)
O(3)	-1536(18)	2500	9990(14)	4c	40(30)
O(4)	1971(15)	4849(7)	5098(18)	8d	40(40)
O(5)	-970(20)	5456(9)	8143(15)	8d	40(50)
O(6)	-2537(18)	3256(8)	7768(15)	8d	40(30)
O(7)	1120(30)	3993(9)	9410(20)	8d	40(80)
B(1)	5110(30)	2500	11350(20)	4c	18(4)
B(2)	2790(20)	2500	6470(20)	4c	13(3)
B(3)	3267(19)	5404(9)	5432(15)	8d	21(3)

U(eq) is defined as one third of the trace of the orthogonalized Uij tensor. The Ca^{2+} and Y^{3+} ions coexist in M1, M2 and M3 positions statistically.

Table 3.1. Atomic coordinates (× 10^4) and thermal parameters (Å^2× 10^3)

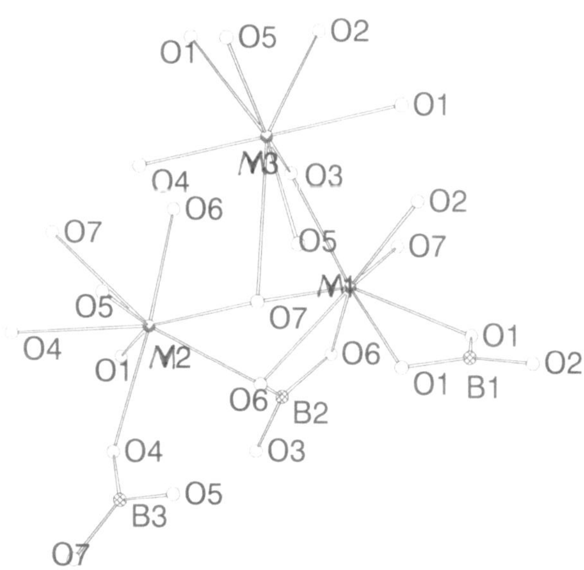

Fig. 3.1a. A structure fragment of crystal Yb:$Ca_3Y_2(BO_3)_4$

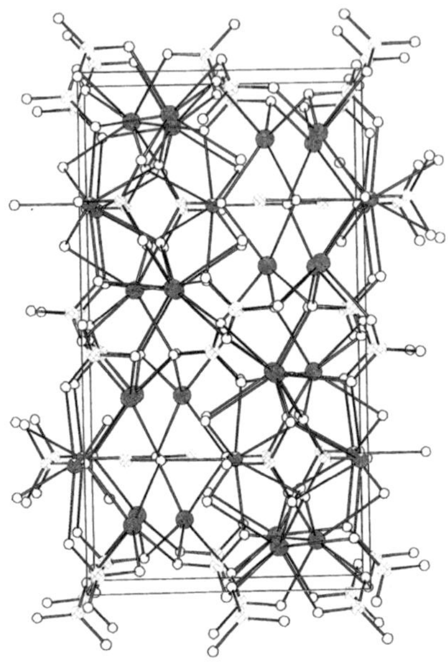

Fig. 3.1b. Packing diagram of cell units of Yb:$Ca_3Y_2(BO_3)_4$

3.2 The crystal growth

The crystal was grown by Czochralski method. The crystal growth was carried out in a DGL-400 furnace with KGPF25-0.3-2.5 power supply of intermediate frequency. An Ir-crucible of 60mm diameter by 35 mm high was used.

The initial compounds for the synthesis were AR grade $CaCO_3$, H_3BO_3 and 4N Y_2O_3,Yb_2O_3. A $Ca_3Y_2(BO_3)_4$ compound with 5mol%Yb_2O_3-doped was synthesized according to the following reactions:

$$6CaCO_3+0.05Yb_2O_3+1.95Y_2O_3+8H_3BO_3 \rightarrow 2Ca_3Yb_{0.05}Y_{1.95}(BO_3)_4+6CO_2\uparrow+12H_2O\uparrow$$

Thoroughly mixed and pressed mixtures of the stoichiometric composition were slowly heated to 500^{0}C at a rate of 50^{0}C/h and further to the synthesis temperature at the rate of 150^{0}C/h in a Pt-crucible. Then the sintered compound was melt in the Ir-crucible under N_2 atmosphere, at a temperature which was 50^{0}C higher than the crystallization temperature, and was kept at this temperature for one hour. Seeding was performed on the Ir-wire. The nitrogen gas pressure is 0.04MPa. Pulling rate and the rotation rate was 1.3~1.5mm/h and 12-20 r.p.m, respectively. When the growth process was ended, the crystal was drawn out of the melt surface and cooled down to room temperature at a rate of 10~30^{0}C/h. The transparent single crystals with a size up to φ20 mm×55 mm was obtained (as shown in Fig.3.2). Fig.3.3 shows the interference fringe of the grown Yb^{3+}:$Ca_3Y_2(BO_3)_4$ crystal, the optical homogeneity is 4×10^{-5}, it means the crystal has excellent quality. Table 3-2 presents the parameters of crystal growth. In order to estimate the solubility of the doping ion in the crystal, it is customary to use an "effective segregation coefficient" defined as [41]: $k_e=C_s/C_l$, C_s is the doped-ion concentration in the crystal and C_l is the doped-ion concentration in the melt, since the concentration of Yb^{3+} ion in Yb^{3+}:CYB was measured to be 1.56wt% by electron probe microanalysis method ,the effective segregation coefficient of Yb^{3+} ion in CYB crystal was calculated to be 0.97.

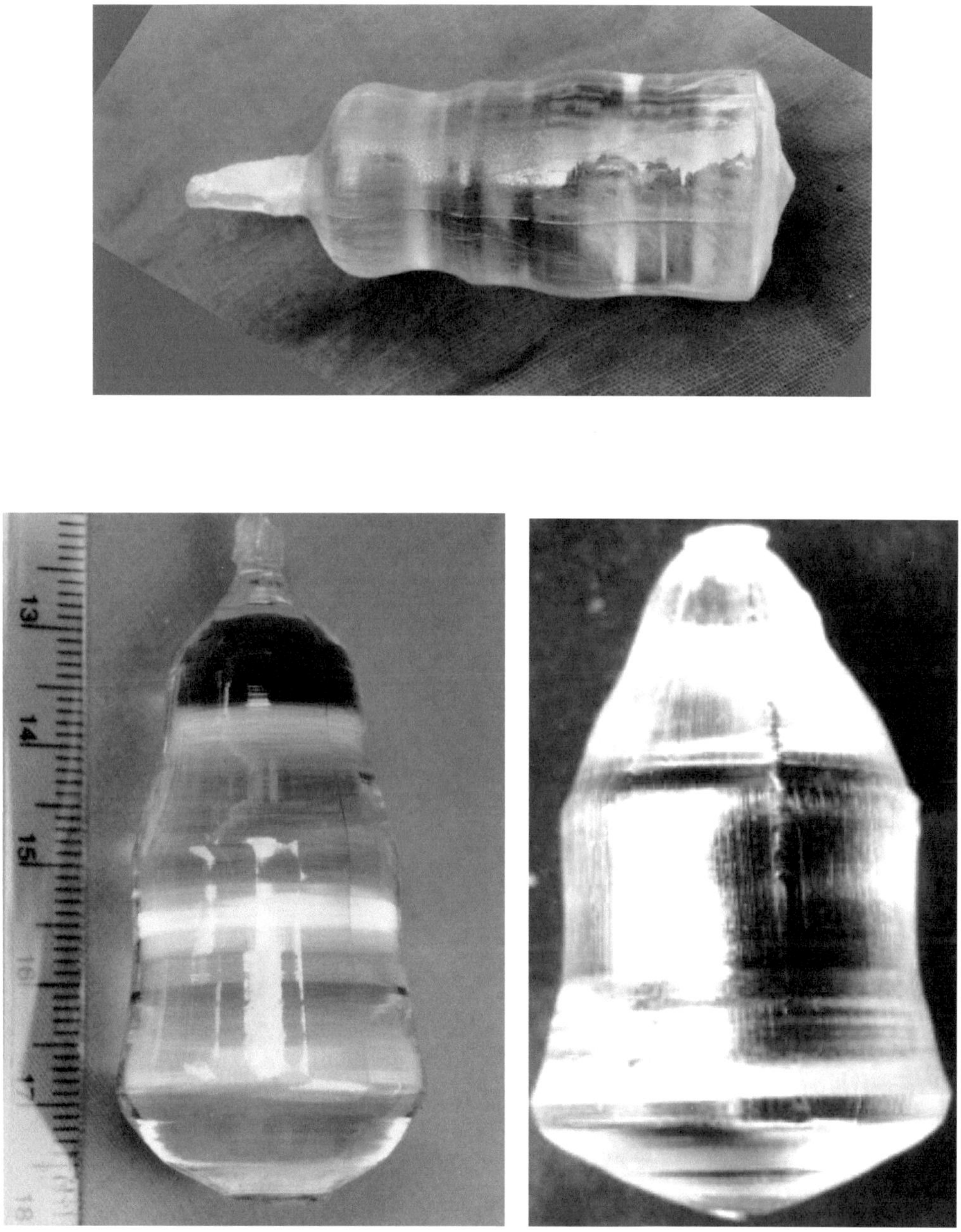

Fig. 3.2. The grown Yb^{3+}:$Ca_3Y_2(BO_3)_4$ crystal

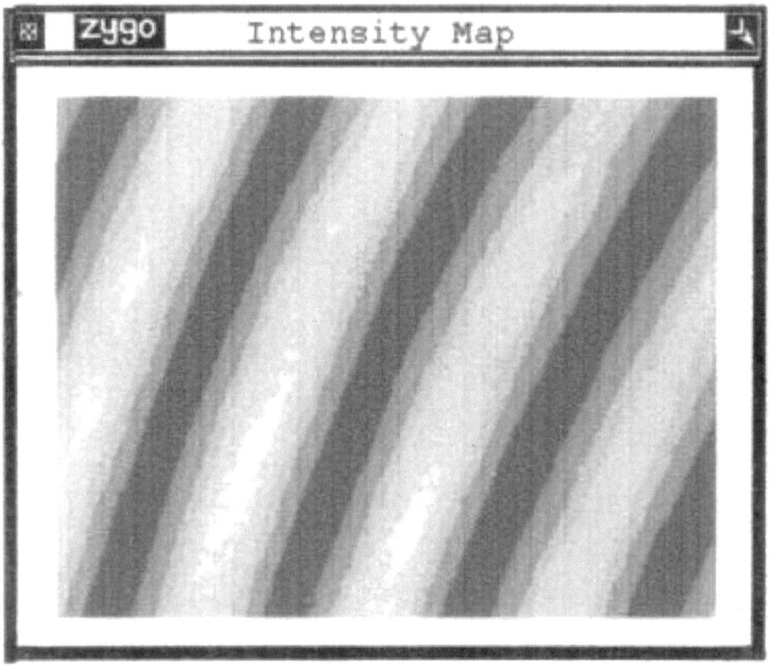

Fig. 3.3. Interference fringe of Yb^{3+}:$Ca_3Y_2(BO_3)_4$ crystal.

Nitrogen gas (MPa)	0.04	Rotate rate(rpm)	12-20
Soaping temp. (^{0}C)	1460	Decreasing rate of temp.(^{0}C/day)	2~10
Soaping time (h)	0.5	Pulling rate(mm/h)	1.3~1.5
Crucible size(mm)	φ60mm×35mm	Annealing rate(^{0}C/h)	10~50
Seeding temp. (^{0}C)	1410	Crystal size(mm)	φ20 mm×55 mm

Table 3.2. The parameters of crystal growth.

3.3 The spectrum characteristics

3.3.1 The spectrum characteristics of Tm^{3+}:$Ca_3Y_2(BO_3)_4$ crystal

Fig.3.4 presents the absorption spectrum of $Ca_3Y_2(BO_3)_4$:Tm^{3+}crystal, in which there are seven peaks centered at 1684nm, 1215 nm, 792.4 nm, 687.9nm, 475.6 nm, 359.9 nm, 262.9nm, corresponding to the transitions from $^3H_6$ to$^3F_4$, $^3H_5$, $^3H_4$, $^3F_3$, $^1G_4$, $^1D_2$, $^1I_6$, respectively[42]. The FWHM at 792.4 nm is about 33.9 nm and the cross-section is about 1.5×10^{-21} cm^2, which is benefit to the pumping of commercial laser diode. The room temperature emission spectrum of $Ca_3Y_2(BO_3)_4$:Tm^{3+} crystal excited at 792.4 nm is presented in Fig.3.5, in which there is a broad emission band ranged from 1332.6 nm to 1429.1 nm. The FWHM of this emission band is 63.6 nm, which is resulted from the statistical distribution of Ca^{2+}, $Y^{3+}$and Tm^{3+} ions.

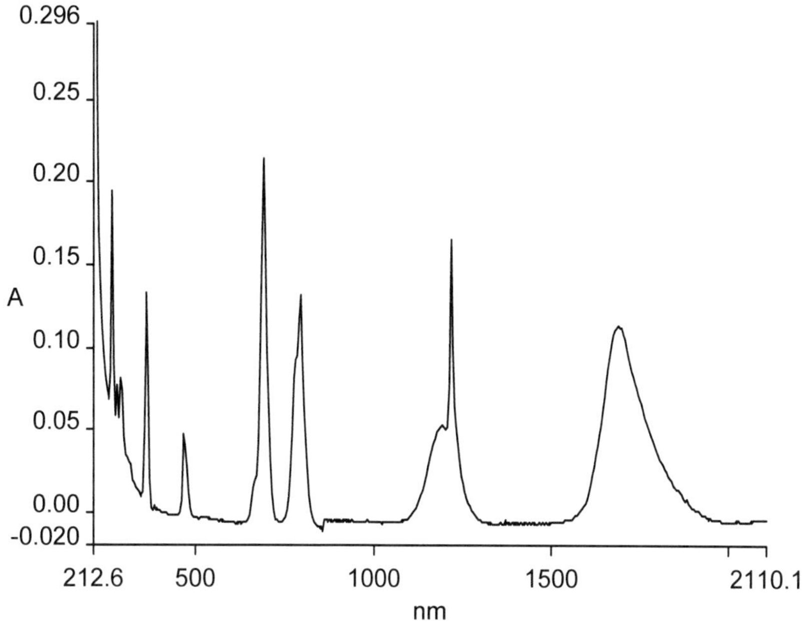

Fig. 3.4. The absorption spectrum of $Ca_3Y_2(BO_3)_4$:Tm^{3+}crystal

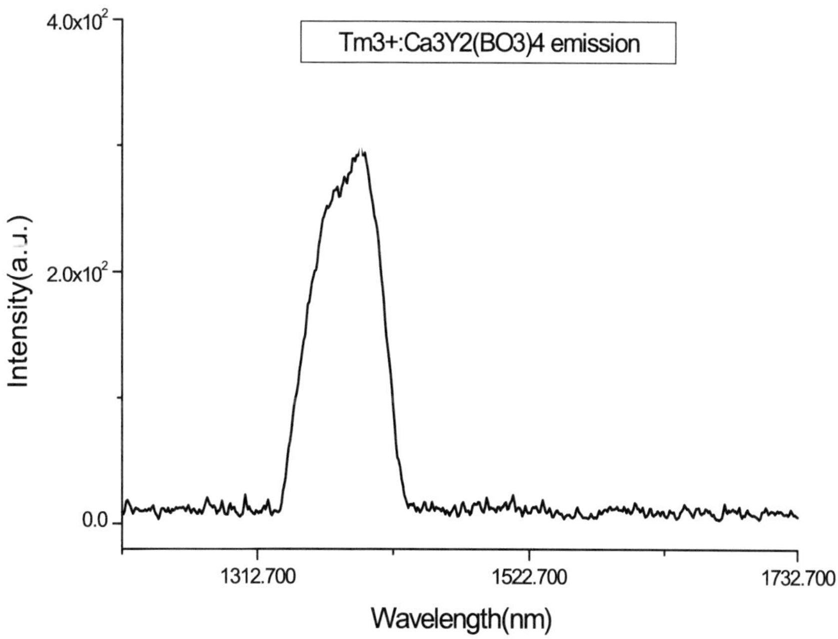

Fig. 3.5. The room temperature emission spectrum of $Ca_3Y_2(BO_3)_4$:Tm^{3+} crystal excited at 792nm

3.3.2 The spectrum characteristic of Er^{3+}:$Ca_3Y_2(BO_3)_4$ crystal

Fig.3.6 presents the absorption spectrum of $Ca_3Y_2(BO_3)_4$:Er^{3+}crystal, in which there are twelve peaks centered at 244.6 nm, 257.3 nm, 366.4 nm, 379.9 nm, 408.1 nm, 452 nm, 522.7 nm, 544.2 nm, 654.3 nm,799.3 nm,976.6 nm,1518.4nm, corresponding to the transitions from $^4I_{15/2}$ *to* $^2I_{11/2}$, $^4D_{5/2}$+$^4D_{7/2}$, $^4G_{9/2}$, $^2G_{9/2}$+$^4F_{9/2}$+$^2H_{9/2}$, $^4F_{5/2}$, $^2H_{11/2}$, $^4S_{3/2}$, $^4F_{9/2}$, $^4I_{9/2}$, $^4I_{13/2}$, respectively[43]. The room temperature emission spectrum of $Ca_3Y_2(BO_3)_4$:Er^{3+} crystal excited at 530 nm is presented in Fig.3.7, in which there is a broad emission band ranged from 1429.4 nm to 1662.8 nm. The FWHM of this emission band is 126 nm. The factor contributing to this broad emission is the disordered structure of the crystal, namely, Ca^{2+} and Y^{3+} ions are statistically situated in three different lattices in the crystal structure determined by us. This broad emission will benefit the energy storage. Therefore, this crystal should be useful as a tunable infrared (in the eye-safe region at 1.54 μm) laser crystal.

Based on the measured absorption spectra of three commutative perpendicularity directions and J-O theory, the J-O parameters are calculated to be Ω_2=1.214×10^{-20} cm^2, Ω_4=1.585×10^{-20} cm^2, Ω_6=1.837×10^{-20} cm^2. The oscillator strength, radiative transition probability A, radiative lifetime τ_{rad} and the fluorescent branching ratio β are also calculated, which are shown in Table 3-3~4. The stimulated emission cross-section at *1535nm* is 6.4×10^{-21} cm^2 and the integral cross-section at 1535 nm is 2.7×10^{-18} cm^2. The lifetime measured is about *792μs* with the luminescent quantum efficiency of *20.8%*.

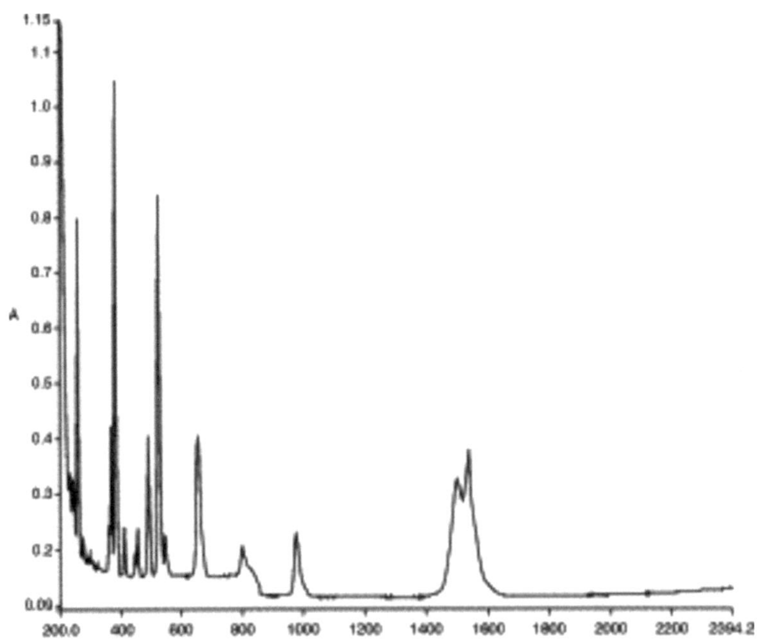

Fig. 3.6. The absorption spectrum of $Ca_3Y_2(BO_3)_4$:Er^{3+}crystal

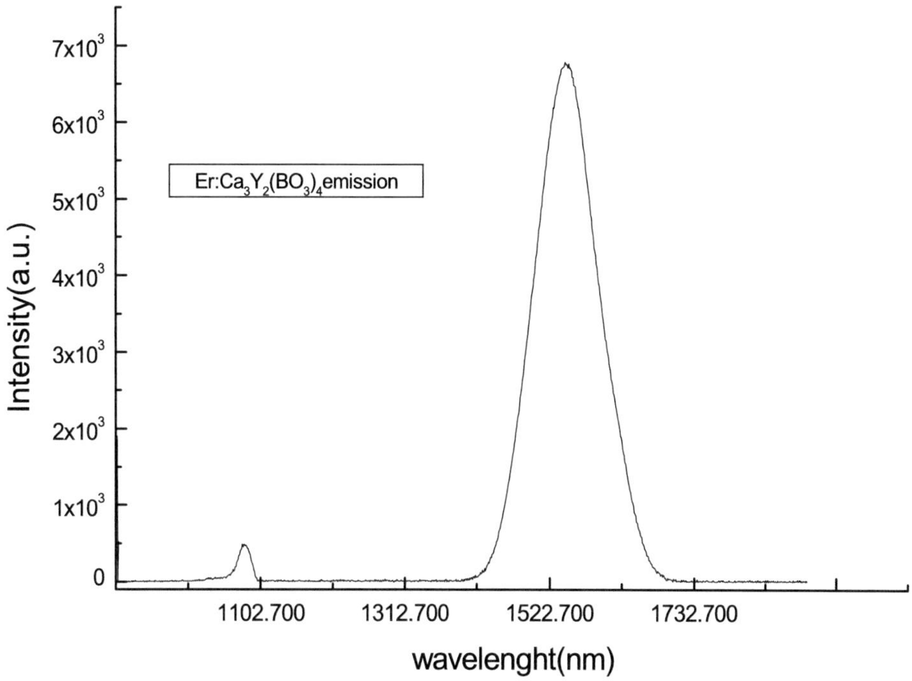

Fig. 3.7. The room temperature emission spectrum of $Ca_3Y_2(BO_3)_4{:}Er^{3+}$ crystal excited at 530 nm

Wavelength λ(nm)	$f_{exp}\times10^6$	$f_{cal}\times10^6$
258	14.380	0.213
368	3.276	6.595
381	11.034	0.018
453	1.261	1.015
491	1.895	3.161
524	6.618	3.614
654	2.926	2.910
800	0.681	0.411
974	0.728	0.876
1534	2.219	2.076
rms$\Delta f=5.864\times10^{-6}$		

Table 3.3. Oscillator strengths of 7 at% Er^{3+} in $Ca_3Y_2(BO_3)_4$ crystal

Transtion J→J'	λ(nm)	$A^{ed}(s^{-1})$	$A^{md}(s^{-1})$	$A^{total}(s^{-1})$	β	τ_{rad}(ms)
$^4I_{13/2}\rightarrow {}^4I_{15/2}$	1535	216.5	46.2	262.7	1	3.807
$^4I_{11/2}\rightarrow {}^4I_{13/2}$	2822	27.489	9.507	362.52	0.102	2.758
$^4I_{15/2}$	976	325.526			0.898	
$^4I_{9/2}\rightarrow {}^4I_{13/2}$	1694	85.45		401.96	0.213	2.488
$^4I_{15/2}$	800	316.512			0.787	
$^4F_{9/2}\rightarrow {}^4I_{9/2}$	3554	3.436	3.03	2519.73	0.003	0.397
$^4I_{11/2}$	1933	126.801	7.718		0.053	
$^4I_{13/2}$	1147	84.96			0.034	
$^4I_{15/2}$	653	2293			0.91	
$^4S_{3/2}\rightarrow {}^4I_{9/2}$	1693	98.665		3366	0.029	0.297
$^4I_{11/2}$	1210	73.585			0.022	
$^4I_{13/2}$	847	921.885			0.274	
$^4I_{15/2}$	545	2272			0.675	
$^2H_{9/2}\rightarrow {}^4F_{9/2}$	1063	53.286	51.595	5412	0.019	0.185
$^4I_{9/2}$	818	27.509	1.162		0.005	
$^4I_{11/2}$	686	674.786	42.768		0.133	
$^4I_{13/2}$	552	2483			0.459	
$^4I_{15/2}$	405	2078			0.384	
$^4G_{11/2}\rightarrow {}^2H_{11/2}$	1369	42.342	14.366	24416.8	0.002	0.041
$^4F_{9/2}$	894	801.92	3.505		0.022	
$^4I_{9/2}$	714	300.563	0.851		0.009	
$^4I_{11/2}$	611	695.246	0.106		0.019	
$^4I_{13/2}$	502	2720	49.515		0.077	
$^4I_{15/2}$	380	19786			0.81	

Table 3.4. Radiative transition probability A, radiative lifetime τ_{rad} and fluorescent branching ratio β of Er^{3+} in 7 at%Er^{3+}:$Ca_3Y_2(BO_3)_4$ crystal at room temperature.

3.3.3 The spectrum characteristic of Yb^{3+}:$Ca_3Y_2(BO_3)_4$ crystal

Fig.3.8 presents the absorption spectrum of $Ca_3Y_2(BO_3)_4$:Yb^{3+}crystal, in which there is a broad absorption band ranged from 850 nm to 1000 nm, corresponding to the transition from $^2F_{7/2}\rightarrow{}^2F_{5/2}$. [6] The FWHM at 977 nm is about 12 nm and the cross-section is about $1.9\times10^{-20}cm^2$, which is benefit to the pumping of commercial laser diode. Table.3-5 shows the absorption properties of some ytterbium-doped compounds. The room temperature emission spectrum of $Ca_3Y_2(BO_3)_4$:Yb^{3+} crystal excited at 977 nm is presented in Fig.3.9, in which there is a broad emission band ranged from 927.95 nm to 1102.7 nm. The FWHM of this emission band is 98 nm and its peak value is located at 1025 nm, which is resulted from the statistical distribution of Ca^{2+}, $Y^{3+}$and Yb^{3+} ions.

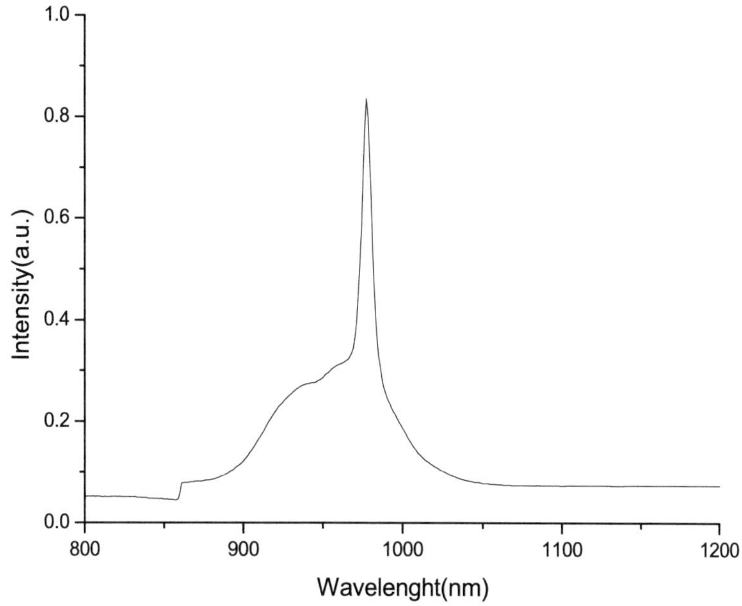

Fig. 3.8. The Absorption spectrum of Yb^{3+} $Ca_3Y_2(BO_3)_4$ crystal at room temperature

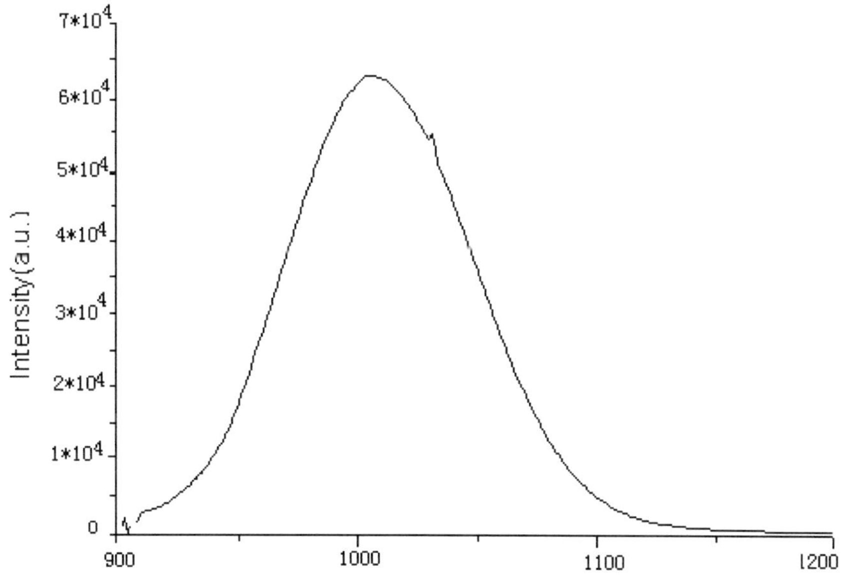

Fig. 3.9. The room temperature emission spectrum of $Ca_3Y_2(BO_3)_4$:Yb^{3+} crystal excited at 977 nm

Compound	λ_p(nm)	σ_a (×10-20 cm2)	FWHM(nm)	References
5at%Yb:CYB	977	1.9	12	[6]
2at%Yb:YCOB	976.1	0.94	3	[44]
1at%Yb:Y_2SiO_5	977.4	2.1	7	[45]
1at%Yb:Sc_2SiO_5	979.5	1.9	4	[45]
Yb:GdCOB	976	0.87	2.6	[46]
15%Yb:YAG	968	0.94	2.6	[47]
15%Yb:BLuB	966	0.29	6.4	[47]
Yb:YAB	975	3.4	3	[48]

Table 3.5. Absorption properties of the ytterbium-doped compounds

3.3.4 The spectrum characteristic of Nd^{3+}:$Ca_3Gd_2(BO_3)_4$ crystal

Based on the absorption spectra of Nd^{3+}:$Ca_3Gd_2(BO_3)_4$ crystal and the measured absorption spectra in three commutative perpendicularity directions and J-O theory, the J-O parameters are calculated to be Ω_2= 2.076×10^{-20} cm^2, Ω_4=4.252×10^{-20} cm^2, Ω_6=5.342×10^{-20} cm^2 [55]. The emission spectra of Nd^{3+}:$Ca_3Gd_2(BO_3)_4$ crystal demonstrates that there are three main emission peaks under excited of 808nm, centered at 912nm, 1064nm and 1337nm, respectively. The stimulated emission cross-section at 1064 nm corresponding to $^4F_{3/2} \rightarrow ^4I_{11/2}$ transition is 2.28×10^{-20} cm^2. The fluorescence decay curve displays that the measured lifetime of $^4F_{3/2}$ level is 175 μs, and the quantum efficiency is estimated to be 72%.

3.3.5 The spectrum characteristic of Er^{3+}:$Ca_3Gd_2(BO_3)_4$ crystal

Fig.3.10 presents the absorption spectrum of $Ca_3Gd_2(BO_3)_4$:Er^{3+}crystal, in which there are twelve peaks centered at 244.6nm, 257.3 nm, 366.4 nm, 379.9nm, 408.1 nm, 452 nm, 522.7nm, 544.2 nm, 654.3 nm, 799.3 nm, 976.6 nm, 1518.4nm, corresponding to the transitions from $^4I_{15/2}$ to $^2I_{11/2}$, $^4D_{5/2}$+$^4D_{7/2}$, $^4G_{9/2}$, $^2G_{9/2}$+$^4F_{9/2}$+$^2H_{9/2}$, $^4F_{5/2}$, $^2H_{11/2}$, $^4S_{3/2}$, $^4F_{9/2}$, $^4I_{9/2}$, $^4I_{13/2}$, respectively[56]. The room temperature emission spectrum of $Ca_3Gd_2(BO_3)_4$:Er^{3+} crystal excited at 530nm is presented in Fig.3.11, in which there is a broad emission band ranged from 1460 nm to 1600 nm. The FWHM of this emission band is 126 nm, which is resulted from the statistical distribution of Ca^{2+}, $Gd^{3+}$and Er^{3+} ions. Based on the measured absorption spectra of three commutative perpendicularity directions and J-O theory, the J-O parameters are calculated to be Ω_2= 4.01×$10^{-20}cm^2$, Ω_4=0.98×$10^{-20}cm^2$, Ω_6=1.72×$10^{-19}cm^2$. Comparing the parameters with those of the other Er^{3+} doped crystal, we found that the parameters are larger. The oscillator strength, radiative transition probability A, radiative lifetime τ_{rad} and the fluorescent branching ratio β are also calculated, which are shown in Tables 3-6~3-8. The stimulated emission cross-section at 1535nm is calculated to be 6.0×10^{-21}cm^2. Fig.3.12 shows the fluorescence lifetime of Er:CGB crystal under the excitation of 530 nm. The lifetime measured is about 792μs, so the luminescent quantum efficiency of the $^4I_{13/2}$ manifold is estimated to be 20%.

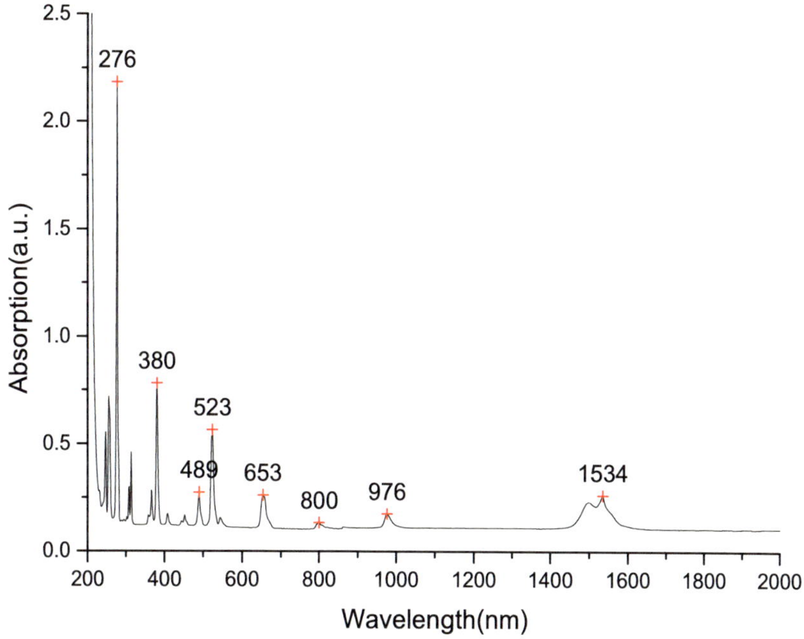

Fig. 3.10. Absorption spectrum of Er:CGB crystal in random direction at RT

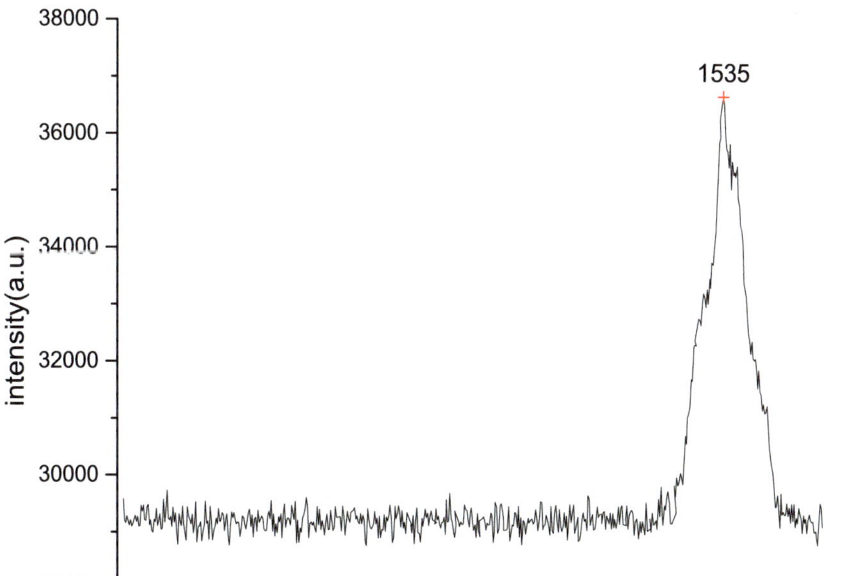

Fig. 3.11. The fluorescence spectrum of Er^{3+}:CGB under the excitation of 530nm

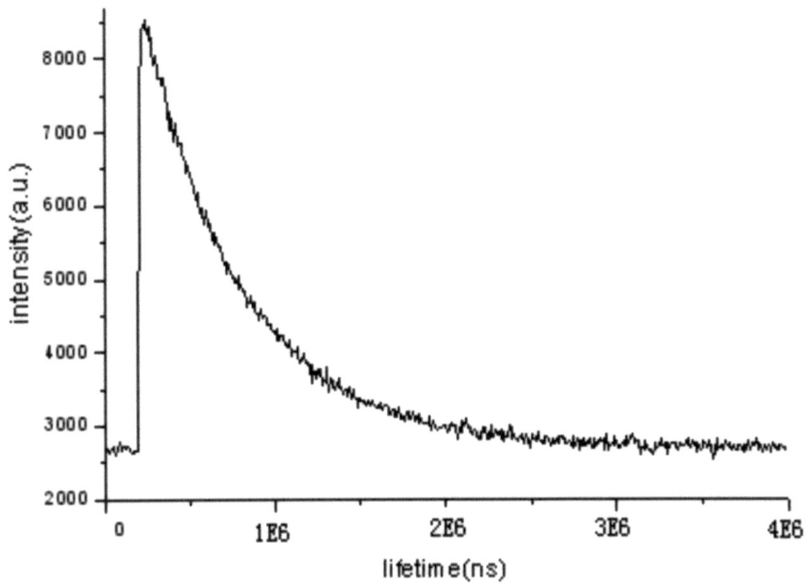

Fig. 3.12. The fluorescence lifetime of Er:CGB crystal under the excitation of 530nm

Wavelength (nm)	$f_{exp}\times10^6$	$f_{cal}\times10^6$
380	13.969	12.976
489	2.343	2.797
523	6.415	7.33
653	2.3	2.261
800	0.241	0.261
976	0.587	0.91
1534	2.053	1.938 ed(a)
		0.54 md(a)
	rmsΔf =5.54×10^{-7}	

(a): ed and md denote electric-dipole and magnetic-dipole transitions respectively.

Table 3.6. Oscillator strengths of Er^{3+} in $Ca_3Gd_2(BO_3)_4$ crystal

Crystals	$\Omega_2\times10^{20}$ cm^2	$\Omega_4\times10^{20}$ cm^2	$\Omega_6\times10^{20}$ cm^2	References
Er^{3+}:$Ca_3Gd_2(BO_3)_4$	4.01	0.98	1.72	[55]
Er^{3+}:YAG	0.19	1.68	0.62	[57]
Er^{3+}:$YAlO_3$	1.06	2.63	0.78	[58]
Er^{3+}:YVO_4	1.25	1.69	0.61	[59]
Er^{3+}:$YLiF_4$	0.97	1.21	1.37	[60]

Table 3.7. The Judd-Ofelt parameters for $Ca_3Gd_2(BO_3)_4$: Er^{3+} compared with other Er-doped crystals

Transition J→J'	λ(nm)	$A^{ed}(s^{-1})$	$A^{md}(s^{-1})$	$A^{total}(s^{-1})$	β	τ_{rad}(ms)
$^4I_{13/2}$→ $^4I_{15/2}$	1534	200.303	46.234	246.537	1	4.056
$^4I_{11/2}$→ $^4I_{13/2}$	2822	27.489	9.507	299.474	0.124	3.339
$^4I_{15/2}$	976	262.478			0.876	
$^4I_{9/2}$→ $^4I_{13/2}$	1694	85.45		197.461	0.433	5.064
$^4I_{15/2}$	800	112.011			0.567	
$^4F_{9/2}$→ $^4I_{9/2}$	3554	3.436	3.03	2029	0.003	0.493
$^4I_{11/2}$	1933	126.801	7.718		0.066	
$^4I_{13/2}$	1147	84.96			0.042	
$^4I_{15/2}$	653	1803			0.889	
$^4S_{3/2}$→ $^4I_{9/2}$	1693	98.665		3366	0.029	0.297
$^4I_{11/2}$	1210	73.585			0.022	
$^4I_{13/2}$	847	921.885			0.274	
$^4I_{15/2}$	545	2272			0.675	
$^2H_{9/2}$→ $^4F_{9/2}$	1063	53.286	51.595	5412	0.019	0.185
$^4I_{9/2}$	818	27.509	1.162		0.005	
$^4I_{11/2}$	686	674.786	42.768		0.133	
$^4I_{13/2}$	552	2483			0.459	
$^4I_{15/2}$	405	2078			0.384	
$^4G_{11}$→$^2H_{11/2}$	1369	42.342	14.366	29680	0.002	0.034
$^4F_{9/2}$	894	801.92	3.505		0.027	
$^4I_{9/2}$	714	300.563	0.851		0.011	
$^4I_{11/2}$	611	695.246	0.106		0.023	
$^4I_{13/2}$	502	2720	49.515		0.093	
$^4I_{15/2}$	380	25050			0.844	

Table 3.8. Radiative transition probability A, radiative lifetime τ_{rad} andfluorescent branching ratio β of Er^{3+} in $Ca_3Gd_2(BO_3)_4$ crystal at room temperature.

3.3.6 The spectrum characteristic of Yb^{3+}:$Ca_3Gd_2(BO_3)_4$ crystal

Fig.3.13 presents the absorption spectrum of $Ca_3Gd_2(BO_3)_4$:Yb^{3+}crystal, in which there is a broad absorption band ranged from 850 to 1000nm, corresponding to the transition from $^2F_{7/2}$ to $^2F_{5/2}$. The FWHM at 980 nm is about 12 nm and the cross-section is about 5.9×10^{-20} cm^2, which is benefit to the pumping of commercial laser diode. Table 3-9 shows the absorption properties of some ytterbium-doped compounds. The room temperature emission spectrum of $Ca_3Gd_2(BO_3)_4$:Yb^{3+} crystal excited at 895 nm is presented in Fig.3.14, in which there is a broad emission band ranged from 930 nm to 1100.7 nm. The FWHM of this emission band is 72.6 nm and its peak is located at 1020nm, which is resulted from the statistical distribution of Ca^{2+}, $Gd^{3+}$and Yb^{3+} ions.

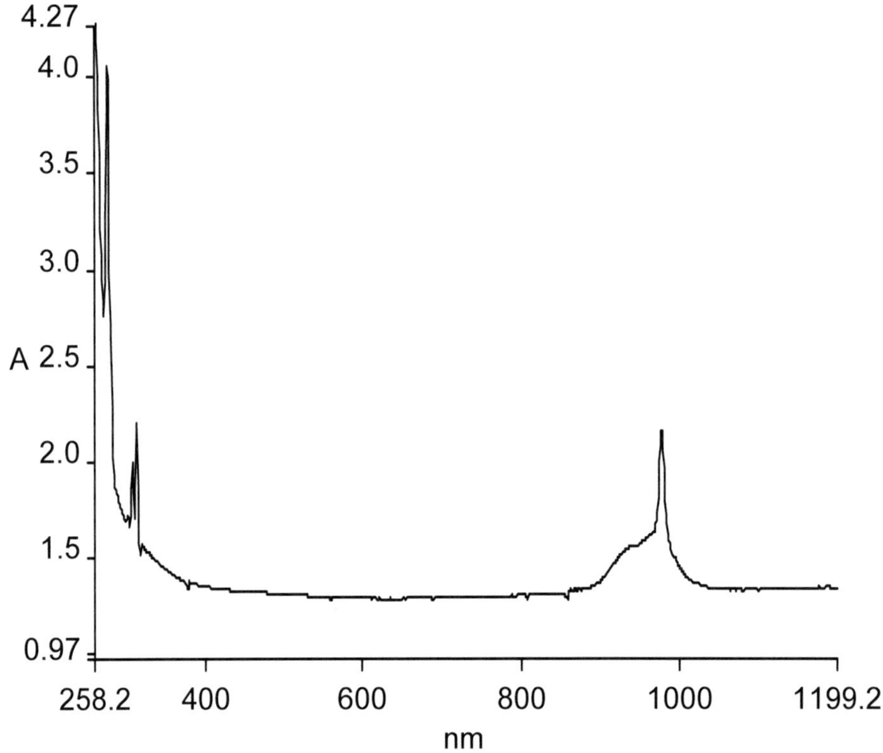

Fig. 3.13. The absorption spectrum of $Ca_3Gd_2(BO_3)_4$:Yb^{3+}crystal

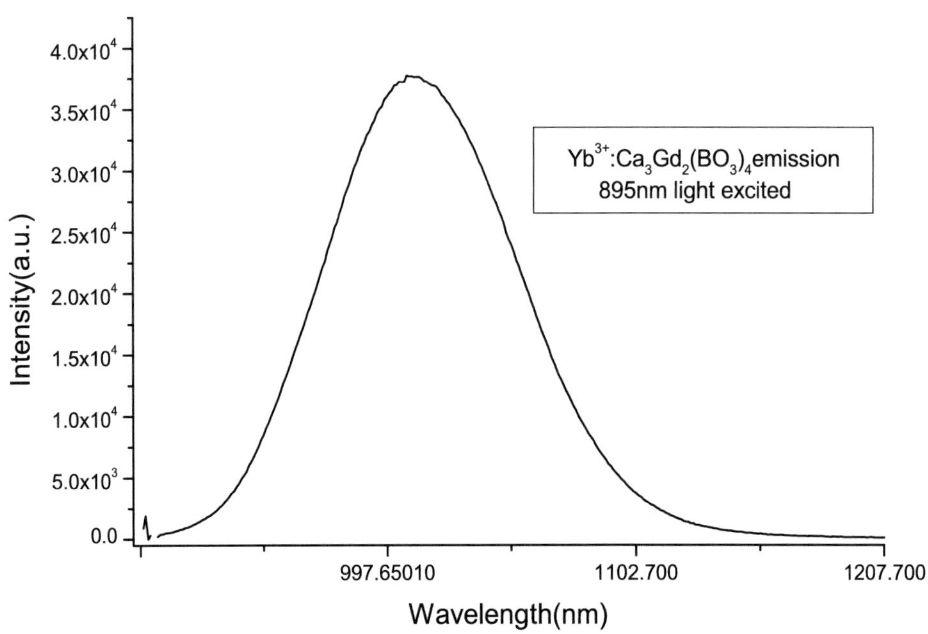

Fig. 3.14. Room temperature emission spectra of $Ca_3Gd_2(BO_3)_4$:Yb^{3+} crystal excited at 895 nm

Compound	λ_P(nm)	σ_a ($\times 10^{-20}$ cm^2)	FWHM(nm)	References
5at%Yb:CGB	980	5.95	12	[5]
2at%Yb:YCOB	976.1	0.94	3	[61]
1at%Yb:Y_2SiO_5	977.4	2.1	7	[62]
1at%Yb:Sc_2SiO_5	979.5	1.9	4	[62]
Yb:GdCOB	976	0.87	2.6	[63]
15%Yb:YAG	968	0.94	2.6	[64]
15%Yb:BLuB	966	0.29	6.4	[64]
Yb:YAB	975	3.4	3	[65]

Table 3.9. Absorption properties of the ytterbium-doped compounds

3.4 The Laser characteristics of Yb^{3+}:$Ca_3Re_2(BO_3)_4$[Re=Y,Gd] crystals

3.4.1 The Laser characteristics of Yb^{3+}:$Ca_3Y_2(BO_3)_4$ crystal[66~69]

An passively mode-locked Yb:$Y_2Ca_3(BO_3)_4$ (Yb:CYB) laser with a partially reflective semiconductor saturable-absorber mirror was achieved. The 244 fs pulses with a repetition rate of ~55 MHz were obtained at the central wavelength of 1044.7 nm. The measured average output power amounted to 261 mW. This was the first demonstration of femtosecond laser in Yb:CYB crystal. Fig.3.15 shows the experimental setup of the laser oscillator. The input mirror M1 was a flat mirror coated with high reflection (HR) in a broad band from 1010 to 1060 nm and high transmission (HT) at 976 nm. The two folding mirrors, M2 and M3, were concave and had the radii of curvature of 1000 and 500 mm, respectively. Both of which were HR-coated placed near normal incidence (~3°). A SESAM with a reflection of 96% at 1040 nm was employed, which had a modulation depth of 1.6% and saturation fluence of 70 μJ/cm^2. Fig.3.16 presents the continuous wave and mode locking average output power versus the absorbed pump power. We can see that the threshold absorbed pump power was 1.9 W and a maximum output power of 783 mW was obtained under the absorbed pump power of 7.0W. The laser oscillation was achieved with the threshold absorbed pump power of 2.8 W when the output coupler was replaced by the SESAM. Within the range of absorbed pump power from threshold to 4.6W, a metastable regime rapidly alter-nating between Q-switched mode locking and continuous wave (CW) mode locking was observed.

Fig.3.17 presents the central wavelength and FWHM of the emission spectrum for mode locking operation. The spectrum was red-shifted obviously at a range from 1041.5 to 1044.7 nm with the absorbed pump power increased from threshold to 7.0 W, which was possibly attributed to the reabsorption effect for quasi-three-level system as the short wavelength part of the absorption spectrum overlaps the emission spectrum. Fig.3.18 shows the pulse train of the cw mode-locked laser with the repetition rate of ~55MHz. Fig.3.19 presents the autocorrelation trace of the 244 fs pulse with the average output power of 261 mW at the central wavelength of 1044.7 nm. The corresponding spectrum had a FWHM of 8.1 nm centered at 1044.7 nm, with a time bandwidth product of 0.54. In this job, a partially reflective SESAM was used as the output coupler that would lower the positive dispersion. When the absorbed pump power was fixed at 7.0 W, the adjustment of the Yb:CYB crystal and SESAM in such a resonator (by either moving or rotating that could vary the amount of material that the light went through) is critical for the stability of mode-locking operation and pulse duration: an average output power of 375 mW could be obtained but the mode locking was unstable; the duration also fluctuated in a wide range from ~1000 to 244 fs.

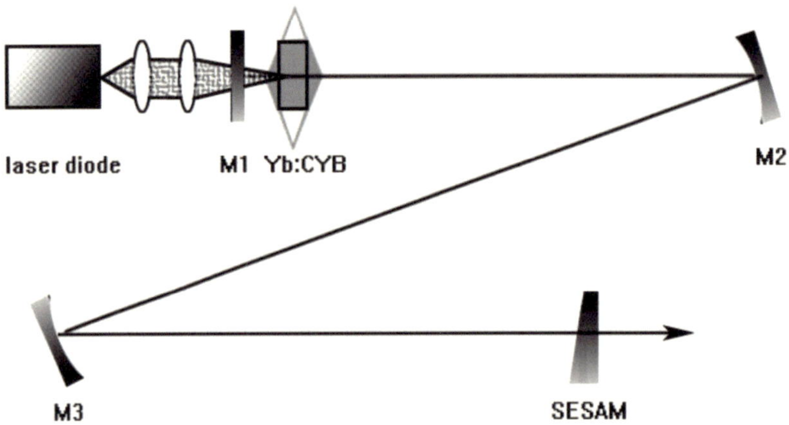

Fig. 3.15. Experimental setup of the laser oscillator. The shaded part in the position of the gain medium refers to the change in shape resulting from the thermal expansion

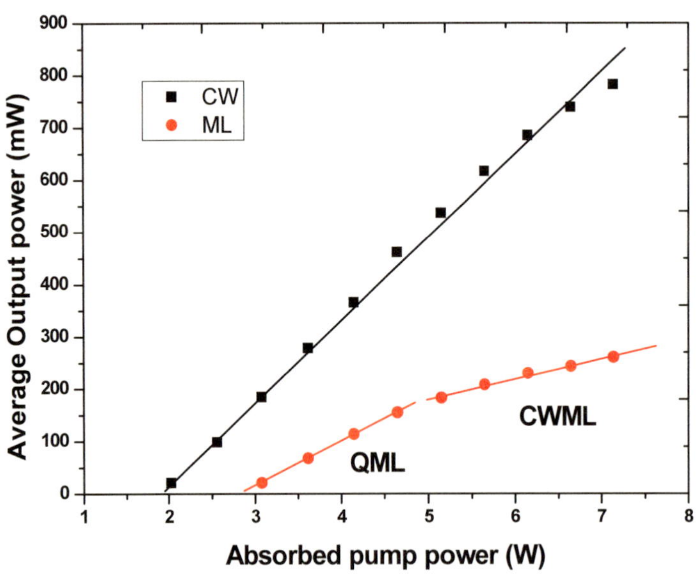

Fig. 3.16. The continuous wave and mode locking average output power versus the absorbed pump power

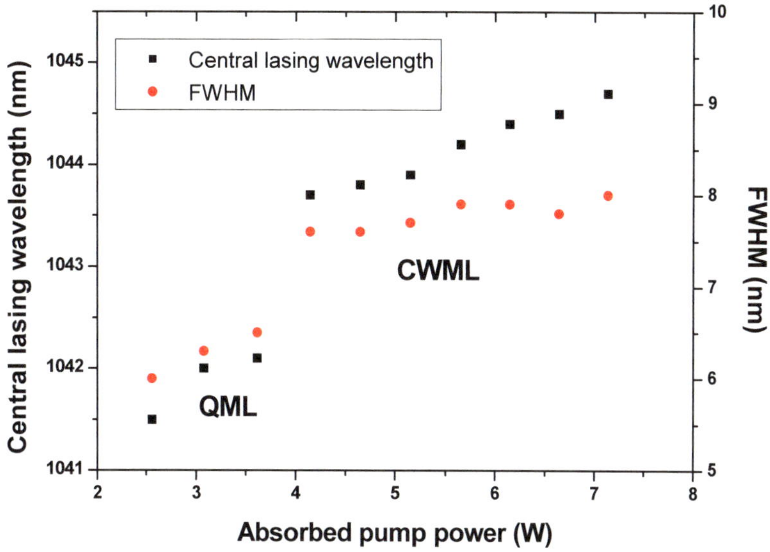

Fig. 3.17. The central wavelength and FWHM of the emission spectrum for mode locking operation

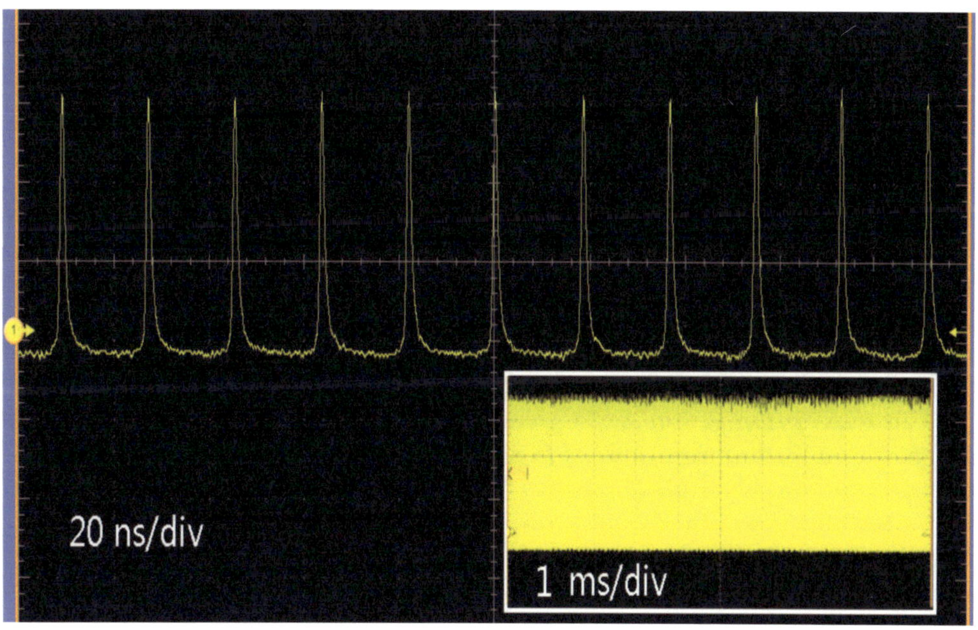

Fig. 3.18. The pulse train of the cw mode-locked laser with the repetition rate of ~55MHz

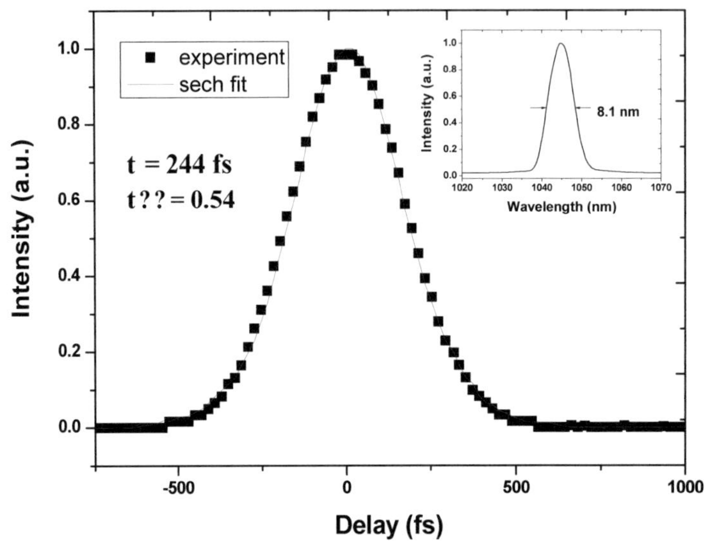

Fig. 3.19. The autocorrelation trace of the 244 fs pulse with the average output power of 261 mW at the central wavelength of 1044.7 nm. And the inset corresponding to the spectrum.

Q-switching and Q-switched mode-locked Yb:$Y_2Ca_3B_4O_{12}$ lasers with an acousto-optic switch are demonstrated. In the Q-switching case, an average output power of 530 mW is obtained at the pulse repetition rate of 10.0 kHz under the absorbed pump power of 6.1 W. The minimum pulse width is 79 ns at the repetition rate of 1.7 kHz. The pulse energy and peak energy are calculated to be 231 μJ and 2.03 kW, respectively. In Q-switched mode-locking case, the average output power of 64 mW with a mode-locked pulse repetition rate of 118 MHz and Q-switched pulse energy of 48 μJ is generated under the absorbed pump power of 6.1W. Fig.3.20 presents the CW and Q-switched average output power versus absorbed pump power. The CW lasers were operated with the absorbed pump power of up to 0.8 W and 1.3 W, respectively, for T = 1% and T = 5% output coupler. When the absorbed pump power reaches 6.1W, the T = 5% output coupler provides the best performance with an output power of 992 mW, which is much higher than 760 mW by using T = 1% output coupler. Fig.3.21 gives the emission spectra of the Yb:CYB laser with plano-concave cavity configuration. (a) is in the CW situation with T = 5% output coupler. (b)-(d) are in the Q-switching situation with T = 5% output coupler, (e) is in the Q-switching situation with T = 1% output coupler the emission spectra of the Yb:CYB laser. Fig.3.22 shows the pulse width versus absorbed pump power. At the pulse repetition rate of 1.7 kHz and the absorbed pump power 6.1 W, the pulse widths of 79 ns and 114 ns are detected by used of T = 1% and 5% coupler, respectively. Fig.3.23 presents the pulse energy and pulse peak power versus the absorbed pump power at the pulse repetition rate of 1.7kHz. The pulse energy of 231 μJ and peak energy of 2.03 kW can be obtained with the T = 5% output coupler. Fig.3.24 gives the single pulse profile of the A-O Q-switched Yb:CYB lasers with the pulse width of 76 ns at the absorbed pump power of 6.1 W, and the inset corresponding to the temporal pulse trains with the repetition rate of 1.7 kHz. The beam quality M2 is measured to be about 1.4 by using the knife-edge scanning method.

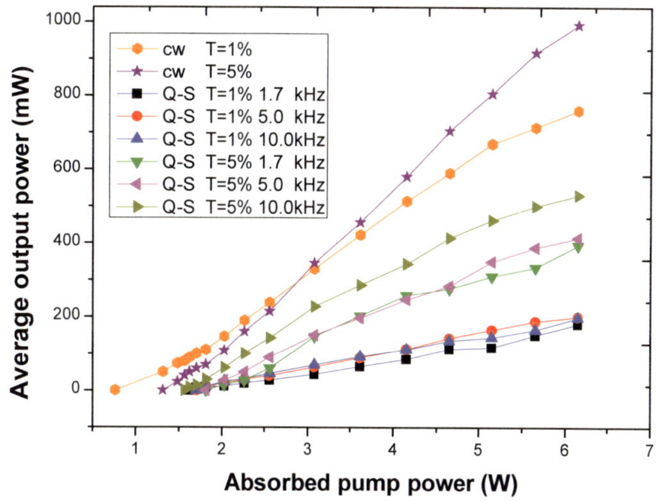

Fig. 3.20. The CW and Q-switched average output power versus absorbed pump power for $T = 1\%$ and 5%

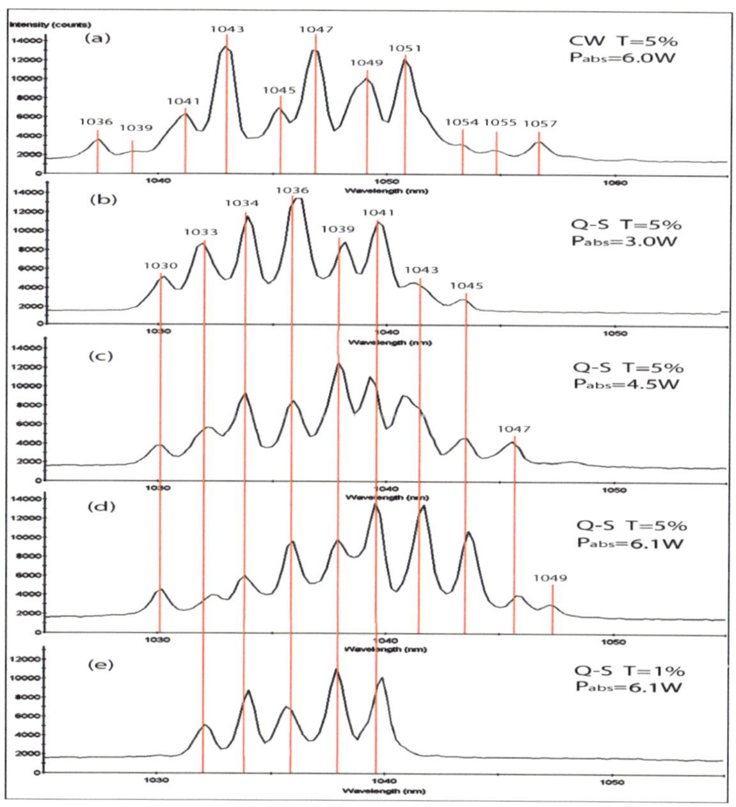

Fig. 3.21. Emission spectra of the Yb:CYB laser.(a) is in CW situation with $T = 5\%$. (b)-(d) are in Q-switching situation with $T = 5\%$, showing the absorbed pump power dependence. (e) is in Q-switching situation with $T = 1\%$.

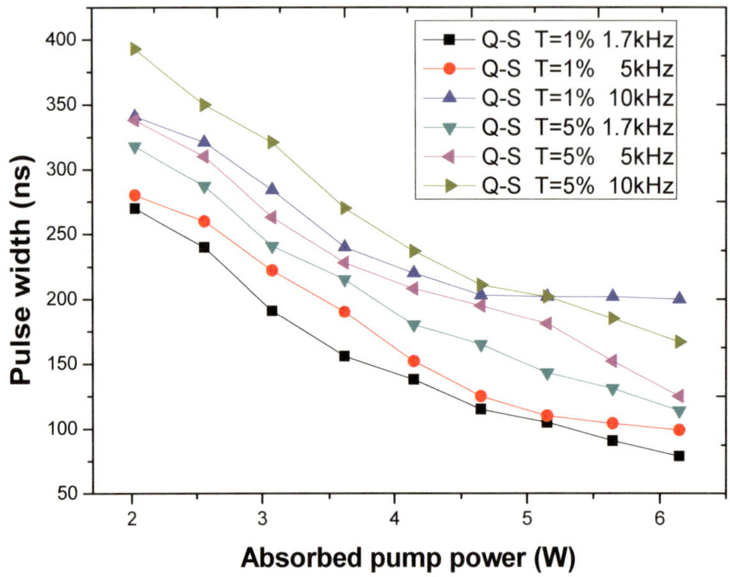

Fig. 3.22. Pulse width versus absorbed pump power

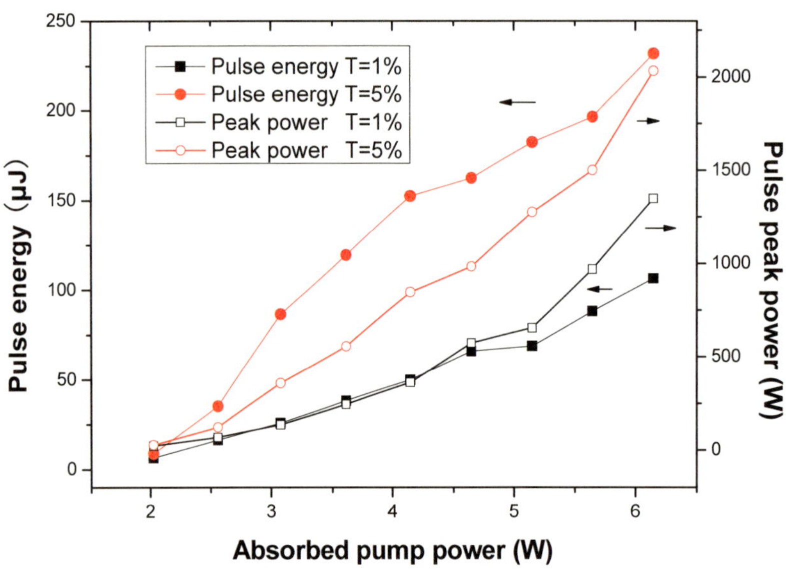

Fig. 3.23. Pulse energy and pulse peak power versus the absorbed pump power at the pulse repetition rate of 1.7kHz

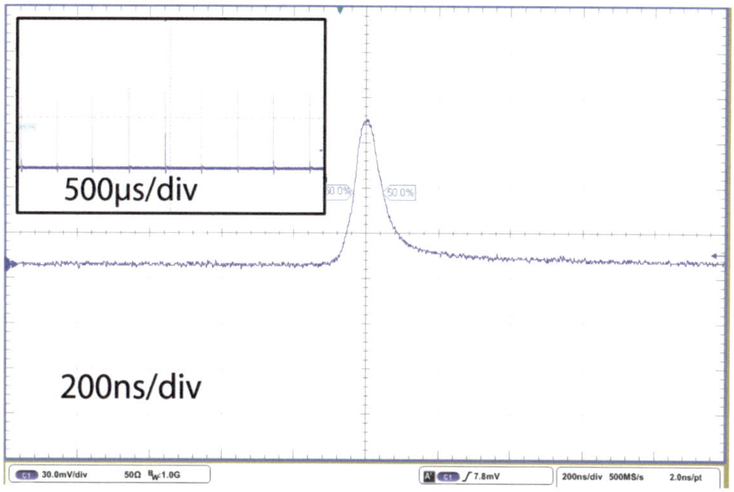

Fig. 3.24. The single pulse profile of the A-O Q-switched Yb:CYB lasers with the pulse width of 76 ns at the absorbed pump power of 6.1 W

Fig.3.25 depicts the schematic arrangement of the four-mirror resonate cavity for Q-switched mode-locking operation. M1 is the same plane mirror as that in Q-switching case described above. M2 and M3 are two spherical concave mirrors with curvature of 500 mm and 100 mm, respectively, and HR coated from 1010 nm to 1060 nm. The two different concave couplers (r = 75 mm) with transmissions of 1% and 5% are also used as the output coupler M4. The distances L1 from M1 to M2, L2 from M2 to M3, L3 from M3 to M4 are set as 440 mm, 770 mm and 60 mm, respectively. Fig.3.26 shows the emission spectra of the Yb:CYB laser. (a)-(c) is in CW situation with T = 5% showing the absorbed pump power dependence. (d) is in Q-switched mode-locking situation with T = 5%, (e) is in Q-switched mode-locking situation with T = 1%. Compared with the spectra of the above plano-concave resonator, there are only three or four emission branches in Figs. 3.26 a–c, since the combined etalon effects of the resonant cavity and uncoated Yb:CYB are different between the two-mirror and four-mirror cavity. Under the absorbed pump power of 3.0 W, there are three nonoverlapping emission branches located at 1042 nm, 1047 nm and 1051 nm. Similar to that in the Q-switched mode, the spectrum shifts to the long side with the increase of the absorbed pump power owing to the reabsorption effect. When the absorbed pump power reaches 6.1 W, a new branch in the short side (1040 nm) appears. The CW and Q-switched average output power versus absorbed pump power for T = 1% and 5% are plotted in Fig.3.27, (a) is T = 1% and (b) is T = 5%. The average output powers of 64 mW and 87 mW are obtained for T = 1% and T = 5% output coupler, respectively. Fig.3.28 shows the pulse energy of the Q-switched envelope versus the absorbed pump power at repetition rate of 1.7 kHz. In Fig.3.29, (a) is the oscilloscope traces of Q-switched pulse train with the T = 5% output coupler and the repetition rate of 1.7 kHz under the absorbed pump power of 6.1 W in the same situation. (b) is the typical QML pulse envelope of T = 5% in the same situation. (c) is the expanded traces of mode-locked train. The repetition rate of the periodic mode-locked pulses is about 118 MHz, which matches exactly with the axial mode interval. The output beam density distribution is close to the fundamental transverse mode (TEM00) and the quality parameter M2 factor is about 1.6.

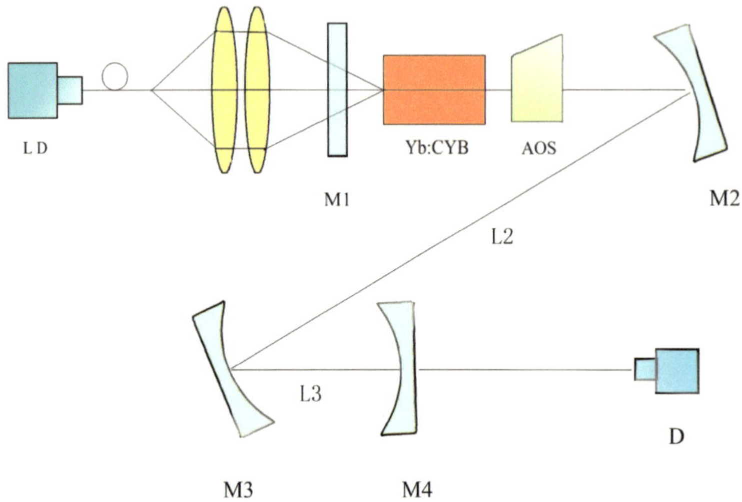

Fig. 3.25. The schematic arrangement of the four-mirror resonate cavity for Q-switched mode-locking operation

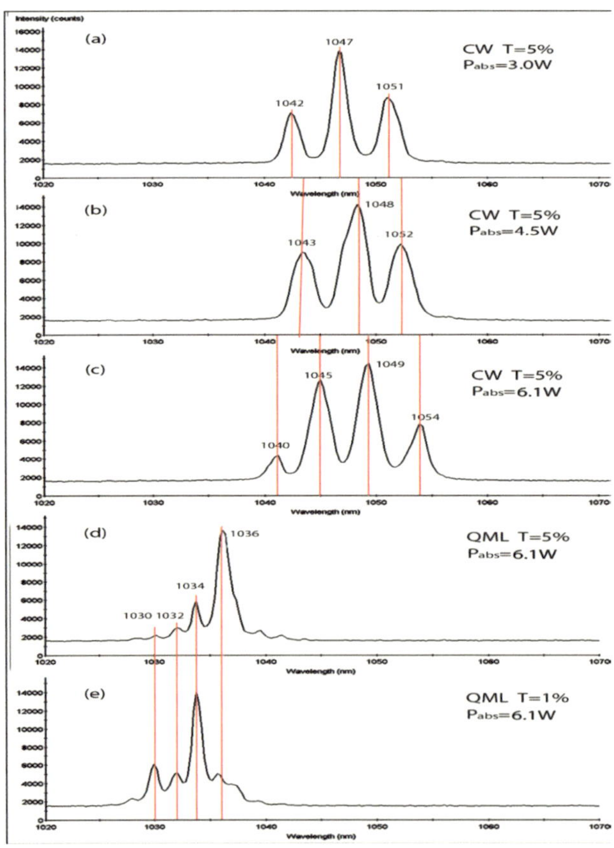

Fig. 3.26. Emission spectra of the Yb:CYB laser. (a)-(c) is in CW situation with $T = 5\%$ showing the absorbed pump power dependence.

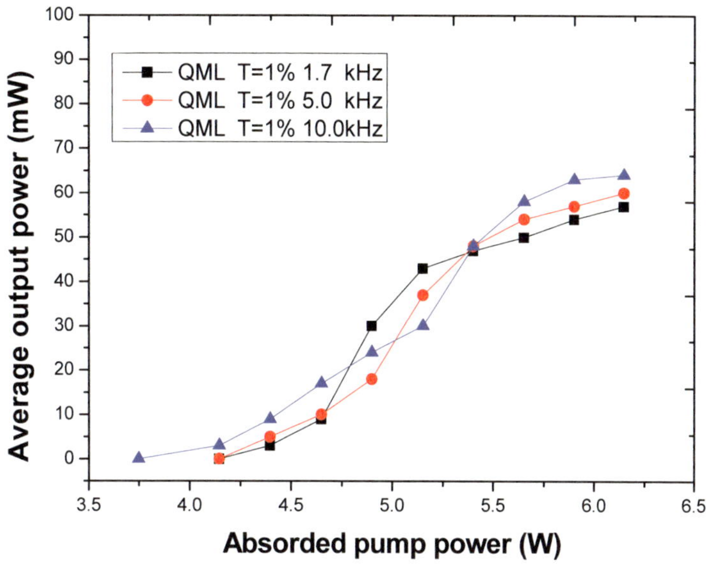

(a)

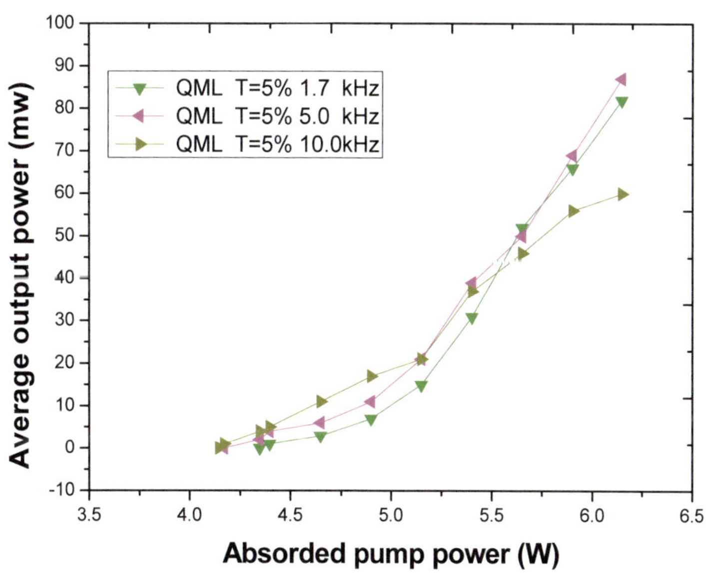

(b)

Fig. 3.27. The CW and Q-switched average output power versus absorbed pump power for $T = 1\%$ and 5%. (a) is $T = 1\%$ and (b) is $T = 5\%$.

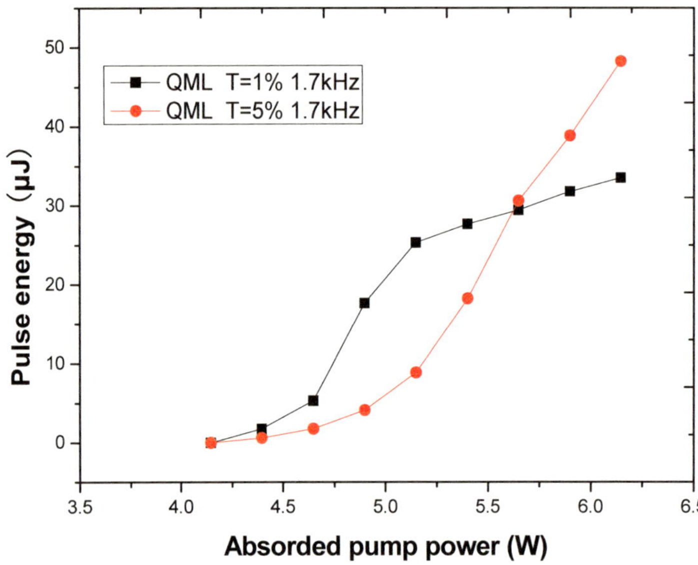

Fig. 3.28. Pulse energy of the Q-switched envelope versus the absorbed pump power at repetition rate of 1.7 kHz.

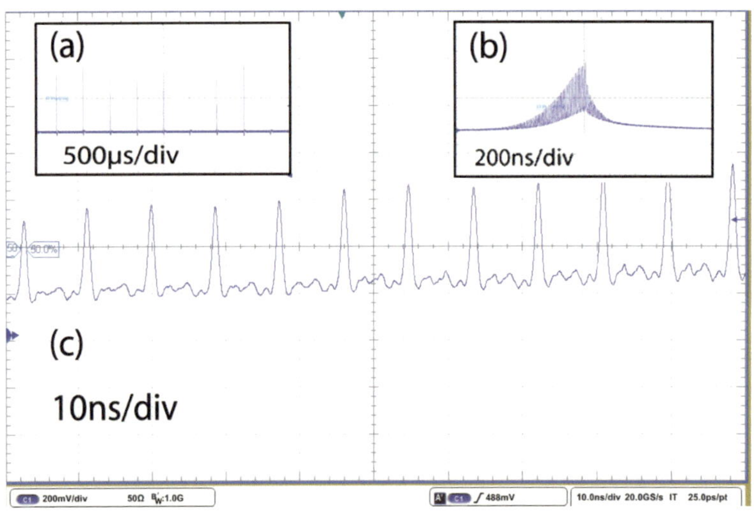

Fig. 3.29. (a) is the oscilloscope traces of Q-switched pulse train with the $T = 5\%$ output coupler and the repetition rate of 1.7 kHz under the absorbed pump power of 6.1 W in the same situation. (b) is the typical QML pulse envelope of $T = 5\%$ in the same situation. (c) is the expanded traces of mode-locked pulses.

The Yb^{3+}-doped $Y_2Ca_3B_4O_{12}$ diode-pumped laser operation in both continuous-wave (CW) and passively Q-switched modes was reached. The differential slopes of the CW output power are in the 22-40 % range under different experimental conditions. Continuous tuning of the laser wavelength is obtained in the 1020-1057 nm range, in agreement with the broad emission spectra. In pulsed regime the repetition rate occurs up to 1.6 kHz and pulse energies of 30-75 μJ with about 40 ns duration are obtained. Fig.3.30 demonstrates the polarized emission spectra of the CYB:Yb^{3+} crystal used for laser experiments. The two orthogonal polarizations of the eigenstates are labeled as *H* and *V*. The main peak at 976.3 nm has 6 nm full width at half maximum (FWHM) and is suitable for diode pumping. Fig.3.31 displays the spectral distribution of the laser emission in CW and passive *Q*-switch modes. We can see that the more intense one corresponds to the H polarization with a broadband peaking near 1040 nm. The time evolution of the fluorescence was displayed with a 9410 Lecroy oscilloscope. The decay time was found to be 1 ms. Fig.3.32 presents repetition rates and pulse energies obtained in passive *Q* switching the $Ca_3Y_2(BO_3)_4$:Yb^{3+} laser. Lasing was obtained in H polarization in agreement with the polarized emission spectra, near 1045 nm (with the 97.5% transmission coupler) and up to 1 W power. Fig.3.33 shows the tunability of the laser emission obtained from rotation of a birefringent filter. Fig.3.34 presents the laser output power versus pump power obtained under different experimental conditions. The obtained slope efficiencies were in the 29%–40% range with the 5 cm radius curvature output coupler and 22% with the 7.5 cm coupler. Table 3-10 shows the pulse energy obtained in passive Q-switching different Yb^{3+}-doped hosts. In particular, we can see that the performances for Yb^{3+} doped GGG and GAB crystals obtained with similar experimental conditions were better than for Yb^{3+}:CYB, with no instability of the pulsed regime and less thermal problems. A plausible explanation is the lower laser emission cross section in Yb^{3+}:CYB and a too low absorbed pump power (61% absorption) of our sample.

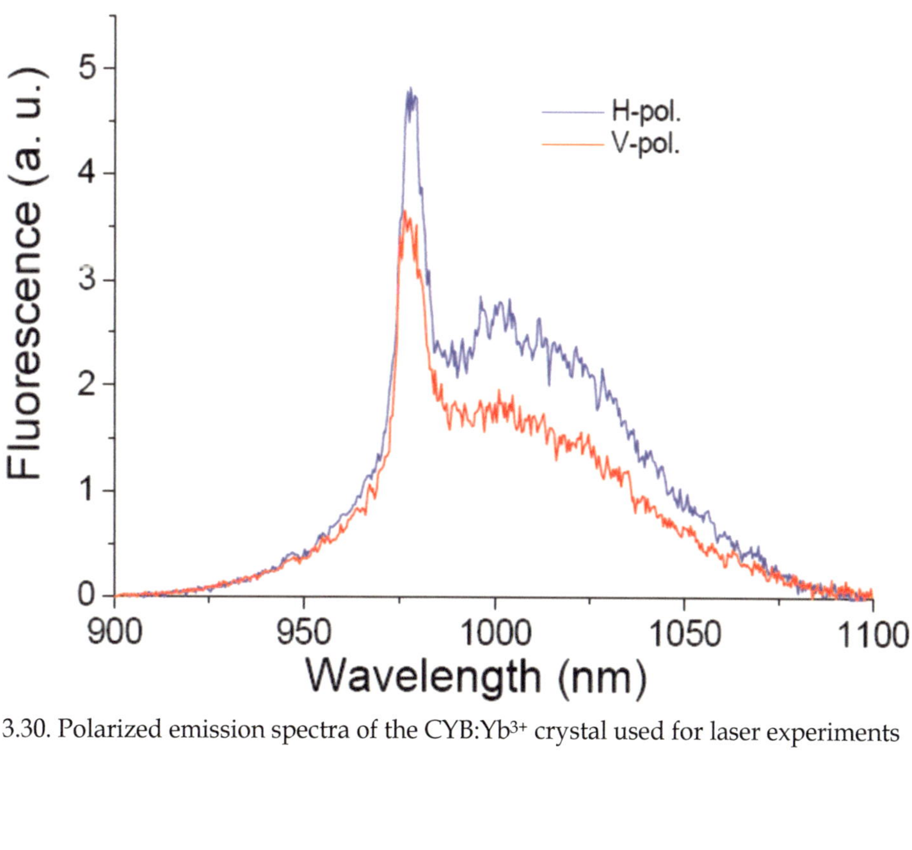

Fig. 3.30. Polarized emission spectra of the CYB:Yb^{3+} crystal used for laser experiments

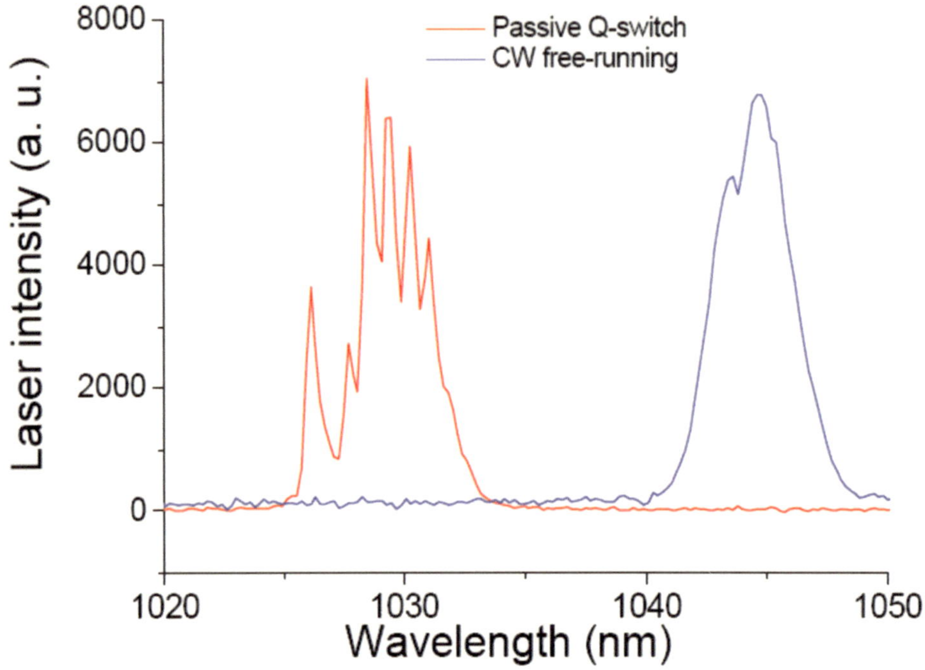

Fig. 3.31. Spectral distribution of the laser emission in cw and passive Q-switch modes

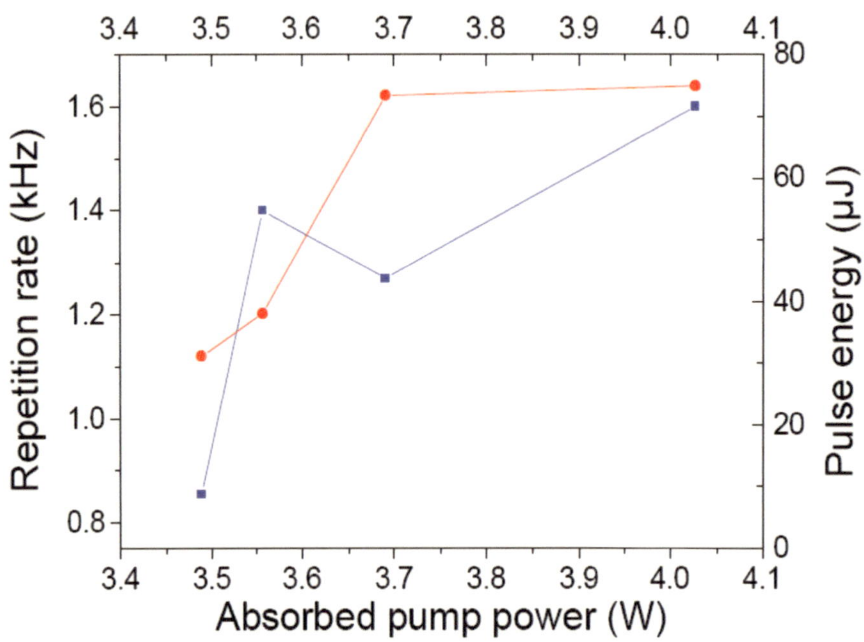

Fig. 3.32. Repetition rates and pulse energies obtained in passive Q switching the $Ca_3Y_2(BO_3)_4:Yb^{3+}$ laser. The inset represents a typical time evolution of the laser pulse

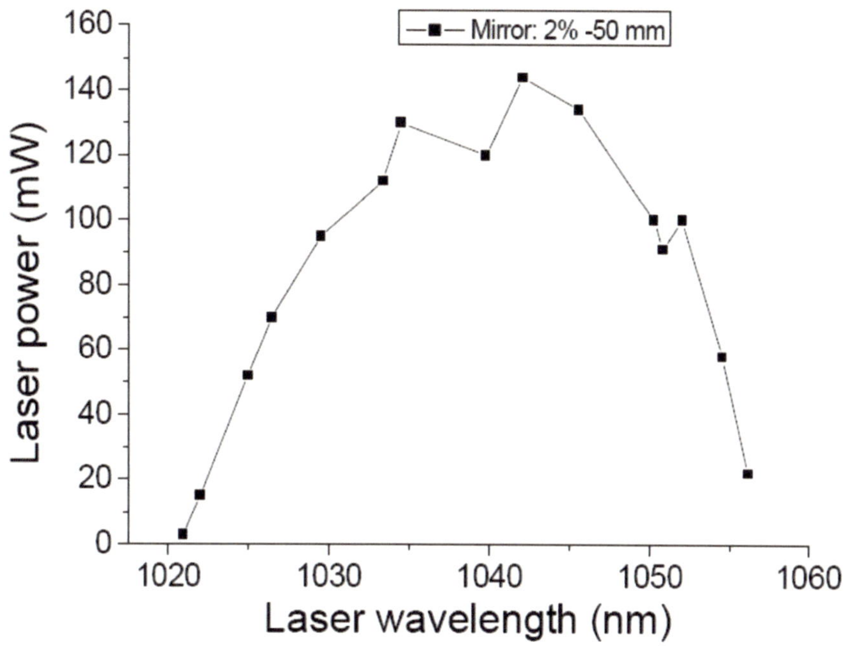

Fig. 3.33. Tunability of the laser emission obtained from rotation of a birefringent filter

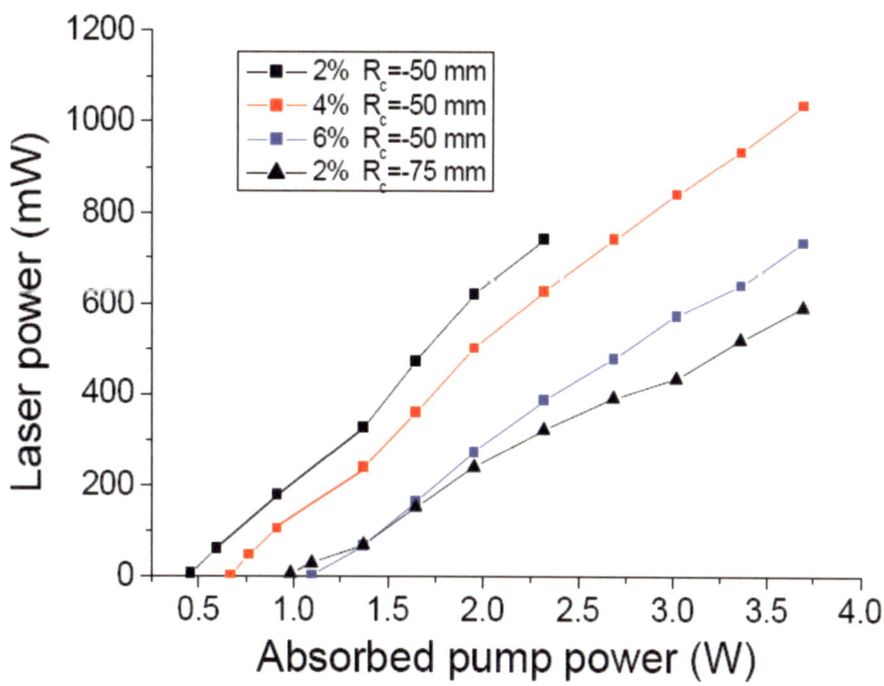

Fig. 3.34. Laser output power versus pump power obtained under different experimental conditions.

Crystal	cw emission wavelength (nm)	Lifetime τ (μs)	Emission cross section ($10^{-20}cm^2$) at cw lasing wavelength	Pulse energy (μJ)
Yb^{3+}:KGdW	1029	951	2.7	3.4[70]
Yb^{3+}:KLuW	1031.7	...	...	32.4[71]
Yb^{3+}:YAG	1023	600	2.03	54,100,179,500[72]
Yb^{3+}:GGG	1025	800	2.0	40,48,87,96[73]
Yb^{3+}:GAB	1045	800	0.83	125,165[74]
Yb^{3+}:CYB	1044	1000	0.60	30-70[69]

Table 3.10. Pulse energy obtained in passive Q-switching different Yb^{3+}-doped hosts

3.4.2 The laser characteristics of Yb^{3+}:$Ca_3Gd_2(BO_3)_4$ crystal[75]

The diode-pumped multi-wavelength continuous-wave laser operation of the disordered Yb:$Ca_3Gd_2(BO_3)_4$ (Yb:CGB) crystal was investigated. The number of the oscillating wavelengths varied from two to five in the range from 1045.4 to1063.6 nm with the absorbed pump power and Yb:CGB crystal length. An output power of 1.4 W was obtained when quadruple wavelengths were emitted simultaneously, corresponding to a slope efficiency of 23.7%. The quintuple-wavelength oscillation at 1049.4, 1051.3, 1053.4, 1055.6 and 1057.4 nm was realized with an output power of 1.0 W. The experiment results exhibited the further possible application of Yb:CGB crystal in terahertz-wave generation. Fig.3.35 represents the experimental setup of the CW Yb:CGB laser oscillator. M1 was a plane mirror with antireflection coating at the pump wavelength and high-reflection coating at a broad band from 1040 to 1070 nm. A concave mirror with 75-mm curvature radius and ~99% reflectance from 1040 to 1070 nm was used as the output coupler M2. Fig.3.36 shows the absorbed pump power and absorption efficiency versus incident pump power for the two Yb:CGB samples, with the same cross section of 3×3 mm^2 but different lengths of 2 and 5 mm (described as sample 1 and 2, respectively). It can be seen that the absorption efficiency of sample 1 was around 40% if the incident pump power was below 4.0 W. But the efficiency decreased dramatically from 40% to 30% when the incident pump power was increased from 4.0 to 8.0 W. Then the efficiency was stable again, varying within a narrow range of 30% ~ 32 %. That was possibly attributed to the saturation of pump absorption to some extent, resulting from the depletion of the population in ground state. The similar phenomenon was observed when sample 2 was tested. Fig.3.37 depicts the relationship between the output power (P_{out}) and absorbed pump power for the two samples. The laser operation was realized with threshold absorbed pump powers of 0.4 and 0.9 W for sample 1 and 2, respectively. The maximum output power was 1.4 W by using sample 2 under the absorbed pump power of 6.8 W, with a slope efficiency of 23.7% and an optical conversion efficiency of 20.6%. The sample 1 exhibited higher slope efficiency of 30.3% and optical conversion efficiency of 27.0% with the output power of 1.0 W under the absorbed pump power of 3.7 W. Fig.3.38 gives the emission wavelengths versus absorbed pump power for the two Yb:CGB samples. Fig.3.39 and Fig.3.40 shows the emission spectra of the simultaneous multi-wavelength Yb:CGB laser with sample 1 and sample 2, repectively. It can seen that the emission wavelengths at each stage were almost same in intensity. That is advantageous to the practice terahertz-wave generation. If the quintuple-wavelength simultaneous emission is employed, multiple terahertz waves can be generated theoretically through difference frequency nonlinear interaction. Furthermore, it is interesting to note the separation of emission peaks varied from 1.0 to 2.0 nm with the absorbed pump power and Yb:CGB crystals, which means the multi-wavelength CW Yb:CGB laser could support the tunable terahertz-wave generation from 0.27 to 2.16 THz as calculated from Fig.3.38. Our experiment also showed that the reabsorption effect in quasi-three-level laser systems depended on the length of laser medium and the level of pump intensity. In addition, this effect had a great influence on the laser characteristics such as output power, optical efficiency and emission wavelength.

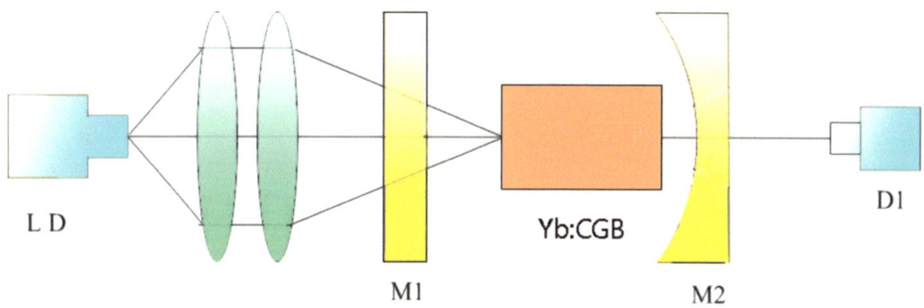

Fig. 3.35. Experimental setup of the CW Yb:CGB laser oscillator

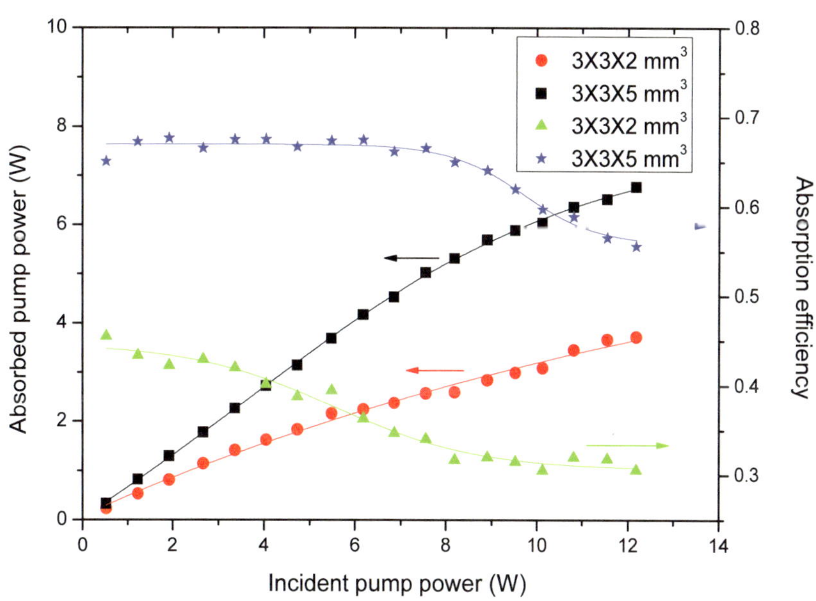

Fig. 3.36. Absorbed pump power and absorption efficiency versus incident pump power for the two Yb:CGB samples

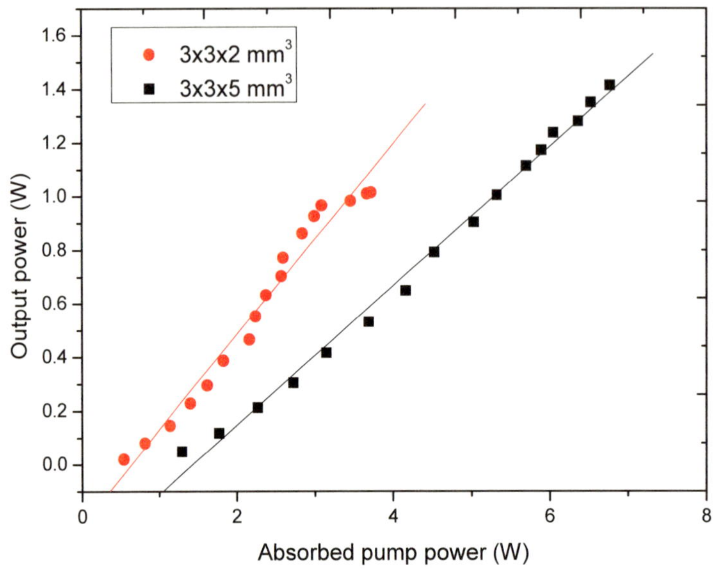

Fig. 3.37. Continuous-wave output power versus absorbed pump power of the Yb:CGB laser

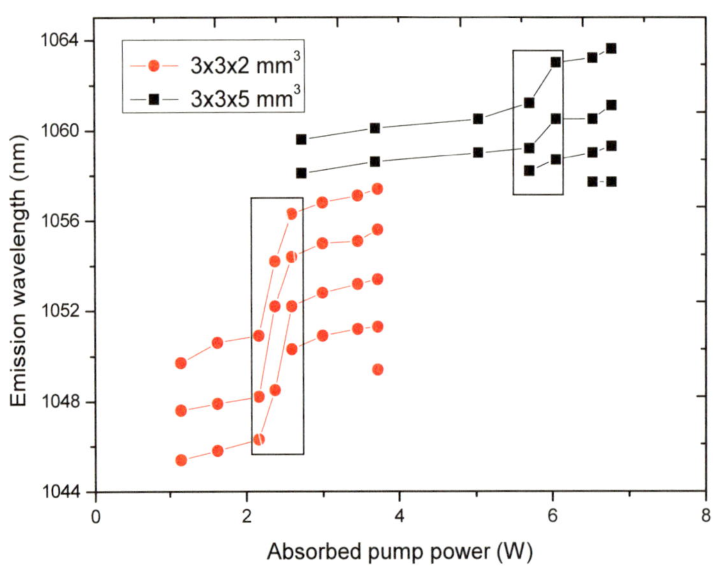

Fig. 3.38. Emission wavelengths versus absorbed pump power for the two Yb:CGB samples. The rapid redshift ranges are marked by squared pattern. The oscillating wavelengths varied in the region from 1045.4 to1063.6 nm with the absorbed pump power and Yb:CGB sample length.

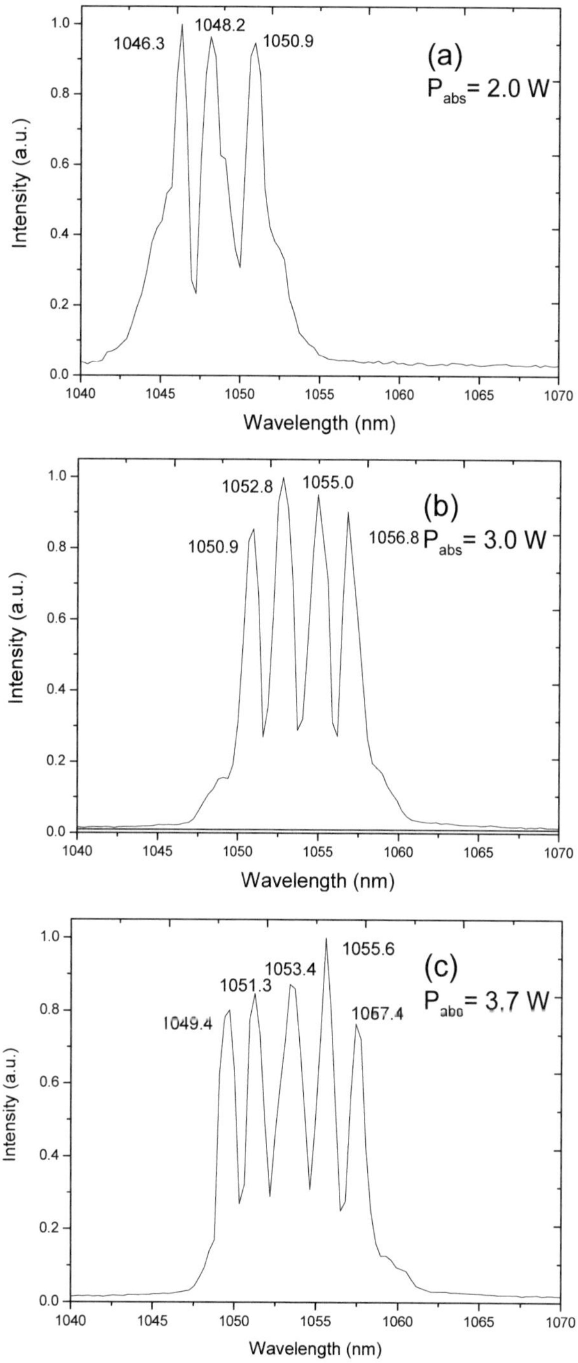

Fig. 3.39. Emission spectra of the simultaneous multi-wavelength Yb:CGB laser with sample 1. (a) triple-wavelength oscillation with P_{out} = 0.4 W and P_{abs} = 2.0 W; (b) quadruple-wavelength oscillation with P_{out} = 0.9 W and P_{abs} = 3.0 W; (c) quintuple-wavelength oscillation with P_{out} = 1.0 W and P_{abs} = 3.7 W.

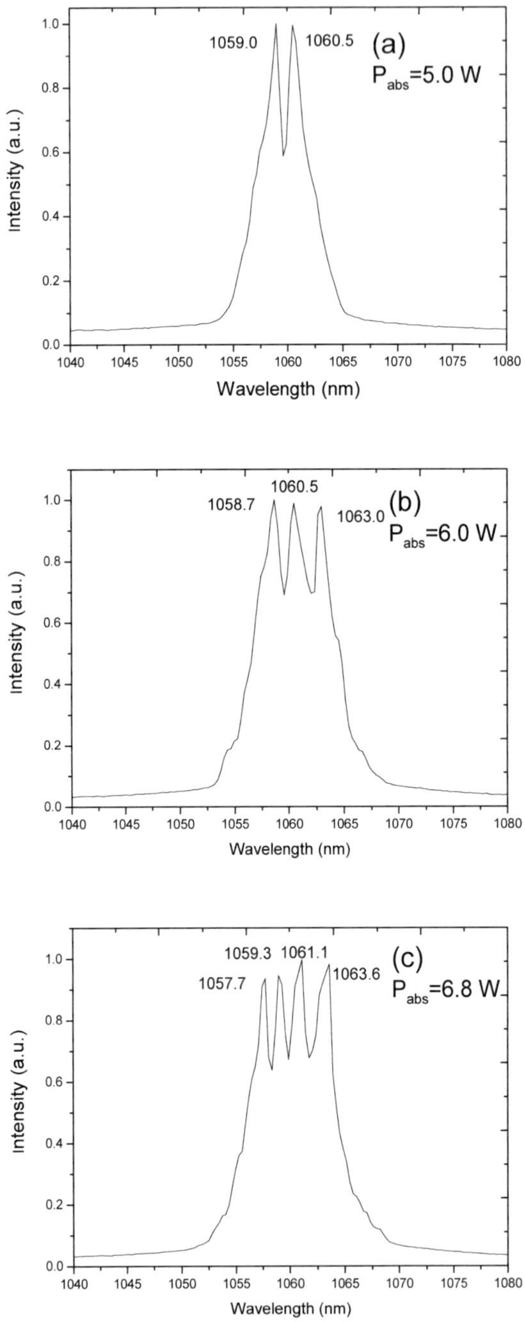

Fig. 3.40. Emission spectra of the simultaneous multi-wavelength Yb:CGB laser with sample 2. (a) is dual-wavelength oscillation with $P_{out} = 0.9$ W and $P_{abs} = 5.0$ W; (b) is triple-wavelength oscillation with $P_{out} = 1.2$ W and $P_{abs} = 6.0$ W; (c) is quadruple-wavelength oscillation with $P_{out} = 1.4$ W and $P_{abs} = 6.8$ W.

4. Nd^{3+}-doped LaB_3O_6 crystals

The LaB_3O_6 [LaBO] single crystal belongs to monoclinic system with the space group of I2/c [76,77]. The cell parameters are as follows[77]: a=9.946(1) Å, b=8.164(1) Å, c=6.4965(5) Å, β=127.06(1)°, [78]. V=420.9 Å^3, Dc=4.219 g/cm^{-3}. It melts congruently at 1145°C [76]. Therefore, this crystal can be obtained with large size by the Czochralski technique. Since the ionic radius of La^{3+} in LaB_3O_6 single crystal is about 1.04 Å [78], it can be substituted by laser exciting ions of lanthanide such as Nd^{3+}, Yb^{3+} ions. Therefore LaB_3O_6 crystal may be a new potential host for laser crystal. Furthermore, rare earth-doped LaB_3O_6 crystal can serve as a microchip laser crystal without any processing because of the cleavage of LaB_3O_6 crystal.

4.1 The crystal growth

Nd^{3+}-doped LaBO crystal with size up to ф20 mm×35 mm was grown using the Czochralski technique by Dr.Guohua Jia[4]. When the crystal was cut into laser bulk, it split into the cleavage crystal with the size of 2.5 mm×9 mm×35 mm as shown in Fig.4.1.

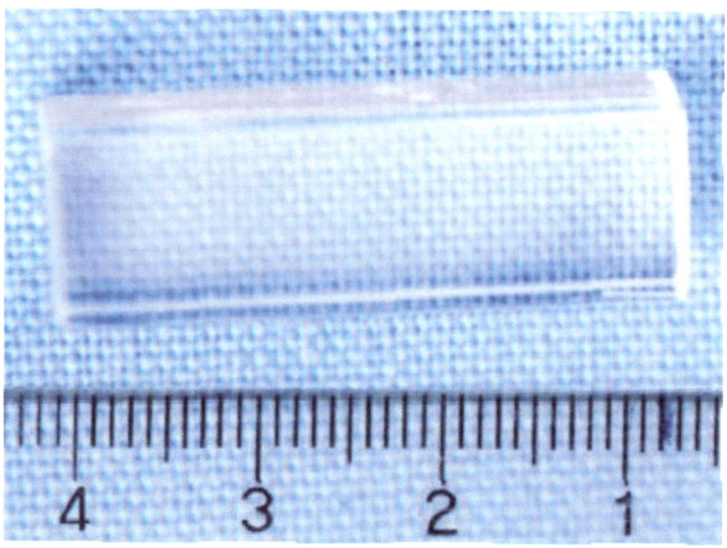

Fig. 4.1. Image of the unprocessed cleavage microchip along the (101) planes directly obtained from the as-grown crystal

4.2 The spectrum characteristics of Nd^{3+}-doped LaB_3O_6 crystal[80]

The room temperature absorption spectrum (Fig.4.2) consists of 10 groups of bands, which are associated with the observed transitions from the $^4I_{9/2}$ ground state. The absorption spectrum of the LaB_3O_6:Nd^{3+} crystal reaches its maximal value at about 800 nm and its FWHM is about 16 nm. The absorption cross-section was measured to be $\sigma_{abs} = 3.37 \times 10^{-20}$ cm^2. [7] This stronger absorption band corresponding to the transition $^4I_{9/2} \rightarrow {}^2H_{9/2}$ is very favorable for commercial GaAlAs diode pumping[4]. The room temperature emission spectrum with the light perpendicular to <1 1 1> planes is presented in Fig.4.3. The $^4F_{3/2} \rightarrow {}^4I_J$ (J = 9/2, 11/2, 13/2) transitions corresponding to the bands centered at 890.7, 1062 and 1329 nm, respectively, are the most important for laser applications. The value of the emission cross-section at 1062 nm of $LaB_3O_6:Nd^{3+}$ is 3.46×10^{-20} cm^2, which is a little smaller than that of other Nd^{3+} doped crystals. The emission cross-section and branching ratio (β) of the $^4F_{3/2} \rightarrow {}^4I_{9/2}$ transition are centered at 891 nm. The values of the emission cross-section at 891 nm and the branching ratio of this transition are 4.07×10^{-21} cm^2 and 0.336, respectively. Fig. 4.4 shows the room temperature fluorescence decay curve of $LaB_3O_6:Nd^{3+}$ crystal from which the fitting result of single exponential decay is 44.465 ns.

Table 4-1 shows the integrated absorbance, the experimental and calculated line and oscillator strengths of $Nd^{3+}:LaB_3O_6$ crystal (note: rms $f = 0.744 \times 10^{-6}$), and Table 4-2 presents the intensity parameters of $Nd^{3+}:LaB_3O_6$ crystal and the comparison of the intensity parameters of other Nd^{3+} doped crystals. Table 4-3 presents the calculated radiative probabilities, radiative branching ratios and radiative time for the emissions from the $^4F_{3/2}$ level of $LaB_3O_6:Nd^{3+}$. Table 4-4 shows the comparison of the $^4F_{3/2} \rightarrow {}^4I_{11/2}$ emission cross-section and radiative branching ratios of Nd^{3+} doped crystals.

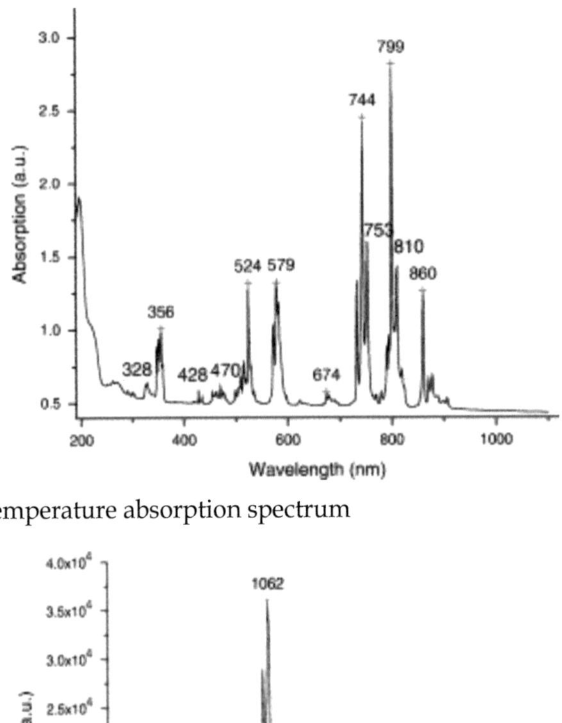

Fig. 4.2. The room temperature absorption spectrum

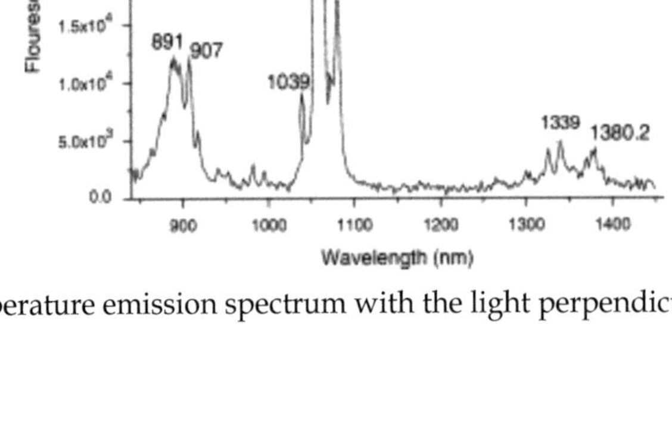

Fig. 4.3. Room temperature emission spectrum with the light perpendicular to <1 1 1> plane

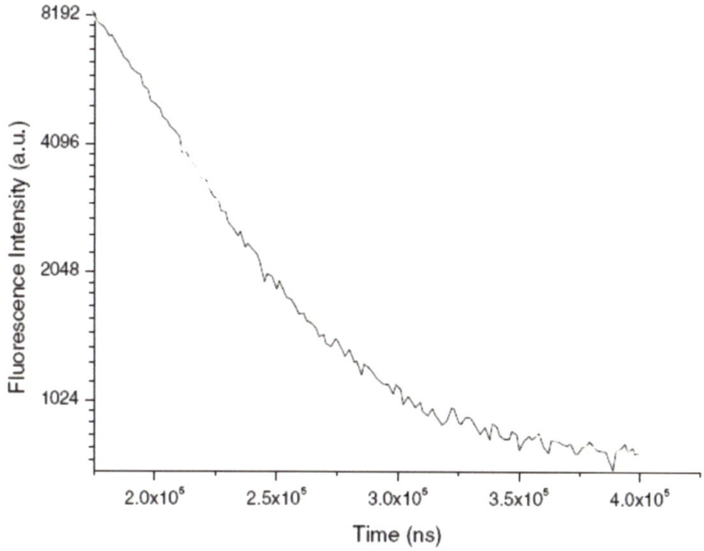

Fig. 4.4. Room temperature fluorescence decay curve of $LaB_3O_6:Nd^{3+}$ crystal

Excited state	Wavelength (nm)	I (nm/cm)	S_{mea} (10^{-20} cm^2)	S_{cal} (10^{-20} cm^2)	f_{exp} (10^{-6})	f_{cal} (10^{-6})
$^4F_{3/2}$	860	35.86	0.411	0.778	1.656	0.329
$^4F_{5/2}$,$^2H_{9/2}$	799	240.81	2.969	2.885	6.810	6.891
$^4F_{7/2}$,$^4S_{3/2}$	744	221.79	2.937	3.081	0.163	0.379
$^4F_{9/2}$	674	41.89	0.612	0.210	1.595	0.970
$^4G_{5/2}$,$^2G_{7/2}$	579	127.94	2.177	2.204	4.900	3.633
$^2K_{13/2}$,$^4G_{7/2}$,$^4G_{9/2}$	524	79.91	1.502	1.114	6.426	6.507
$^2K_{13/2}$,$^2G_{9/2}$,$^2P_{3/2}$,$^4G_{11/2}$	470	20.92	0.483	0.267	1.533	0.532
$^2P_{1/2}$,$^2D_{5/2}$	428	1.77	0.041	0.095	6.746	7.708
$^4D_{3/2}$,$^4D_{5/2}$,$^2I_{11/2}$,$^4D_{1/2}$	356	51.26	1.418	1.435	6.351	6.171
$^4D_{7/2}$,$^2I_{17/2}$	328	10.58	0.318	0.063	0.816	1.547

Table 4.1. The integrated absorbance, the experimental and calculated line and oscillator strengths of $Nd^{3+}:LaB_3O_6$ crystal (note: rms $f = 0.744 \times 10^{-6}$)

Crystals	Ω_2 (10^{-20} cm^2)	Ω_4 (10^{-20} cm^2)	Ω_6 (10^{-20} cm^2)	References
LaB_3O_6	0.54	2.31	4.51	[80]
YVO_4	5.88	4.08	5.11	[81]
$CaZn_2Y_2Ge_3O_{12}$	0.94	3.25	3.68	[82]
$Sr_2GdGa_3O_7$	2.94	6.93	6.96	[83]
$Gd_3Ga_5O_{12}$	0.05	2.9	9.3	[81]
YAG	1.0	2.9	9.3	[81]
$Ca_3Sc_2Ge_3O_{12}$	0.99	4.24	7.14	[90]
$Lu_3ScGa_3O_{12}$	0.082	2.844	3.137	[84]

Table 4.2. The intensity parameters of $Nd^{3+}:LaB_3O_6$ crystal and the comparison of the intensity parameters of other Nd^{3+} doped crystals

Start levels	Wavelength (nm)	A	β (s^{-1})	τ (μs)
$^4I_{9/2}$	891	917.29	0.336	366.03
$^4I_{11/2}$	1062	1.47×10^3	0.538	
$^4I_{13/2}$	1329	326.55	0.12	
$^4I_{15/2}$	1852	17.05	6.24×10^{-3}	

Table 4.3. The calculated radiative probabilities,radiative branching ratios and radiative time for the emissions from the $^4F_{3/2}$ level of LaB_3O_6:Nd^{3+}

Crystals	Wavelength (nm)	σ_p (10^{-20} cm^2)	β	References
$NdAl_3$ $(BO_3)_4$	1060	28.4	0.518	[85]
$CaSc_2Ge_3O_{12}$:Nd^{3+}	1060	2.6	0.37	[24]
$NdSc_3$ $(BO_3)_4$:Nd^{3+}	1061	2.0×10^2	0.38	[86]
La_2 $(WO_4)_3$:Nd^{3+}	1058	11.2	0.5098	[87]
$NdAl_3$ $(BO_3)_4$	1063 (σ polarized)	16.0	0.52	[88]
	1063.5 (π polarized)	14.3	0.52	[88]
$Gd_{0.8}La_{0.2}VO_4$:Nd^{3+}	1063	33.2	0.498	[89]
LaB_3O_6:Nd^{3+}	1062	3.46	0.538	[80]

Table 4.4. The comparison of the $^4F_{3/2} \rightarrow {^4I_{11/2}}$ emission cross-section and radiative branching ratios of Nd^{3+} doped crystals.

4.3 The laser characteristics of Nd^{3+}-doped LaB_3O_6 crystal

A method utilizing an unprocessed Nd^{3+}-doped LaB_3O_6 crystal cleavage microchip as the solid-state laser gain medium was proposed by Prof. Huang[10]. Pumped by a Ti:sapphire laser at 871 nm, 1060 nm continuous-wave laser emission with slope efficiency of 23% has been achieved in an unprocessed microchip directly obtained from a cleavage Nd^{3+}:LaB_3O_6 crystal. Fig.4.5 shows the infrared laser output power at 1060 nm as a function of absorbed pump power at 871 nm. A maximum output power of 112 mW was obtained when the absorbed pump power was 580 mW. The laser performance of the unprocessed cleavage Nd^{3+}:LaBO microchip cannot compare with those of other microchip lasers yet, such as widely investigated Nd^{3+}:YAG and Nd^{3+}:YVO_4.[91~94]

5. Summary

The growth, thermal, optical and spectrum characteristics and laser characteristics of rare earth-doped $Ln_2Ca_3B_4O_{12}$ (Ln = La, Gd, or Y) double borate family laser crystals, $Ca_3(BO_3)_2$ and LaB_3O_6 laser crystals were reviewed.

From a passively mode-locked Yb:$Y_2Ca_3(BO_3)_4$ (Yb:CYB) laser, the 244 *fs* pulses with a repetition rate of ~55 MHz were obtained at the central wavelength of 1044.7 nm. The measured average output power amounted to 261 mW. Q-switching and Q-switched mode-locked Yb:$Y_2Ca_3B_4O_{12}$ lasers with an acousto-optic switch were also demonstrated. In the Q-switching case, an average output power of 530 mW was obtained at the pulse repetition rate of 10.0 kHz under the absorbed pump power of 6.1 W. The minimum pulse width is 79 ns at the repetition rate of 1.7 kHz. The pulse energy and peak energy are calculated to be 231 μJ and 2.03 kW, respectively. In Q-switched mode-locking case, the average output

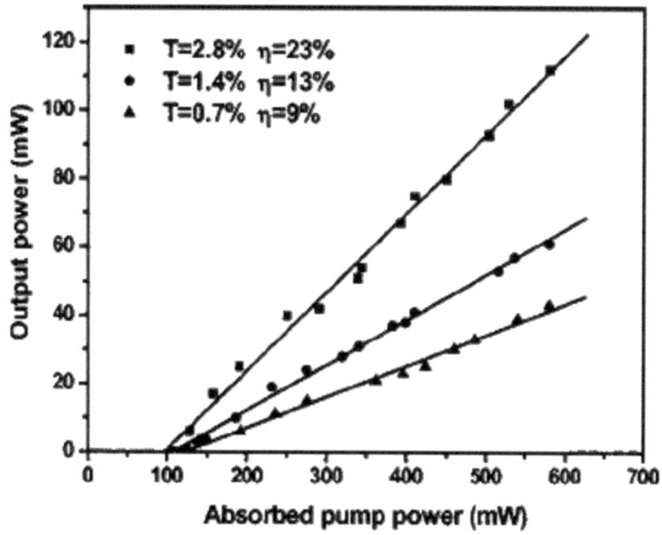

Fig. 4.5. Infrared laser output power at 1060 nm as a function of absorbed pump power at 871 nm

power of 64 mW with a mode-locked pulse repetition rate of 118 MHz and Q-switched pulse energy of 48 μJ was generated under the absorbed pump power of 6.1W. Meanwhile, Yb^{3+}-doped $Y_2Ca_3B_4O_{12}$ diode-pumped laser operation in continuous-wave (CW) was reached. The differential slopes of the CW output power are in the 22-40% range under different experimental conditions. Continuous tuning of the laser wavelength is obtained in the 1020-1057 nm range, in agreement with the broad emission spectra. Also, the diode-pumped multi-wavelength continuous-wave laser operation of the disordered $Yb:Ca_3Gd_2(BO_3)_4$ (Yb:CGB) crystal was reached. An output power of 1.4 W was obtained when quadruple wavelengths were emitted simultaneously, corresponding to a slope efficiency of 23.7%.

Finally, the laser property of microchip $Nd^{3+}:LaB_3O_6$ crystal are reviewed. Pumped by a Ti:sapphire laser at 871 nm, 1060 nm continuous-wave laser emission with slope efficiency of 23% has been achieved in an unprocessed microchip directly obtained from a cleavage $Nd^{3+}:LaB_3O_6$ crystal. A maximum output power of 112 mW was obtained when the absorbed pump power was 580 mW.

6. Acknowledgements

Some works of this chapter were supported by National Nature Science Foundation of China (50902129, 61078076, 91122033), Major Projects from FJIRSM (SZD09001), the Knowledge Innovation Program of the Chinese Academy of Sciences (KJCX2-EW-H03), Science and Technology Plan Major Project of Fujian Province of China (2010I0015).

7. References

[1] Xiuai Lu,Zhenyu You,Jianfu Li,et al., Journal of Crystal Growth,281(2-4)(2005) 416;

[2] Xiuai Lu, Master dissertation, The study of new－style rare earth doped $Ca_3(BO_3)_2$and $NaY(MoO_4)_2$ laser crystals, Graduated School of Chinese Academy of Sciences, 2006.
[3] P.H. Haumesser and R. Gaume et al., J. Crystal Growth 233 (2001), 233.
[4] Guohua Jia, Chaoyang Tu*,Zhenyu You, et al., J. Cryst. Growth, 266 (4) (2004)492;
[5] Chaoyang Tu, Yan Wang, Zhenyu You, Journal of Crystal Growth, 265(1-2)(2004) 154-158;
[6] Yan Wang，Chaoyang Tu*, ChangChang Huang, Journal of Materials Research 19 (4) (2004) :1204;
[7] E. Cavalli, E. Bovero, and A. Belletti, J. Phys.: Condens. Matter 14, 5221 (2002).
[8] Y. J. Chen, X. H. Gong, Y. F. Lin, et al., J. Appl. Phys. 99(2006)103101.
[9] Y. J. Chen, X. H. Gong, Y. F. Lin, et al., Appl. Phys. B 83(2006)195.
[10] Y. J. Chen, Y. D. Huang*, X. Q. Lin,et al., Appl. Phys. Lett. 86(2005)021115.
[11] A. Vegas, F.H. Cano and S. Carcia-Blanco, Acta Crystallographica B 31 (1975), 1416;
[12] H. Choosuwan, R. Guo, A.S. Bhalla, U. Balachandran, J. Appl. Phys. 91(2002):5051
[13] J.J. Carvajal, R. Sole, Jna. Gavalsa, et al. Chem. Matt. 15 (2003):2730;
[14] Xiuai Lu, Zhenyu You, Jianfu Li, et al., Journal of Crystal Growth, 281(2-4)(2005)416-425;
[15] Xiuai Lu, Zhenyu You, Jianfu Li, et al., Physica Status Solidi (a) 203 (2006)551;
[16] E. Cavalli, E. Bovero, and A. Belletti, J. Phys.: Condens. Matter 14(2002)5221;
[17] A. A. Kaminskii, in: Crystalline Lasers: Physical Processes and Operating Schemes (Boca Raton, FL: Chemical Rubber Company, 1996);
[18] H. R. Xia, X. L. Meng, M. Guo, L. Zhu, H. J. Zhang, and J. Y. Wang, J. Appl. Phys. 88, 5134 (2000);
[19] M. H. Randles, J. E. Creamer, and R. F. Bclt, Opt. Soc. Am. B 10(1998)289;
[20] T. Jansen, V. G. Ostroumov, J.-P. Meyn, G. Huber, A. I. Zagumennyi, and I. A. Scherbakov, Appl. Phys. B 58(1994)373;
[21] G. H. Jia, C. Y. Tu, J. F. Li, et al., Opt. Commun. 242 (2004)79;
[22] H. D. Jiang, H. J. Zhang, J. Y. Wang, et al., Opt. Commun. 198(2001)447;
[23] E. Cavalli, E. Zannoni, A. Belletti, V. Carozzo, A. Toncelli, M. Toncelli, and M. Bettinelli, Appl. Phys. B 68 (1999)677;
[24] A. A. Kaminskii, Zh. Tekh. Fiz. Pis. 1 (1975)256;
[25] R. C. Powell, in: Physics of Solid-State Laser Materials (Springer, New York, 1998)
[26] D. Jaque, O. Enguita, U. Caldino, G. M. O. Ramirez, J. Garcia Sole, C. Zaldo, J. E. Munoz-Santiuste, A. D. Jiang, and Z. D. Luo, J. Appl. Phys. 90(2001)561;
[27] Y. J. Chen, X. Q. Lin, and Y. D. Huang, Chem. Phys. Lett. 381(2003)598;
[28] Xiuai Lu, Zhenyu You, Jianfu Li, et al., J. Appl. Phys ,100(2006)033103;
[29] I. Sokólska, Appl. Phys. B: Lasers Opt. 71(2000)157;
[30] S. A. Payne, L. L. Chase, L.-K. Smith, L. Kway, and W. F. Krupke, IEEE J. Quantum Electron. 28(1992) 2619;
[31] G. S. Ofelt, J.Chem.Phys.37(1962)511
[32] S. Zhang, X. Wu, Y. Song, D. Ni, et al., J. Cryst. Growth 252(2003) 246;
[33] W. T. Carnall, P. R. Field, and K. Rajnak, J. Chem. Phys. 49(1968)4424;
[34] M. J. Weber, Phys. Rev. 157(1967)262;

[35] F. S. Ermeneux, R. Moncorge, P. Kabro, et al., OSA Trends Opt. Photon. Series 1(1996)498;
[36] Xiuai Lu, Zhenyu You, Jianfu Li, et al., Journal of Physics and Chemistry of Solids, 66(10)(2005)1801;
[37] S. Zhang, X. Wu, Y. Song, et al., J. Cryst. Growth 252 (2003)246.
[38] B.R. Judd, Phys. Rev. 127 (1962)750.
[39] B.V.Mill,A.M.Tkachuk et al. Opt.& Spec.84(1)(1998)65;
[40] Khamaganova,T.N.,Trunov.V.K.et al.Kristallografiva, 35(1990)856;
[41] W.bardsley,D.T.J.hurle & J.B.Mullin.(ed.)Crystal Growth: a tutorial Approach, North-Holland Publishing Company,1979:160;
[42] Chaoyang Tu, Yan Wang, Zhenyu You, et al.,Journal of Alloys and Compounds, 368(1-2)(2004)123;
[43] Chaoyang Tu, Yan Wang, Zhenyu You, et al.,Journal of Crystal Growth, 260(3-4)(2004)410;
[44] H.Jiang,J.Wang, H.Zhang et al.Chem.Phys.Lett. 361(2002)499;
[45] Gaume,B.Viana et al.Opt.Mater. 22(2003)107;
[46] W.F.Krrupke. Curr. Opin. Solid State Mater.Sci.4(1999)197;
[47] P.H.Haumesser,R.Gaume et al. J.Opt.Soc.Am.B, 19(10) (2002)2365;
[48] .H.Jiang, J.Li,J.Wang et al. J.Cryst.Growth. 233(2001)248;
[49] Chaoyang Tu, Doctor Dissertation,The study on the growth,structure,spectra and laser characteristics of new rare earth-actived laser crystals, Graduated School of Chinese Academy of Sciences,2005.
[50] Bruesselbach H, Umida D S. Optics Letter , 21(1996)480;
[51] Shimokozono M, Sugimoto N, Tate A and Katoh Y.Appl.Phys., 68(1996)2177;
[52] Payne S A, Deloash L D, Smith L K et al. J , . Appl. Phys., 76(1994)497;
[53] Deloach L D , Payne S A, Smith L K et al. J. Opt . Soc . Am ., B, 11(1994) 269;
[54] Kuleshov N V, Lagatsky A A, Shcherbitsky V G, et al . Appl. Phys., B64(1997)409;
[55] Yan Wang, Chaoyang Tu*, Zhenyu You, et al.,Optical spectroscopy of $Ca_3Gd_2(BO_3)_4$:Nd^{3+} laser crystal", Journal of Modern Optics, 53(8) (2006):1141-1148;
[56] Chaoyang Tu, Yan Wang, Zhenyu You, et al.,Optical Materials, 29(2-3)(2006)257;
[57] Hart D W, Jani M, Barnes N P. Optical Letters , 21(1996)728 Communications, 132(1996)107;
[58] Jojmson L F, Geusic J E & Van Uitert L G, Appl. Phys, Lett.,7(5)(1965)127;
[59] Bums G. Phys .Rev., 167(2)(1968)31;
[60] Reinberg A.R. Riseberg L A. Brown R M,et al .GaAs:Si IED Pue
[61] Chen Y , Majior L , and Kushawaha V. Applied Optics, 35(1996) 3203;
[62] Shen D, Wang C, Shao Z, Meng X. and Jian M. Appiled Optics, 35(1996)3203;
[63] Meyn J-P, Jensen T and Huber G.IEEE, Quantum ;
[64] Ostroumov V G, Heine F, K? ck S, et al. Appl. Pyhs ., B64(1997)301;
[65] Wang Guofu, Doctor Dissertation, in Thesis of Growth and optical Characterisation of Cr^{3+}, Ti^{3+}; and Nd^{3+} doped $RX_3(BO_3)_4$ Borate Crystals, 1996, University of Strathclyde UK;
[66] J.-L. Xu, J.-L. He, H.-T. Huang,et al., Laser Phys. Lett., 1–4 (2010) / DOI 10.1002/lapl.201010089 1;
[67] J.-L. Xu, J.-L. He, H.-T. Huang, et al.,Laser Phys. Lett. 7 (3) (2010)198;

[68] C.-Y. Tu ·Y.Wang ·J.-L. Xu et al., Appl Phys B 101(2010) 855;
[69] A.Brenier, Chaoyang Tu, Yan Wang,et al., Journal of applied physics,104(2008)013102;
[70] A. A. Lagatsky, A. Abdolvand, and N. V. Kuleshov, Opt. Lett. 25(2000)616;
[71] J. Liu, U. Griebner, V. Petrov, H. Zhang, J. Zhang, and J. Wang, Opt. Lett. 30(2005) 2427;.
[72] Y. Kalisky, C. Labbe, K. Waichman, L. Kravchik, U. Rachum, P. Deng, J. Xu, J. Dong, and W. Chan, Opt. Mater. (Amsterdam, Neth.) 19(2002)403;
[73] X. Zhang, A. Brenier, Q. Wang, Z. Wang, J. Chang, P. Li, S. Zhang, S. Ding, and S. Li, Opt. Express 13(2005) 7708;
[74] A. Brenier, C. Tu, Z. Zhu, and J. Li, Appl. Phys. Lett. 90(2007) 071103;
[75] Jin-Long Xu, Chao-Yang Tu, Yan Wang, Jing-Liang He, Optical Materials, In Press, Corrected Proof, Available online 5 July 2011;
[76] G.K. Abdullayev, Kh.S. Mamedov and G.G. Dzhafarov. Kristallografiya 26 (1981) 837.
[77] J.S. Ysker and W. Hofmann. Naturwissenschaften 57 (1970) 129.
[78] A.N. Shekhovtsov, A.V. Tolmachev, M.F. Dubocik et al.. Nucl. Instrum. Methods A 456 (2001) 280.
[79] E.F. Dolzhenkova, A.N. Shekhovtsov, A.V. Tolmachev et al.. J. Crystal Growth 233 (2001) 473.
[80] Guohua Jia, Chaoyang Tu, Jianfu Li, et al.,Optics Communications, 242(1-3)(2004) 79-85;
[81] A.A. Kaminskii, Crystalline Lasers, Physical processes and Operating Schemes, CRC Press, New York (1996).
[82] D.K. Sardar, S. Vizcarra, M.A. Islam, et al., J. Lumin. 60 (1994) 97.
[83] W. Koechner, Solid State Laser Engineering (third ed.), Springer, Berlin, Heidelberg (1992).
[84] A.A. Kaminskii, G. Boulon, M. Buoncristiani, et al., Phys. Status Solidi A 141 (1994) 471.
[85] Xueyuan Chen, Zundu Luo, D. Jaque, et al., J. Phys.: Condens. Mat. 13 (2001) 1171
[86] John B. Gruber, Thomas A. Reynolds, Douglas A. Keszler and Bahram Zandi, J. Appl. Phys. 87 (2000) 7159.
[87] Yujin Chen, Xiuqin Lin, Xiuqin Lin and Yidong Huang, Chem. Phys. Lett. 381 (2003)598.
[88] D. Jaque, O. Enguita, U. Caldino, et al., J. Appl. Phys. 90 (2001)561.
[89] H.D. Jiang, J.Y. Wang, H.J. Zhang, et al., J. Appl. Phys. 92 (2002)3647.
[90] K.A. Gschneidner and Le Roy Eyring, Handbook on the Physics and Chemistry of Rare Earths vol. 9, North Holland (1987) pp. 11.
[91] T. Sasaki, T. Kojima, A. Yokotani, O. Oguri, and S. Nakai, Opt. Lett. 16(1991)1665;
[92] G. J. Spuhler, R. Paschotta, M. P. Kulloerg, M. Graf, M. Moser, E. Mix, G. Huber, C. Harder, and U. Keller, Appl. Phys. B: Lasers Opt. 72(2001)285;
[93] V. Lupei, N. Pavel, and T. Taira, Appl. Phys. Lett. 80(2002)4309;
[94] Y. Sato, T. Taira, N. Pavel, and V. Lupei, Appl. Phys. Lett. 82(2003)844;

4

Rare-Earth-Doped Low Phonon Energy Halide Crystals for Mid-Infrared Laser Sources

M. Velázquez[1], A. Ferrier[2], J.-L. Doualan[3] and R. Moncorgé[3]
[1]*CNRS, Université de Bordeaux, ICMCB, Pessac*
[2]*LCMCP,CNRS-Université de Paris 6-Collège de France-Paris Tech,Paris*
[3]*CIMAP, CEA-CNRS- ENSICaen-Université de Caen, Caen*
France

1. Introduction

Since ~15 years, solid state lasers emitting in bands II (2.7-4.3, 4.5-5.2 μm) and III (8-14 μm) of the atmosphere transparency spectral range are being developed for imaging, polluting species detection as well as military NRBC detection and optronic countermeasures. Because most of these applications require highly brilliant and/or important peak power laser sources, several RE^{3+}-doped (RE=rare earth) low phonon energy ($\hbar\omega$<400 cm^{-1}) chloride and bromide crystals, such as APb_2X_5 (A=K,Rb;X=Cl,Br) or $CsCdBr_3$, stand out as promising laser gain media in the mid-infrared (MIR) spectral range [Doualan & Moncorgé, 2003; Isaenko et al., 2008]. Indeed, these bulk crystals, transparent up to more than 18 μm, could be used at room temperature and their emission lifetime-emission cross section product ($\sigma_{EM}\tau_R$) at the laser wavelength is high enough to allow for energy storage and subsequent pulsed regime laser operation. Moreover, in these systems, the laser beam quality could be high even at high output powers.

This chapter is composed of three parts. The first one reviews all the successful laser operations ever demonstrated. The basic thermal, mechanical, optical and spectroscopical properties characterizations are presented for a series of halide crystals : single crystal Raman spectroscopy, Fourier-transformed infrared (FTIR) spectroscopy, X-Ray diffraction (XRD) and thermal conductivity data, showing at a glance the transparency range, the highest phonon energies, the site symmetry of RE^{3+} ions as well as mechanical hardness, among other laser-related characteristics. The second part deals with RE^{3+} ions spectroscopy related to the pumping strategy for a better inversion population kinetics at play during laser operation. Such mechanisms as upconversion energy transfers (ETU) and excited-state absorption (ESA) are detailed in the case of Er^{3+} and Pr^{3+} ions. Spectroscopic data on more exotic Tl_3PbX_5 (X=Cl,Br) crystals, which are optically non linear in addition to the general propensity to MIR laser operation is presented only to illustrate crystal field strength trends affecting the absorption and emission bands, as well as energy transfer mechanisms between Er^{3+} ions and ultimately gain cross sections. The third part of this chapter addresses the synthesis and crystal growth of pure and RE-doped chlorides and bromides in relation to their spectroscopical and laser operation properties. The choice of such growth parameters as the nature and shape of the crucible, the nature of the gas and its pressure, the growth rate, is fundamental to avoid bubble formation, stabilize RE^{3+} oxidation state, minimize the complications arising from crystallographic phase transition and the mechanical stresses upon cooling, control as much as possible RE^{3+}-ion segregation, and so on. All these aspects ultimately affect laser operation and the relationships between growth conditions, growth defects and laser performances have scarcely been discussed in the literature on these laser materials, which is surprising if these crystals are to be produced on a large industrial scale by the well spread Bridgman-Stockbarger method.

2. MIR solid state laser operation

To date, the longest laser wavelength ever achieved, 7.15 μm [Bowman et al., 1994, 1996], has been obtained with a 4-mm long $LaCl_3:Pr^{3+}$ (0.7 % at.) single crystal operated in the pulsed regime. This pioneering work by Bowman and his collaborators has not been reproduced since 1996, probably because the crystals were so hygroscopic that they had to be kept in a cryostat during all the measurements to avoid their deliquescence. In these crystals, the most energetic phonon vibrates at 210 cm^{-1} and the transparency range extends up to 15 μm. The pumping scheme of this laser emission involves an ETU mechanism by photon addition from the thermalized $^3H_6$ and $^3F_2$ levels according to $2{*}^3H_6 \rightarrow {}^3H_5 + {}^3F_3$ (figure 1). The laser crystal is pumped at 2.02 μm on the $^3F_2$ level by means of a Tm:YAG laser (being itself diode- or flashlamp-pumped) delivering free-running pulse trains at frequencies from 2 to 100 kHz with an average power 70 W. In spite of this "awkward and expensive 2 μm pumping scheme", as Howse *et al.* recently put it [Howse et al., 2010], the crystal exhibits a high absorption cross section around 2 μm (10 times higher, for instance, than the absorption one around 800 nm, which corresponds to the $^4I_{9/2}$ pump level of $Er^{3+}:KPb_2Cl_5$, figures 2 and 3). The laser emission centered at 7.15 μm occurs between levels $^3F_3$ (the experimental lifetime of which τ_2, measured by direct pumping with an Er^{3+} laser at 1.6 μm, equals 58 μs) and $^3F_2$. At 20 °C, the laser slope efficiency reaches 3.9 %, the absorbed power conversion yield 2.3 % and the pump threshold ~4 mJ. Bowman *et al.* suggested that thermalization of levels $^3F_4$ and $^3F_3$ would make efficient a direct pumping at 1.5 μm on level

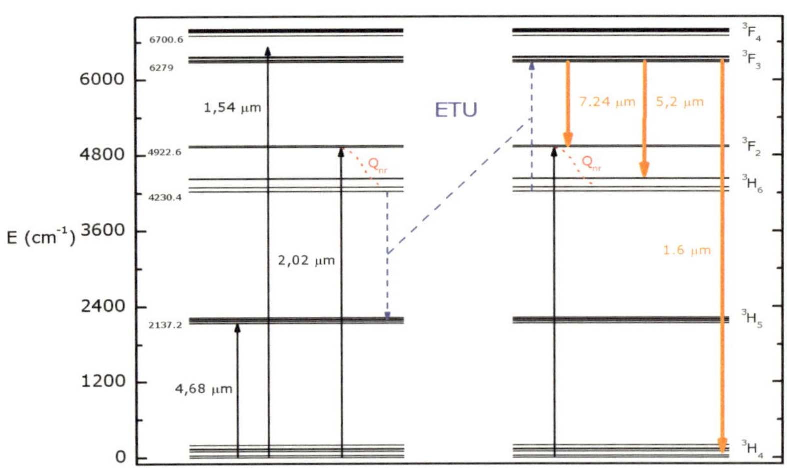

Fig. 1. Energy level diagram of Pr^{3+} ions in the $LaCl_3$ host crystal. Blue arrows indicate the phonon assisted non radiative ETU mechanism.

$^3F_4$ (figure 1). They were also the first to tune laser operation from 4.42 to 4.70 µm on the $^4I_{9/2}\rightarrow^4I_{11/2}$ transition of Er^{3+} ions in KPb_2Cl_5 single crystals, with diode pumping at 800 nm and an absorbed power conversion yield at room temperature of 7.6 % (to be compared with the quantum yield of 17.4 %) [Bowman et al., 1999, 2001; Condon et al., 2006a] (see in Fig. 2).

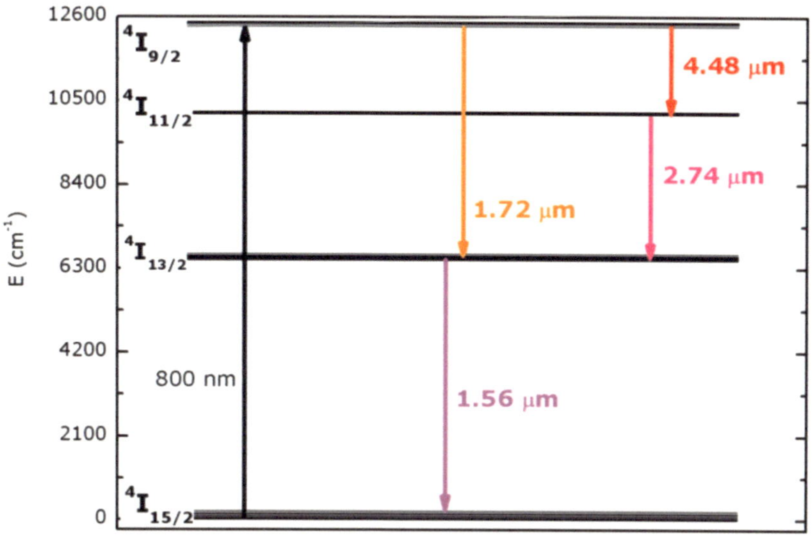

Fig. 2. Four first energy levels diagram of Er^{3+} ions in Tl_3PbBr_5 and their experimental lifetimes.

The main reason why the APb_2X_5 (A=K,Rb;X=Cl,Br) family of laser hosts has triggered a continuous breed of publications since 2001 lies in the non hygroscopicity of the crystals, which turns out to be unusual among chlorides and bromides host crystals with luminescent properties [Egger at al., 1999; Kaminska et al., 2011; Nitsch et al., 1993, 1995a, 1995b, 2004; Nitsch & Rodová 1999; Riedener et al., 1997; Rodová et al., 1995; Vinogradova et al., 2005; Zhou et al., 2000]. The interest of diode pumping lies in the simple and compact cavity design (cavity length of 8 mm with KPb_2Cl_5:Er^{3+} crystals), and that of KPb_2Cl_5 crystals, as previously stated, to manipulate a non hygroscopic and air stable gain medium. Even if a 19 W power diode had to be used (absorption rate ~4 %) with a beam waist ~150 µm, it should be noted that this crystal was efficiently operated at ~50 Hz, without any cooling system, clearly establishing its satisfying thermomechanical properties (table 1). Laser operation was demonstrated at 5.5 µm in $RbPb_2Cl_5$ crystals doped with Dy^{3+} ions (2.10^{19} cm^{-3}, $^6H_{9/2}+^6F_{11/2}\rightarrow^6H_{11/2}$), with flash-lamp pumped YAG:Nd^{3+} laser operating in free multimode simultaneously at 1.32 and 1.34 µm, with a repetition rate of 2.5 Hz and a beam waist in the crystal ~300 µm [Okhrimchuk et al., 2007]. Preliminary laser tests on unoriented crystals obviously full of scattering defects gave a laser slope ~0.1 % and a threshold of 25 mJ. Although Dy^{3+} ion spectroscopy were investigated in $CaGaS_4$ and KPb_2Cl_5 crystals, because of their seemingly promising laser transitions $^6H_{11/2}\rightarrow^6H_{13/2}$ at 4.31 µm, we shall not insist on that since they can not be currently diode pumped at 1.32 µm [Nostrand et al., 1998, 1999; Okhrimchuk, 2008].

Properties	KPb_2Cl_5	$RbPb_2Cl_5$	KPb_2Br_5	Tl_3PbBr_5	$CsCdBr_3$
Structure, space group	Monoclinic, $P2_1/c$	Monoclinic, $P2_1/c$	Monoclinic, $P2_1/c$	Orthorhombic, $P2_12_12_1$	Hexagonal, $P6_3/mmc$
Cell parameters (Å)	a=7.919 Å b=8.851 Å c=12.474 Å β=90.13 °	a=8.959 Å b=7.973 Å c=12.492 Å β=90.12 °	a=9.256 Å b=8.365 Å c=13.025 Å β=90.00(3) °	a=15.395 Å b=9.055 Å c=8.544 Å	a=7.681 Å c=6.726 Å
Z	4	4	4	4	2
Volumic mass (g.cm^{-3})	4.629	5.041	5.619	6.80	4.77
RE^{3+} substitution site density (cm^{-3})	Pb^{2+}, C_1, 4.6×10^{21} (*)	unknown, C_1, 4.5×10^{21} (*)	unknown, C_1, 4.0×10^{21} (*)	unknown, C_1, 3.4×10^{21}	Cd^{2+}, D_{3d} (C_{3v} with charge compensation) 5.8×10^{21}
Thermal conductivity at room temperature (W.m^{-1}.K^{-1})	4.62	4			0.3
Young moduli (GPa)	E_1~29.9 E_2~24.9 E_3~27.0				E_1=E_2~22.2 E_3~29.4
Shear moduli (GPa)	G_{44}~11 G_{55}~11.1 G_{66}~14.3				G_{44}=G_{55}~4.5 G_{66}~7.7
Hardness (Mohs)	2.5	2.5	2.5	2 502 MPa	294 Mpa
Thermal expansion coefficients (10^{-6} K^{-1})	α_a=38.6 α_b=42.6 α_c=39.3	α_a=26 α_b=36 α_c=33	α_a=40 α_b=36 α_c=28	α_a=41 α_c=62	
Maximum phonon frequency (cm^{-1})	204.8	203	138	137.5	161
Transparency range (μm)	0.3-20	0.37-20	0.4-30	0.4->24	0.28-35
Energy bandgap (eV)	4.79	4.83	4.12		~3.5-4
Refractive index	n_x=1.9406 n_y=1.9466 n_z=1.9724 (@ 1 μm)	n=2.019 (@ 0.63 μm)	n_x=2.191 n_y=2.189 n_z=2.247 (@ 0.63 μm)	n~2.23 (@ 0.63 μm)	1.76 (@ 1 μm)
Thermo-optic coefficient (10^{-5} K^{-1}, @ 1 μm)	-7.0 -10.0 -10.5		-13.0 -14.1 -14.9		
Er^{3+}-doped crystals (figure 2)					
Absorption cross section around 800-810 nm (10^{-21} cm^2)	2.4			2.9	

$^4I_{9/2}$ experimental lifetime (ms)	2.62		1.2 or 1.9	1.8	10.7
$^4I_{9/2} \rightarrow ^4I_{11/2}$ branching ratio (%)	1.2				
$^4I_{11/2}$ experimental lifetime (ms)	3.23		2.1	3.1	
Emission cross section around 4.4-4.6 μm (10^{-21} cm^2)	2			2.1	
$\sigma_{EM}\tau_{exp}$ around 4.4-4.6 μm (10^{-24} $cm^2.s$)	5.2			3.6	
$C_{DD}(^4I_{9/2},^4I_{15/2}) \rightarrow (^4I_{15/2},^4I_{9/2})$ (10^{-40} $cm^6.s^{-1}$)	1.19			0.95	
$C_{DA}(^4I_{9/2},^4I_{11/2}) \rightarrow (^4I_{15/2},^4F_{3/2})$ (10^{-40} $cm^6.s^{-1}$)	2.54			0.48	P_{DA}=1942 s^{-1}
$C_{DA}(^4I_{9/2},^4I_{13/2}) \rightarrow (^4I_{15/2},^4S_{3/2})$ (10^{-40} $cm^6.s^{-1}$)	1			0.44	
Auzel crystal field parameter (cm^{-1})	1771 (C_s/C_2)			1327 (C_s/C_2)	1502 (C_{3v})
Pr^{3+}-doped crystals (figure 1)					
Absorption cross section around 2 μm (10^{-20} cm^2)	3.5		4.6	1.9	5.5
$^3F_3$ experimental lifetime (ms)	0.362	0.11	0.3	0.89	1.57
$(^3F_4,^3F_3) \rightarrow ^3F_2$ branching ratio (%)	ε			ε	ε
$^3F_2$ experimental lifetime (ms)	0.95	1	1.5	3.1	3.6
$^3H_5$ experimental lifetime (ms)	4.5	5	26.5	42.2	27.5
Emission cross section around 4.6-4.8 μm (10^{-21} cm^2)	4.8		4	5.4	3.2
$\sigma_{EM}\tau_{exp}$ around 4.6-4.8 μm (10^{-23} $cm^2.s$)	2.2		10.6	22.8	8.8
Auzel crystal field parameter (cm^{-1})	1746 (C_s/C_2)				4968 (C_{3v})

Table 1. Laser-related crystallographic, thermal, mechanical, optical and spectroscopical properties of selected low phonon energy halide crystals [Aleksandrov et al., 2005; Atuchin et al., 2011; Cockroft et al., 1992; Doualan & Moncorgé, 2003; Ferrier et al., 2006a, 2007, 2008a, 2008b, 2009a, 2009b; Heber et al., 2001; Isaenko et al., 2008, 2009a, 2009b; Malkin et al., 2001; Mel'nikova et al., 2005; Merkulov et al., 2005; Neukum et al., 1994; Nostrand et al., 2001; Okhrimchuk et al., 2006; Popova et al., 2001; Quagliano et al., 1996; Ren et al., 2003; Singh et al., 2005; Velázquez et al., 2006a, 2006b; Virey et al., 1998; Vtyurin et al., 2004, 2006]. (*) per lead type of crystallographic site.

3. Spectroscopic parameters for laser operation

3.1 Er^{3+}-doped single crystals

The great number of close energy levels within the 4f configuration explains the interest for RE ions as optically active species for MIR applications. When dissolved in low phonon energy crystals such as halides and bromides, nonradiative multiphonon emission probabilities between the two levels of the laser transition are significantly reduced. Hence, they allow for reaching long emitting level lifetimes, on the order of a few tens of µs to tens of ms. Consequently, RE-doped APb_2X_5 crystals are likely to insure sufficient energy storage for amplification. The shape and magnitude of the absorption bands around 800 and 980 nm (figure 3), where efficient, compact, rugged, high-powered and cheap laser diodes are easily available as pumping sources, has been widely characterized. The forced electric dipole emission bands from the Er^{3+} ions $^4I_{9/2}$ multiplet, which are never obtained in the MIR range (1.7 µm, 4.5 µm) in oxides and fluorides, was also exhaustively discussed (figure 4). On the other hand, energy transfers between Pr^{3+} ions (figure 1), on which the world record in terms of laser wavelength is based, has been investigated with an emphasis put on ion pairing effects, by comparing the efficiency of energy transfers in KPb_2Cl_5, Tl_3PbBr_5 and $CsCdBr_3$, all of which being non hygroscopic. Exactly as in the case of Er^{3+} ions, the shape, magnitude and possible polarization effects of the absorption and emission bands involved in laser operation around 5 µm have been commented and compared to well established laser systems in the near infrared such as Nd^{3+}:YAG.

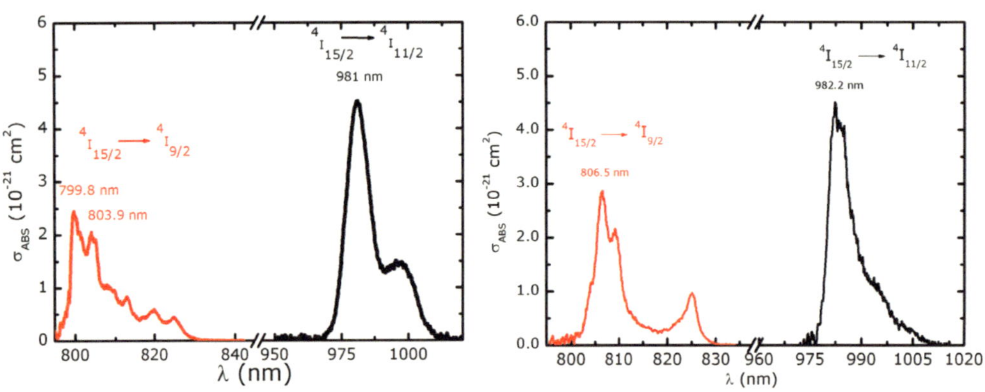

Fig. 3. Room temperature absorption cross section around 800 and 980 nm, left : KPb_2Cl_5:Er^{3+} ; right : Tl_3PbBr_5:Er^{3+}.

Absorption (or transmission), emission and excitation spectra recorded as a function of temperature at low temperature (10-50 K) over a broad spectral range allow for accurately determining (±3 cm^{-1}) the crystal-field energy sublevels of Er^{3+} ions dissolved into KPb_2Cl_5 and Tl_3PbBr_5 and fitting the energy level structure with parameterized crystal-field model Hamiltonians. This makes possible, among other things :

- to calculate the electronic sum over states function of the first 11 multiplets (from $^4I_{15/2}$ to $^2H_{9/2}$)
- to characterize the possible existence of several incorporation sites for RE ions

The knowledge of these data is mandatory not only to estimate absorption and emission cross sections of the optical lines exploited for laser operation, but also to assign the bands observed on the excitation and anti-Stokes emission spectra. This has given a strong impetus to spectroscopic investigations in Russia, in Europe and in the USA [Balda et al., 2004; Gruber et al., 2006; Jenkins et al., 2003; Quimby et al., 2008; Tkachuk et al., 2005, 2007]. The peak-by-peak assignment of crystal-field sublevels also permits to check the number of peaks expected for a complete degeneracy lift of each multiplet (J+1/2). This demonstrates the occurrence of only one symmetry-type site for Er^{3+} ions in these two host compounds, *a priori* C_1 (table 1). However, as a great number of non equivalent defects is likely to form (for instance, more than one hundred in KPb_2Cl_5 [Velázquez et al., 2006b]) and the point group symmetry is very low for all the atomic positions, the precise characterization of the Er^{3+} incorporation sites in both compounds remains difficult. Crystal-field calculations with C_s/C_2 point groups lead to a satisfactory agreement between experimental and calculated energy levels [Ferrier et al., 2007, 2008a; Gruber et al., 2006].

Absorption cross sections around 800 and 980 nm, spectral range in which diode pumping is commonly available, are shown in figure 3. The absorption bands are large (for a RE^{3+}-doped system) and so suitable for diode pumping. Judd-Ofelt analysis performed with absorption spectra over the first 11 excited states showed that the branching ratio of the $^4I_{9/2} \rightarrow {}^4I_{11/2}$ transition around 4.5 μm amounts to only 1.2 %. But even if this is low as compared to other laser systems, these absorption cross sections are typical of forced electric dipole transitions and sufficient to be exploited under diode pumping [Bowman et al., 1999, 2001; Condon et al., 2006a], provided that a high power is used. Cross section calibrated emission spectra obtained under excitation at 800 nm display a poorly structured and virtually independent on polarization broad band, which exemplifies the interest in chlorides and bromides for MIR laser applications (figure 4).

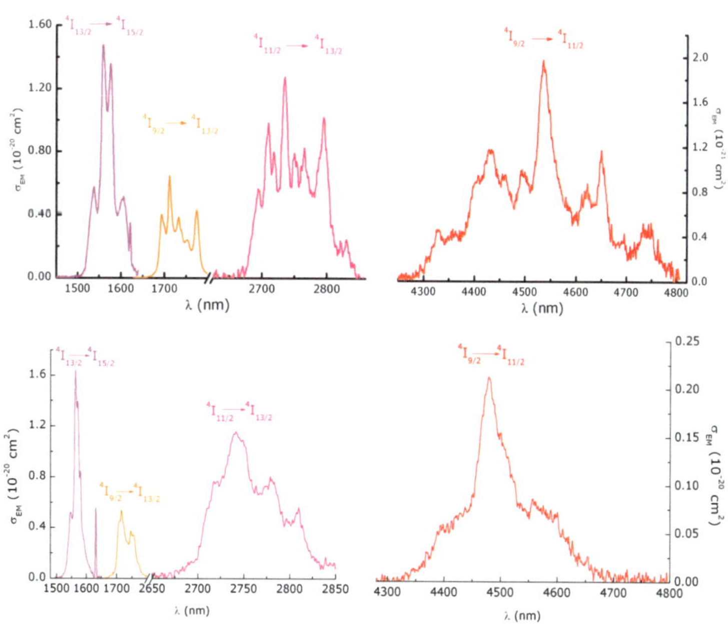

Fig. 4. Er^{3+} ions emission spectra in KPb_2Cl_5 (top) and in Tl_3PbBr_5 (bottom) at room temperature, under cw excitation by Sa:Ti laser at 800 nm.

Indeed, emissions from multiplet $^4I_{9/2}$, to multiplet $^4I_{13/2}$ at 1.7 μm, and to multiplet $^4I_{11/2}$ at 4.5 μm are never observed neither in oxides nor in fluorides. In particular, in $Tl_3PbBr_5:Er^{3+}$, 15 « Raman » phonons of the highest energy are required to match the energy difference between the two multiplets implied in the laser transition $^4I_{9/2}\rightarrow{}^4I_{11/2}$, which entails multiphonon emission kinetics much slower than radiative de-excitation ones (figure 2).

Emission cross-sections, ~2.10^{-21} cm^2 ($\sigma_{EM}\tau_R$~(3-5).10^{-24} cm^2.s at λ_{max}~4.5 μm), are as high in Tl_3PbBr_5 as in KPb_2Cl_5, and the experimental lifetimes of the three first excited levels 2 to 4 ms in both compounds, which is favourable to diode pumped pulsed regime laser operation. As the experimental lifetime of the terminal level (more than 3 ms) is longer than that of the emitting one (around 2 ms), we investigated the possibility of observing anti-Stokes emissions at room temperature and recording their excitation spectra likely to unveil parasitic mechanisms depleting $^4I_{9/2}$ and $^4I_{11/2}$ energy levels. Figure 5 shows the anti-Stokes luminescence issued from $^4G_{11/2}$, $^2H_{9/2}$ (violet), $^4F_{3/2}/^4F_{5/2}$ (blue), $^2H_{11/2}$, $^4S_{3/2}$ (green), $^4F_{9/2}$ and $^2H_{11/2}$ (red) by Ti:Sa laser cw excitation at 804 nm, in Er^{3+}-doped KPb_2Cl_5 and Tl_3PbBr_5 crystals. In order to understand by which mechanism(s), likely to affect population inversion kinetics during laser operation, all these levels gets populated in spite of the fact that excitation around 800 nm is non resonant, it proved useful to get excitation spectra of these emissions as well as fluorescence decay measurements under pulsed non resonant excitation. References [Balda et al., 2004; Ferrier et al., 2007, 2008a; Quimby et al., 2008; Tkachuk et al., 2007] highlight the relevance of such processes as :

- excited state absorption $^4I_{9/2}\rightarrow{}^2H_{9/2}$ from 816 to 837 nm (σ_{ESA}~(2-3).10^{-21} cm^2), $^4I_{11/2}\rightarrow(^4F_{3/2},^4F_{5/2})$ at 815 and 839 nm (σ_{ESA}~(1-1.5)×10^{-20} cm^2) and $^4I_{13/2}\rightarrow(^2H_{11/2},^4S_{3/2})$ around 800 and 847 nm (σ_{ESA}~(3-4).10^{-21} cm^2) ;
- resonant energy transfers by ETU $2{*}^4I_{9/2}\rightarrow{}^4I_{15/2}+{}^2H_{9/2}$, $^4I_{11/2}+{}^4I_{9/2}\rightarrow{}^4I_{15/2}+(^4F_{3/2},^4F_{5/2})$ and $2{*}^4I_{9/2}\rightarrow{}^4I_{13/2}+{}^4S_{3/2}$ around 800 nm (figure 6), the laser pumping wavelength which is foreseen in this system.

In Tl_3PbBr_5 crystals, excited state absorption cross section are virtually the same in magnitude but with a systematic red shift ~5-6 nm. Time resolved emission spectroscopy is mandatory to quantify the different relaxation rates, since population mechanisms of $^2H_{9/2}$ and $(^4F_{3/2},^4F_{5/2})$ levels coexist more or less as a function of the excitation wavelength. Calibration of the spectra in cross sections units by successive application of Fuchtbauer-Ladenburg and MacCumber relationships permits to estimate the energy transfer microparameters which appear in the rate equations driving the population inversion kinetics during laser operation, and consequently to optimize the wavelength and temporal width of pumping pulses. This investigation demonstrated that the effect of ETU (see in Fig. 6) turns out to be completely negligible until the concentration of 3.3 mol % is achieved in KPb_2Cl_5 crystals [Ferrier et al., 2007] (which corresponds to an average Er^{3+}-Er^{3+} distance of ~12 Å), and that optical pumping at 800 nm, that is, at the absorption peak in $KPb_2Cl_5:Er^{3+}$, does not lead to substantial excited state absorption losses towards the $^2H_{9/2}$ level (see in Fig. 5). As a matter of fact, when the Er^{3+} ion concentration increases, losses by ETU also increase, but less rapidly than the population rate of the emitting $^4I_{9/2}$ level, so that the resulting population inversion also increases. A striking structure-property relationship can be seen in the fact that the $^4F_{7/2}$ experimental lifetime is seven to eight times longer in Tl_3PbBr_5 (70 μs) and in KPb_2Br_5 (85 μs) than in KPb_2Cl_5 (10 μs), which demonstrates that non radiative multiphonon emission kinetics are much slower in bromides than in chlorides [Hömmerich et al., 2005]. Indeed, the $^2H_{11/2}$ level just below the $^4F_{7/2}$ one lies at 1321 (resp. 1296) cm^{-1} in KPb_2Cl_5 (resp. Tl_3PbBr_5), requiring 6.4 (resp. 9.4) phonons of the highest frequency to match this energy difference. Riedener and Güdel [Riedener & Güdel, 1997] have already observed blue and very weak anti-Stokes luminescence in $RbGd_2Br_7:Er^{3+}$ crystals, that was spaced by a few hundreds of cm^{-1} from intense anti-Stokes green luminescence in $RbGd_2Cl_7:Er^{3+}$ crystals. It was obtained under excitation at 977.6 nm in the bromide and at 975.1 nm in the chloride on the $^4I_{11/2}$ level, followed by ESA on the $^4F_{7/2}$ level.

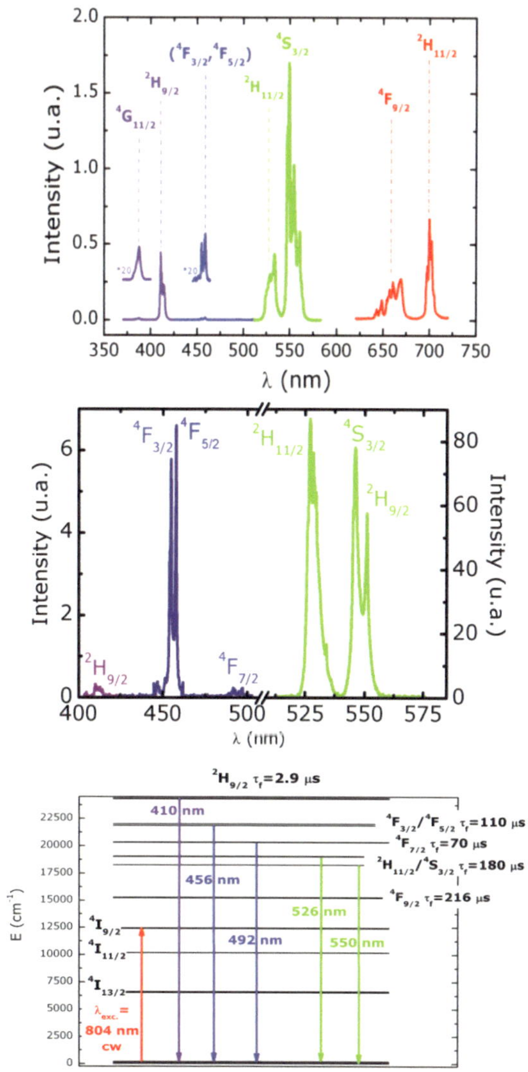

Fig. 5. Top left, anti-Stokes emission spectra of $KPb_2Cl_5:Er^{3+}$ at room temperature under cw excitation at 800 nm ; top right, anti-Stokes luminescence spectra of $Er^{3+}:Tl_3PbBr_5$, transitions from $^2H_{9/2}$ (410 nm), ($^4F_{3/2}$,$^4F_{5/2}$) (456 nm), $^4S_{3/2}$ (550 nm) and $^4I_{13/2}$ (1.55 μm) to $^4I_{15/2}$ under cw excitation at 804 nm. Bottom, energy levels diagram of Er^{3+} ions in Tl_3PbBr_5.

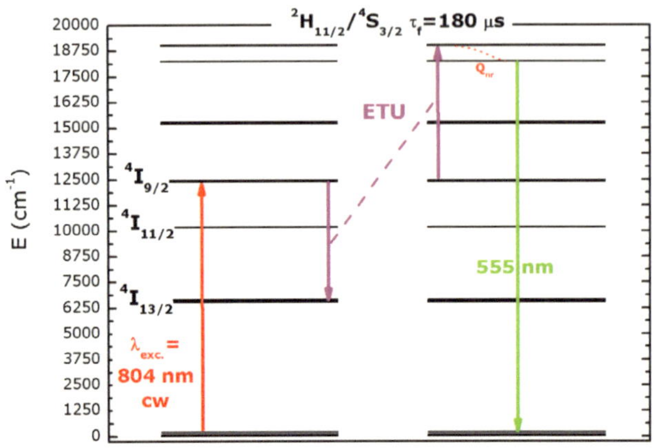

Fig. 6. Green anti-Stokes luminescence of Er^{3+} ions in KPb_2Cl_5 at room temperature, under Sa:Ti laser excitation at 804 nm, and most prominent ETU mechanism.

In addition, the bottleneck effect of the laser effect likely to arise from the long emission lifetime of the $^4I_{13/2}$ level ≈4.6 ms, is unsignificant in pulsed regime because the latter is virtually empty. The gain cross section is displayed in figure 7 and permits the identification of the most probable laser oscillation wavelength at the maxima.

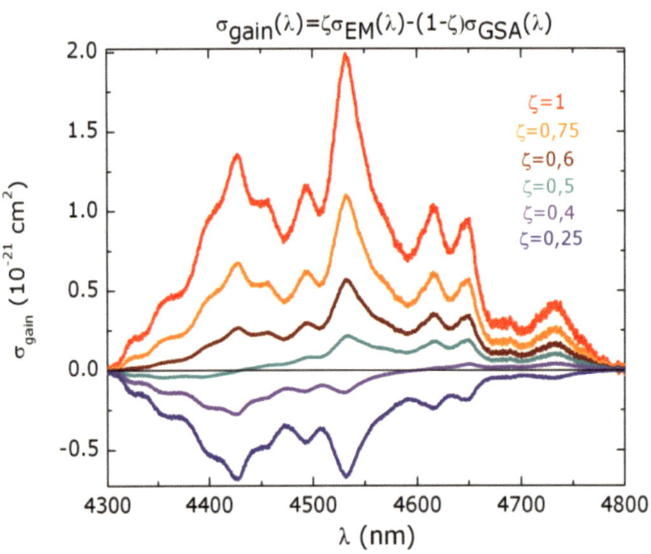

Fig. 7. Gain cross section around 4.5 µm in $KPb_2Cl_5:Er^{3+}$, with $\zeta = n^4I_{9/2}/(n^4I_{9/2}+n^4I_{11/2})$.

3.2 Pr^{3+}-doped single crystals

By adjusting their fluorescence data at 1.6 µm issued from the $^3F_3$ level with the system of population rate equations reported hereafter, Bowman *et al.* could extract the value of the crossed relaxation parameter $\chi \approx 1.3 \times 10^{-16}$ $cm^3.s^{-1}$ and that of the $(^3F_4,^3F_3)\rightarrow(^3F_2,^3H_6)$ transition branching ratio, $\beta \approx 1\%$:

$$\frac{\partial N_{({}^3F_2;{}^3H_6)}}{\partial t} = \frac{\beta}{\tau_2} N_{({}^3F_4;{}^3F_3)} - 2\chi N^2_{({}^3F_2;{}^3H_6)}$$

$$\frac{\partial N_{({}^3F_4;{}^3F_3)}}{\partial t} = \chi N^2_{({}^3F_2;{}^3H_6)} - \frac{N_{({}^3F_4;{}^3F_3)}}{\tau_2}$$

Moreover, by matching the decay rate to an exponential law, they determined the apparent lifetime $"\tau_2" \approx 900\mu s$ (that is, 15.5 times longer than the real τ_2) standing for the electronic feeding of the $^3F_3$ level due to the crossed relaxation mechanism (figure 1). It is worth noting the laser emission centered at 5.2 μm (figure 1) which is obtained at 130 K with the same pumping scheme, a power conversion yield of 23 % and a threshold of 2 mJ that increases strongly with temperature. Homovalent substitution of La^{3+} cations to Pr^{3+} ones modifies very weakly local vibrational frequencies because there is neither vacancies nor interstitial formation in the vicinity of Pr^{3+} ions and the molar masses are very close (only 1.4 % relative difference). It seems that Pr^{3+} ions form pairs, explaining the efficiency of the energy transfer [Guillot-Noël et al., 2004]. This pionnering work of Bowman and his coworkers triggered the interest for a systematic study of Pr^{3+} ion fluorescence kinetics around 7.2 μm, in host compounds as KPb_2Cl_5, Tl_3PbBr_5 and $CsCdBr_3$, all non hygroscopic, by means of two pumping schemes :

- at 2 μm, with a diode pumped Tm^{3+} laser, on the $^3F_2$ level followed by thermalization and ETU ;
- at 1.5 μm, with an Er^{3+} laser, directly on $^3F_4$ level followed by thermalization.

In the near future, such systems will also be pumped at 4.6 μm by means of high power quantum cascade laser (QCL) diodes for the emission at 5 μm from the $^3H_5$ level, particularly in KPb_2Cl_5 because the pumping strategy involving ETU seems more favourable than in $CsCdBr_3$, not only because of lower phonon energies but also because the one dimensional crystal structure of the latter compounds favours ion pairing of Pr^{3+} [Amedzake et al., 2008; An & May, 2006; Balda et al., 2002, 2003; Bluiett et al., 2008; Chukalina, 2004; Ferrier et al., 2009a, 2009b; Gafurov et al., 2002; Guillot-Noël et al., 2004; Neukum et al., 1994; Rana & Kaseta, 1983]. Absorption and emission spectra recorded as a function of temperature at low temperature (10-50 K) over a broad spectral range allows for the determination of Pr^{3+} ions crystal field sublevels dissolved in KPb_2Cl_5, and to match them with the eigenvalues of a parameterized crystal field model hamiltonian taking into account 74 sublevels of the 12 first multiplets. The $^3P_1$ and $^1I_6$ were not included in the refinement procedure because their crystal field sublevels could not be successfully deconvoluted, even at low temperatures. Absorption cross section peaks around 2 μm are virtually twice higher in $CsCdBr_3$ than in Tl_3PbBr_5 crystals, and the absorption line profiles in $CsCdBr_3$:Pr^{3+} differs from those observed in the two other compounds (figure 8). In the former crystals, absorption lines are narrow and structured, whereas in the latter ones they are broad, consistently with point defect disorders expected for the three crystal structures [Ferrier et al., 2006a; Guillot-Noël et al., 2004; Velázquez et al., 2006b]. The absorption cross section at 1.55 μm is of the same magnitude as that at 2 μm in KPb_2Cl_5 and Tl_3PbBr_5, suggesting the possibility of a direct pumping on the ($^3F_4$,$^3F_3$) levels with an Er^{3+} Kigre laser (or any other means : fibre, IR laser diode, etc.). Crystal field calculations with point symmetry C_s/C_2 lead to a good agreement between experimental and simulated energy levels [Ferrier et al., 2008b]. Figure 1 allows for checking the many possible resonances around 2.1 μm, between $^3H_4$, $^3H_5$, ($^3F_2$,$^3H_6$) and ($^3F_4$,$^3F_3$) levels, propicious to efficient energy transfers (even at low Pr^{3+} ions concentration) which make difficult the resolution of the emission spectra and the interpretation of the relaxation kinetics of these complex excitations. Fluorescence decay measurements, from the $^3F_3$ level after excitation at 1.54 μm, realized on weakly (~5.10^{18} ions.cm^{-3}) and « strongly » (~5.31×10^{19} ions.cm^{-3}) Pr^{3+}-doped KPb_2Cl_5 crystals, revealed an exponential decay in the first case, and non exponential in the second one, firmly establishing energy transfers with concentration such as : $^3H_4 + {}^3F_3 \rightarrow {}^3H_5 + {}^3H_6$ and $^3H_5 + {}^3F_3 \rightarrow 2{*}^3H_6$.

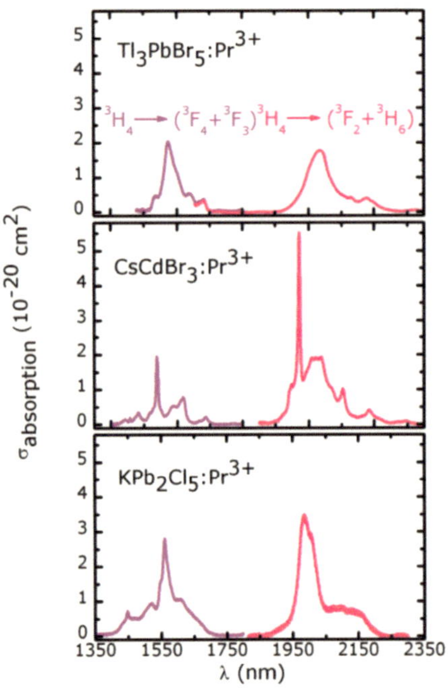

Fig. 8. Absorption cross sections around 1.55 and 2 μm at room temperature. The absorption cross section around 4.6 μm are shown in figure 10.

Judd-Ofelt analysis realized with absorption spectra on the 10 first excited states, among which 3 of them ($^3F_2$, $^3F_4$ and $^1I_6$) were thermalized with their closest lower lying level, showed that the branching ratio of the ($^3F_4$,$^3F_3$)→$^3F_2$ transition is negligibly small, that of the ($^3F_4$,$^3F_3$)→$^3H_6$ transition amounts to ~2.3 % (resp. ~1.6 %) and that the radiative lifetime of ($^3F_4$,$^3F_3$) levels is ~0.5 ms (resp. ~1 ms) in KPb_2Cl_5 (resp. $CsCdBr_3$) (for experimental lifetimes of ~0.096 ms in KPb_2Cl_5, ~1.57 ms in $CsCdBr_3$ and ~0.89 ms in Tl_3PbBr_5). Despite similarly weak branching ratios in $LaCl_3$ crystals, laser operations at 5.2 and 7.2 μm could be evidenced [Bowman et al., 1994, 1996]. Utilizing a flash lamp pumped phosphate glass Kigre laser at 1.54 μm, delivering pulse trains of 30 mJ at a frequency of 2 to 10 Hz, no luminescence issued from ($^3F_4$,$^3F_3$) multiplets at these two wavelengths could be observed in KPb_2Cl_5, $CsCdBr_3$ and Tl_3PbBr_5 crystals. However, luminescence at 1.6 and 2.4 μm associated to ($^3F_4$,$^3F_3$)→$^3H_4$ (β~74 % in KPb_2Cl_5, ~61.3 % in $CsCdBr_3$) and ($^3F_4$,$^3F_3$)→$^3H_5$ (β~24 % in KPb_2Cl_5, ~37 % in $CsCdBr_3$) transitions were observed (figure 9).

Broad band emission spectra are poorly structured and exhibit important emission cross section, ~(2-4).10^{-20} cm^2 ($\sigma_{EM}\tau_R$~(2-4).10^{-23} cm^2.s at the emission peak, virtually as much as in YAG:Nd^{3+} at 1.064 µm), in the two compounds $CsCdBr_3$ and KPb_2Cl_5. Several levels emit around 4.5-5 µm [Rana & Kaseta, 1983] : the ($^3F_4$,$^3F_3$) levels around 5 µm, the ($^3F_2$,$^3H_6$) levels around 4.5 µm and the $^3H_6$ level around 4.9 µm. The former ones already gave rise to the laser operation described in section 1, but in comparison with the other levels, the branching

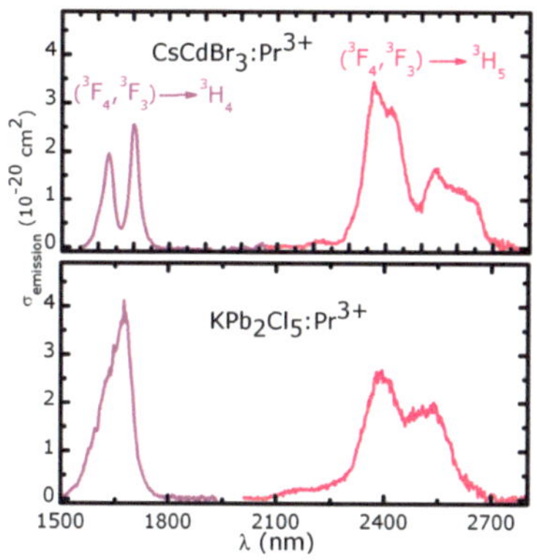

Fig. 9. Room temperature emission spectra of Pr^{3+} ions dissolved into $CsCdBr_3$ and KPb_2Cl_5 crystals, excited by pulses of a Kigre Er^{3+} laser at 1.54 µm.

ratio of the laser transition proves to be rather weak (~2.3 % (resp. ~1.6 %) versus ~14.7 % (resp. ~8.4 %) and ~48.4 % (resp. ~54.4%) in KPb_2Cl_5 (resp. $CsCdBr_3$)). The $^3H_6 \rightarrow ^3H_5$ transition seems interesting not only because of its branching ratio, but also because the experimental lifetime is quite long, 1 ms (3.6 ms in $CsCdBr_3$), and so favourable to energy storage. Nevertheless, this transition competes with ground state absorption and energy transfer processes induced by the excitation at 2 µm, likely to provoke important losses. Finally, the simplest solution remains the $^3H_5 \rightarrow ^3H_4$ transition which, even if it ends on the ground state, has a long lifetime, a quantum yield close to 1 (β=100 %) and broad and high absorption and emission cross sections. Such an emission could be efficiently pumped by means of QCL diodes around 4.6 µm. Although several energy levels (($^3F_4$,$^3F_3$), ($^3F_2$,$^3H_6$) and $^3H_6$) can emit around 4-5 µm, by means of the same excitation scheme as described above, and with a time resolved detection, it was possible to discriminate the IR luminescence associated with the $^3H_5 \rightarrow ^3H_4$ transition, and to calibrate the spectra in cross section units by means of the Fuchtbauer-Ladenburg formula (for Tl_3PbBr_5, we took the experimental lifetime, τ_f($^3H_5$)=30 ms). Absorption and emission spectra around 4.6 µm turn out to be very broad in the three compounds (figure 10). The emission cross section is about twice that found in KPb_2Cl_5:Er^{3+} around 4.5 µm and the radiative lifetimes are several tens of ms ($\sigma_{EM}\tau_R$~(2-3).10^{-22} cm^2.s at the peak). Judd-Ofelt analysis gives τ_R($^3H_5$)=38 ms (versus τ_f=5.65 ms in KPb_2Cl_5), τ_R($^3H_5$)=82 ms (versus τ_f=29 ms dans $CsCdBr_3$), once again favourable to energy storage with a view to short-pulse MIR laser operation. For a total Pr^{3+} ions concentration typically ~10^{20} (resp. 5.10^{19}) cm^{-3} and a crystal length of 5 mm, the absorption rate of a pump beam at 4.6 µm would reach 21.3 (resp. 11.3) %. The emission lines profiles, shown in figure 10 for the two compounds KPb_2Cl_5 and $CsCdBr_3$, were checked by calibrating them by the reciprocity method, and the agreement with the previous spectra is satisfying [Ferrier et al., 2008b, 2009b]. Gain cross section calculated with these spectra are displayed in figure 11. In the case of KPb_2Cl_5 and of Tl_3PbBr_5, the gain cross section is very broad and devoid of maximum, which could give rise to a broad tunability over a large spectral range. In the case of $CsCdBr_3$, the gain cross section, weaker, exhibits a peak at 4.75 μm and a rounded shape ~5 μm, where laser operation is expected. In all cases, we observe that the gain cross section is positive as soon as ζ=0.4, from 4.96 to 5.6 μm for $CsCdBr_3$, and from 4.87 to 5.48 μm for KPb_2Cl_5. As this type of 3-level lasers require a high population inversion ratio (ζ=0.4), an efficient pumping source must be designed, which can excite the strongest absorption bands around 1.6 ($^3H_4 \rightarrow (^3F_4, ^3F_3)$) and 2 μm ($^3H_4 \rightarrow (^3F_2, ^3H_6)$) with solid state lasers or Er^{3+} or Tm^{3+}-doped fibers now widely spread.

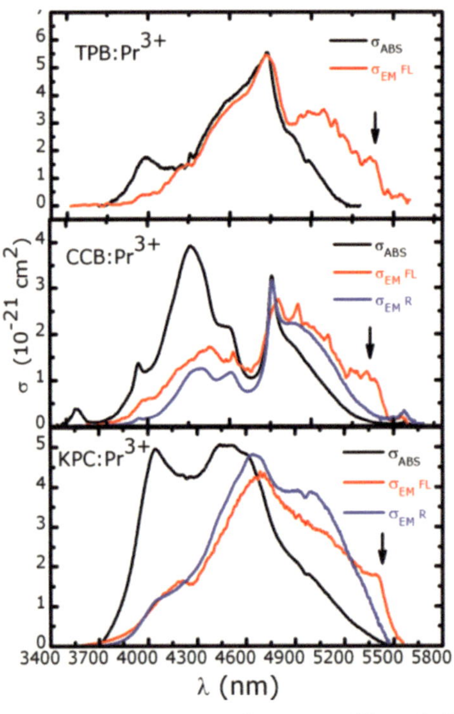

Fig. 10. Absorption and emission cross sections between $^3H_5$ and $^3H_4$ levels at room temperature. The arrow indicates an artefact peak due to the response function of a filter λ>3.5 μm incorporated in the setup in order to eliminate fluorescence harmonics.

Another way would consist in codoping the crystal with Yb^{3+} ions (resp. Tm^{3+} ions), and to exploit Yb^{3+}-Pr^{3+} energy transfer [Balda et al., 2003; Bluiett et al., 2008; Howse et al., 2010] thanks to high-powered low cost 980 nm (resp. 800 nm) diode pumping, but the high concentrations required remain hardly achievable in these compounds on the one hand, and the energy difference between laser and pumping wavelengths which would entail an important thermal load, on the other hand, lead us to consider another promising alternative : QCL diode pumping around 4.6 μm (which delivers 1.8 W at room temperature [Bai et al., 2008]), that is, pumping directly in the emitting level. Preliminary calculations, which do not take into account neither excited state absorption nor energy transfers, suggest that these sources should allow for reaching a sufficient population inversion ratio to obtain laser operation at reasonable pumping powers with power conversion yields on the order ~10 % [Ferrier et al., 2008b, 2009b].

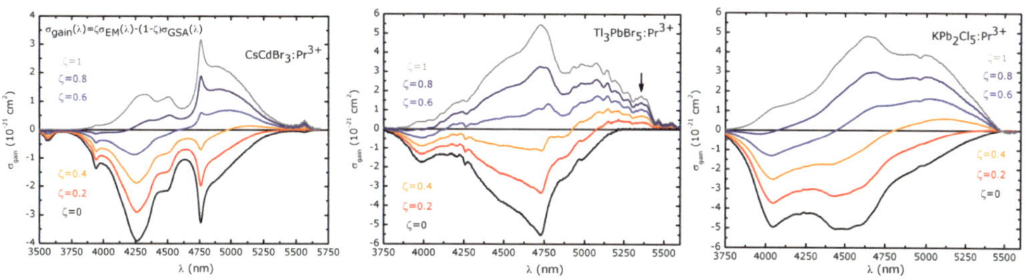

Fig. 11. Effective gain cross section around 4-5 µm in $CsCdBr_3$, Tl_3PbBr_5 and KPb_2Cl_5 doped with Pr^{3+}, with $\zeta=n^3H_5/(n^3H_5+n^3H_4)$.

Let us stress once again the necessity of both obtaining microstructure-related loss-free crystals and designing an efficient pumping scheme. Indeed, even if the gain cross section becomes positive at 40 % population inversion, the latter quickly saturates at 53.5 % in $KPb_2Cl_5:Pr^{3+}$ for a pump beam waist at 4.6 µm of 100 µm, and one expects laser emission centered at 5.05 µm, where the gain cross section reaches a low $\sim1.1\times10^{-21}$ cm^2 (figure 11). For a crystal length of 5 mm and an excited ion concentration of 10^{19} cm^{-3}, the cavity roundtrip gain amounts to 1.1 %, that is, virtually the same as that found in $KPb_2Cl_5:Er^{3+}$. In order to explain the homogeneous and inhomogeneous substantial broadening of the absorption and emission bands in $Pr^{3+}:KPb_2Cl_5$, a scrutinous examination of the absorption line profile on the $^3P_0$ profile at 487.5 nm is mandatory [Ferrier et al., 2008b]. As a matter of fact, for this non degenerate level one expects one single line per substitution site, provided that the second crystal field sublevel of the ground state $^3H_4$ does not contribute to the absorption signal shape : as it lies at 15 cm^{-1}, it should not be significantly populated at temperatures lower than 21.6 K. Figure 12 clearly shows the presence of five absorption lines and consequently as many substitution sites for Pr^{3+} ions in KPb_2Cl_5, giving a beginning of explanation to the origin of the inhomogeneous band broadening, which consequently dominates the homogeneous broadening at low temperature [Ferrier et al., 2008b].

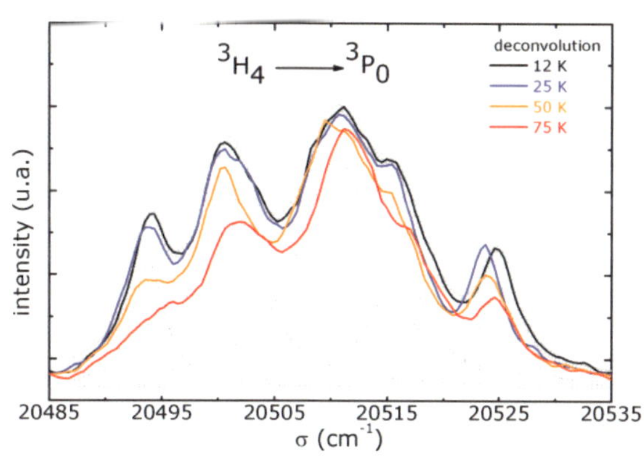

Fig. 12. Ground state absorption on the $^3P_0$ level of Pr^{3+} ions dissolved into KPb_2Cl_5, at low temperature and as a function of temperature. Five lorentzian functions are necessary to deconvolute the whole signal.

This characteristic is consistent with the structural description of the point defects [Velázquez et al., 2006b], but if it is so, why does Er^{3+} or Eu^{3+} substitution lead to majoritarily one type of point defect [Cascales et al., 2005; Ferrier et al., 2007; Gruber et al., 2006; Velázquez et al., 2009] ? Moreover this characteristic confirms the advantage of using KPb_2Cl_5 instead of $LaCl_3$, the lines of which are much narrower and force the experimentalist to find a pumping source emitting precisely, for instance, at 1.54 μm. Fluorescence around 1.6 and 2.4 μm, associated with $(^3F_4,^3F_3)\rightarrow{}^3H_4$ and $(^3F_4,^3F_3)\rightarrow{}^3H_5$ transitions obtained under excitation at 2.012 μm with a YAG:Tm^{3+}, give strong evidence for efficient energy transfers from the $(^3F_2,^3H_6)$ levels.

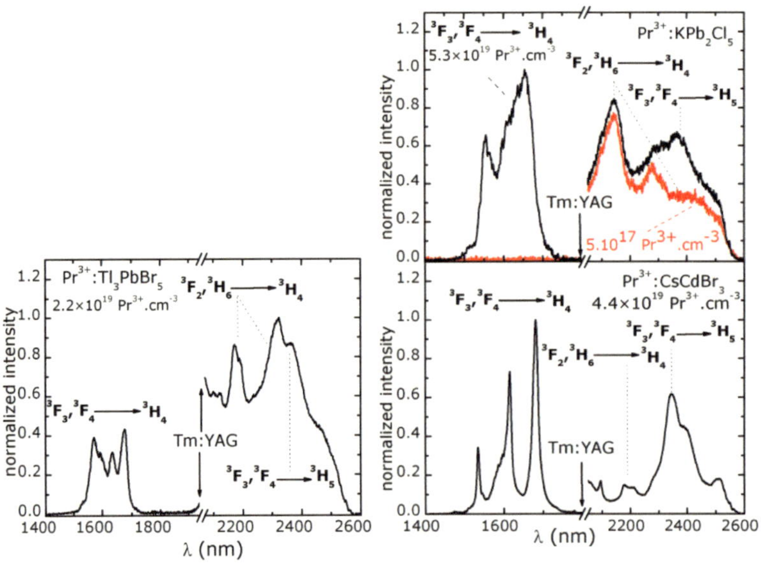

Fig. 13. Emission spectra, recorded at room temperature, under cw excitation at 2.011 μm (YAG:Tm), issued from $(^3F_4,^3F_3)$ and $(^3F_2,^3H_6)$ levels.

In Figure 13, we see that these energy transfers do not only occur undistinctly in KPb_2Cl_5 and in $CsCdBr_3$, but that they also occur in the former compound with Pr^{3+} ions concentrations a hundred times weaker. The Pr^{3+} ion pairing in $CsCdBr_3$, favourable to energy transfers described above, can be related to the decrease by a factor of 4 of the relative intensity of the emission issued from $(^3F_2,^3H_6)$ levels in favour of those issued from $(^3F_4,^3F_3)$ levels.

Finally, another alternative could be found in the works performed on Tb^{3+} and (Tb^{3+},Tm^{3+})-doped KPb_2Cl_5, KPb_2Br_5 and $CsCdBr_3$ single crystals, because of their potentially interesting MIR transitions $^7F_5\rightarrow{}^7F_6$ (4.6-5.5 μm) and $^7F_4\rightarrow{}^7F_5$ (8 μm) [Bluiett et al., 2011; Ganem et al., 2002; Lichkova et al., 2006; Okhrimchuk et al., 2002; Rademaker et al., 2004; Roy et al., 2005; Shakurov et al., 2004]. So far, no laser operation based on these systems could be demonstrated. The $^7F_6$ multiplet is both the terminal level and ground state and the branching ratio of the $^7F_4\rightarrow{}^7F_5$ transition is ~2 % (resp. ~6.6 %) and the experimental lifetimes of several tens of ms (resp. $\tau(^7F_5)\approx 22$ ms and $\tau(^7F_4)\approx 0.36$ ms) in KPb_2Cl_5 (resp. KPb_2Br_5 [Hömmerich et al., 2006]).

4. Loss mechanisms and current challenges in the growth of MIR laser halide crystals

The growth conditions of pure or RE^{3+}-doped APb_2X_5 (A=K,Rb ; X=Cl,Br) and Tl_3PbX_5 (X=Cl,Br) single crystals for MIR laser and/or nonlinear optical applications have been exhaustively detailed in an impressive body of work [Amedzake et al., 2008; Atuchin et al., 2011; Bekenev et al., 2011; Condon et al., 2006b; Ferrier et al., 2006a, 2006b; Gang et al., 2008; Isaenko et al., 2001; Nitsch et al., 1993, 1995a, 1995b, 2004; Nitsch & Rodová 1999; Oyebola et al., 2010; Rodová et al., 1995; Roy et al., 2003; Singh et al., 2005; Tigréat et al., 2001; Velázquez et al., 2006a, 2006b, 2009; Voda et al., 2004; Wang et al., 2007]. Special safety conditions for the manipulation of large amounts of TlCl or TlBr powders must be followed, but this does not entail insuperable difficulties [Peter & Viraraghavan, 2005]. It is now widely established that crystal growth by the Bridgman method must be carried out in sealed silica ampoules under a low pressure (~10^{-2} atm) of Cl_2 or HCl (HBr for bromides) gas to avoid bubble formation, and at low growth rates (~0.1-1 mm.h^{-1}) in axial thermal gradients typically 15-20 K.cm^{-1} with opening angles of the ampoule bottom lower than 120° to avoid shear stress induced cracks. PbX_2 (X=Cl,Br) starting materials must be purified by previous solidification runs by the Bridgman method [Basiev et al., 2004]. In KPb_2Cl_5 crystals, roundtrip losses mechanisms, such as laser beam depolarization and double refraction [Condon et al., 2006b], are likely to be due to a classical kind of twinning, which occurs above a finite stress threshold, consisting in twin planes perpendicular to [100] and [001] directions and leading to a loss of the laser power transmitted within the cavity of up to 50 %. This has seriously hindered further development of highly brilliant laser systems at 4.6 μm (and presumably higher wavelengths), since with an absorption cross section $\sigma_{abs} \approx 2.10^{-21}$ cm^2 at 800 nm, an Er^{3+} ion concentration $n_{^4I_{15/2}}$ ~5.10^{19} cm^{-3} and a typical crystal length of 5 mm, the pump beam absorption rate in a classical two-mirror cavity based on a KPb_2Cl_5:Er^{3+} amplifier is about 5 %. On the other hand, the expected round-trip gain ($g = 2\sigma_{ém} n_{^4I_{9/2}} \ell$, for a 100 % inversion rate), in an otherwise perfect crystal unaffected by thermal lens nor birefringence effects, with $n_{^4I_{9/2}}$ ~5.10^{18} cm^{-3} and $\sigma_{em}(4.53\mu m) \approx 1.9\times10^{-21}$ cm^2, barely reaches 1 %. So, the two main current challenges in terms of laser crystal growth are to increase the Er^{3+} ion content and avoid twinning of the crystals upon cooling. The twinning patterns observed under polarized light are due to the phase transition that occurs upon cooling the crystal, at T_t=255 °C. The mechanism of this phase transition has been definitely unveiled in recent years simultaneously by us and a Russian team [Mel'nikova et al., 2005, 2006; Merkulov et al., 2005; Velázquez et al., 2006b, 2009; Velázquez & Pérez, 2007]. The phase transition is driven by the K^+ and $Pb(2)^{2+}$ cationic ordering that induces a Pmcn to $P2_1/c$ symmetry change upon cooling, at 255 °C. Similar twinning patterns were observed in KPb_2Br_5 crystals [Mel'nikova et al., 2005], and twinning patterns associated with the crossing of another phase transition was also observed in Tl_3PbBr_5 [Singh et al., 2005]. It is possible to produce, by means of slow cooling, small size (~1 to 8 mm^3) untwined single crystals, but any stress of magnitude higher than a certain threshold value will result in the formation of a twinning pattern. In order to get rid of that, one must increase the diffusion activation energy, E_a, by increasing the radius of the A^+ cation in APb_2X_5 (X=Cl,Br) compounds, in such a way that E_a>RT_m, with T_m=432.3 °C for KPb_2Cl_5. On passing from A=K to A=Rb, the crystal does not undergo any phase transition between the melting and room temperatures, which explains the interest of several research groups in developing RE-doped $RbPb_2Cl_5$ crystals. However, according to our preliminary experiments, in rubidium lead chloride crystals, the Er^{3+} solubility seems lower than in potassium lead chlorides. So, it is clear that some innovative codoping of Er^{3+} ions with another ion devoid of optical activity must be found, in order to increase its solubility in rubidium lead halides [Isaenko et al., 2009b; Tarasova et al., 2011; Velázquez et al., 2009].

5. Conclusion

This chapter has emphasized the laser potential of the currently most important rare-earth doped chlorides and bromides laser crystals. Some of them ($LaCl_3:Pr^{3+}$, $KPb_2Cl_5:Er^{3+}$, $KPb_2Cl_5:Dy^{3+}$) have already led to laser systems operating in bands II and III of the atmosphere transmission window. All these crystals were grown exclusively by the Bridgman method in sealed silica ampoules. After 15 years of research on MIR solid state lasers carried out in Europe, USA, Russia and more recently China, lots of improvements remain to be done in the realm of synthesis and crystal growth of the laser materials, as well as on the optimization of the pumping strategy. Both kind of advances will require new and original solid solution crystal engineering in order to both get rid of the phase transition inducing twinned microstructures detrimental to efficient laser operation and increase the suitable RE ions solubilities in the halide host compounds. As the gain cross sections in the spectral range where laser operation was demonstrated are relatively weak ($\sim$(1-5).10^{-21} cm^2), an important effort must be devoted to the annihilation of the losses which could also be achieved by a clever shaping and functionalization of the crystals. In this perspective, 3-level laser operation by QCL diode pumping on the Pr^{3+} $^3H_5$ multiplet in KPb_2Cl_5 crystals could be interesting because of the low thermal load, the substantial absorption cross section and also because the refractive index difference between twin domains should be much smaller in this spectral range.

6. References

Aleksandrov, K. S.; Vtyurin, A. N.; Eliseev, A. P.; Zamkova, N. G.; Isaenko, L. I.; Krylova, S.N.; Pashkov, V. M.; Turchin, P. P. & Shebanin, A. P. (2005). Vibrational spectrum and elastic properties of KPb_2Cl_5 crystals. *Phys. Sol. St.*, Vol. 47, No.3, pp. 531-538.

Amedzake, P.; Brown, E.; Hömmerich, U.; Trivedi, S. B. & Zavada, J. M. (2008). Crystal growth and spectroscopic characterization of Pr-doped KPb_2Cl_5 for mid-infrared laser applications. *J. Cryst. Growth*, Vol. 310, pp. 2015-2019.

An, Y. & May, P. S. (2006). Temperature dependence of excited-state relaxation processes of Pr^{3+} in Pr-doped and Pr,Gd-doped $CsCdBr_3$. *J. Lum.*, Vol. 118, pp. 147-157.

Atuchin, V. V.; Isaenko, L. I.; Kesler, V. G. & Tarasova, A. Yu. (2011). Single crystal growth and surface chemical stability of KPb_2Br_5. *J. Cryst. Growth*, Vol. 318, No.1, pp. 1000-1004.

Bai, Y.; Slivken, S.; Davish, S. R. & Razeghi, M. (2008). Room temperature continuous wave operation of quantum cascade lasers with 12.5% wall plug efficiency. *Appl. Phys. Lett.*, Vol. 93, No.2, pp. 021103/1-3.

Balda, R.; Voda, M.; Al-Saleh, M. & Fernandez, J. (2002). Visible luminescence in $KPb_2Cl_5:Pr^{3+}$ crystal. *J. Lumin.*, Vol. 97, pp. 190-197.

Balda, R.; Fernández, J.; Mendioroz, A.; Voda, M. & Al-Saleh, M. (2003). Infrared-to-visible upconversion processes in Pr^{3+}/Yb^{3+}-codoped KPb_2Cl_5. *Phys. Rev. B*, Vol. 68, pp. 165101/1-7.

Balda, R.; Garcia–Adeva, A. J.; Voda, M. & Fernández, J. (2004). Upconversion processes in Er^{3+}-doped KPb_2Cl_5. *Phys. Rev. B*, Vol. 69, pp. 205203/1-8.

Basiev, T. T. ; Danileiko, Yu. K.; Dmitruk, L. N.; Galagan, B. I.; Moiseeva, L. V.; Osiko, V. V.; Sviridova, E. E. & Vinogradova, N. N. (2004). The purification, crystal growth, and spectral-luminescent properties of $PbCl_2$:RE. *Opt. Mater.*, Vol. 25, pp. 295-299.

Bekenev, V. L.; Khyzhun, O. Yu.; Sinelnichenko, A. K.; Atuchin, V. V.; Parasyuk, O. V.; Yurchenko, O. M.; Bezsmolnyy, Yu.; Kityk, A. V.; Szkutnik, J. & Calus, S. (2011). Crystal growth and the electronic structure of Tl_3PbCl_5. *J. Phys. Chem. Sol.*, Vol. 72, No.6, pp. 705-713.

Bluiett, A. G.; Pinkney, E.; Brown, E. E.; Hömmerich, U.; Amedzake, P.; Trivedi, S. B. & Zavada, J. M. (2008). Energy transfer processes in doubly doped Yb,Pr:KPb_2Cl_5 for mid-infrared laser applications. *Mater. Sci. Engin. B*, Vol. 146, pp. 110-113.

Bluiett, A. G.; Peele, D.; Norman, K.; Brown, E.; Hömmerich, U.; Trivedi, S. B. & Zavada, J. M. (2011). Mid-infrared emission characteristics and energy transfer processes in doubly doped Tm,Tb:KPb_2Br_5 and Tm,Nd:KPb_2Br_5. *Opt. Mater.*, Vol. 33, pp. 985-988.

Bowman, S. R.; Ganem, J.; Feldman, B. J. & Kueney, A. W. (1994). Infrared laser characteristics of praseodymium-doped lanthanum trichloride. *IEEE J. Quant. Elect.*, Vol. 30, No.12, pp. 2925-2928.

Bowman, S. R.; Shaw, L. B.; Feldman, B. J. & Ganem, J. (1996). A 7-μm praseodymium-based solid-state laser. *IEEE J. Quant. Elect.*, Vol. 32, No.4, pp. 646-649.

Bowman, S. R.; Searles, S. K.; Ganem, J. & Smidt, P. (1999). Further investigations of potential 4-μm laser materials. *Trends in Optics and Photonics*; Fejer, M. M., Ingeyan, H., Keller, U., Eds.; Optical Society of America : Washington, DC, Vol. 26, pp. 487-490.

Bowman, S. R.; Searles, S. K.; Jenkins, N. W.; Qadri, S. B. & Skelton, E. F. (2001). Diode pumped room temperature mid-infrared erbium laser. *Trends in Optics and Photonics*; Marshall, C. Ed.; Optical Society of America : Washington, DC, Vol. 50, pp. 151 156.

Cascales, C.; Fernández, J. & Balda, R. (2005). Investigation of site-selective symmetries of Eu^{3+} ions in KPb_2Cl_5 by using optical spectroscopy. *Optics Express*, Vol. 13, No.6, pp. 2141-2152.

Chukalina, E. P. (2004). Study of hyperfine structure in the optical spectra of crystals doped with Pr^{3+} and Er^{3+} ions. *J. Opt. Techn.*, Vol. 71, No.9, pp. 581-585.

Cockroft, N. J.; Jones, G. D. & Nguyen, D. C. (1992). Dynamics and spectroscopy of infrared-to-visible upconversion in erbium-doped cesium cadmium bromide ($CsCdBr_3$:Er^{3+}). *Phys. Rev. B*, Vol. 45, No.10, pp. 5187-5198.

Condon, N. J.; O'Connor, S. & Bowman, S. R. (2006a). Growth and mid-infrared laser performance of Er^{3+}:KPb_2Cl_5. *IEEE LEOS Annual Meeting, Conference Proceedings, 19th*, Montreal, QC, Canada, Oct. 29-Nov. 2, Vol. 2, pp. 743-744.

Condon, N. J.; O'Connor, S. & Bowman, S. R. (2006b). Growth and characterization of single-crystal Er^{3+}:KPb_2Cl_5 as a mid-infrared laser material. *J. Crystal Growth*, Vol. 291, pp. 472-478.

Doualan, J.-L. & Moncorgé, R. (2003). Laser crystals with low phonon frequencies. *Ann. Chim. Sci. Mat.*, Vol. 28, pp. 5-20.

Egger, P.; Burkhalter, R. & Hulliger, J. (1999). Czochralski growth of $Ba_2Y_{1-x}Er_xCl_7$ ($0<x\leq1$) using growth equipment integrated into a dry-box. *J. Cryst. Growth*, Vol. 200, pp. 515-520.

Ferrier, A. ; Velázquez, M.; Portier, X. ; Doualan, J.-L. & Moncorgé, R. (2006a). Tl_3PbBr_5 : a possible crystal candidate for middle infrared nonlinear optics. *J. Cryst. Growth*, Vol. 289, No.1, pp. 357-365. Note that in this paper, a misprint error occurred: we used a KDP and not a KD*P powder as a standard.

Ferrier, A.; Velázquez, M.; Pérez, O.; Grebille, D.; Portier, X. & Moncorgé, R. (2006b). Crystal growth and characterization of the non-centrosymmetric compound Tl_3PbCl_5. *J. Cryst. Growth*, Vol. 291, No.2, pp. 375–384.

Ferrier, A.; Velázquez, M.; Doualan, J.-L.; Moncorgé, R. (2007). Energy level structure and excited-state absorption properties of Er^{3+}-doped KPb_2Cl_5. *J. Opt. Soc. Amer. B*, Vol. 24, pp. 2526-2536.

Ferrier, A.; Velázquez, M. & Moncorgé, R. (2008a). Spectroscopic characterization of Er^{3+}-doped Tl_3PbBr_5 for midinfrared applications. *Phys. Rev. B*, Vol. 77, pp. 075122/1-11.

Ferrier, A.; Velázquez, M.; Doualan, J.-L. & Moncorgé, R. (2008b). Midinfrared luminescence properties and laser potentials of Pr^{3+} doped KPb_2Cl_5 and $CsCdBr_3$. *J. of Applied Physics*, Vol. 104, pp. 123513.

Ferrier, A.; Velázquez, M.; Doualan, J.-L. & Moncorgé, R. (2009a). Pr^{3+}-doped Tl_3PbBr_5 : a non hygroscopic, non-linear and low-energy phonon single crystal for the mid-infrared laser application. *Appl. Phys. B*, Vol. 95, pp. 287-291.

Ferrier, A.; Velázquez, M.; Doualan, J.-L. & Moncorgé, R. (2009b). Spectroscopic investigation and mid-infrared luminescence properties of the Pr^{3+} doped low phonon single crystals $CsCdBr_3$, KPb_2Cl_5 and Tl_3PbBr_5. *J. Lumin.*, Vol. 129, No.12, pp. 1905-1907.

Gafurov, M. R.; Iskhakova, A. I.; Kurzin, I. N.; Kurzin, S. P.; Malkin, B. Z.; Nikitin, S. I.; Orlinskii, S. B.; Rakhmatullin, R. M.; Shakurov, G. S.; Tarasov, V. F.; Demirbilek, R. & Heber, J. (2002). Spectra and relaxation of electronic excitations in $CsCdBr_3:Yb^{3+}$ and $CsCdBr_3:Nd^{3+}$ single crystals. *SPIE Proceedings*, Vol. 4766, pp. 279-291.

Ganem, J.; Crawford, J.; Schmidt, P.; Jenkins, N. W. & Bowman, S. R. (2002). Thulium cross-relaxation in a low phonon energy crystalline host. *Phys. Rev. B*, Vol. 66, pp. 245101/1-15.

Gang, C.; Chun-he, Y.; Jian-rong, C.; Hai-li, W. & Nan-hao, Z. (2008). *Rengong Jingti Xuebao*, Vol. 37, No.6, pp. 1567-1570.

Guillot-Noël, O.; Goldner, P.; Higel, P. & Gourier, D. (2004). A practical analysis of electron paramagnetic resonance spectra of rare earth ion pairs. *J. Phys. : Condens. Matter*, Vol. 16, pp. R1-R24.

Gruber, J. B.; Yow, R. M.; Nijjar, A. S.; Russell, C. C.; Sardar, D. K.; Zandi, B.; Burger, A. & Roy, U. N. (2006). Modeling the crystal-field splitting of energy levels of Er^{3+} ($4f^{11}$) in charge-compensated sites of KPb_2Cl_5. *J. Appl. Phys.*, Vol. 100, pp. 0431081-6.

Heber, J.; Demirbilek, R.; Altwein, M.; Kübler, J.; Bleeker, B. & Meijerink, A. (2001). Electronic states and interactions in pure and rare-earth doped $CsCdBr_3$. *Radiation Effects & Defects in Solids*, Vol. 154, pp. 223-229.

Hömmerich, U.; Nyein, E. E. & Trivedi, S. B. (2005). Crystal growth, upconversion, and infrared emission properties of Er^{3+}-doped KPb_2Br_5. *J. Lum.*, Vol. 113, pp. 100-108.

Hömmerich, U.; Brown, E.; Amedzake, P.; Trivedi, S. B. & Zavada, J. M. (2006). Mid-infrared (4.6 μm) emission properties of Pr^{3+} doped KPb_2Br_5. *J. Appl. Phys.*, Vol. 100, pp. 113507/1-4.

Howse, D.; Logie, M.; Bluiett, A. G.; O'Connor, S.; Condon, N. J.; Ganem, J. & Bowman, S. R. (2010). Optically-pumped mid-IR phosphor using Tm^{3+}-sensitized Pr^{3+}-doped KPb_2Cl_5. *J. Opt. Soc. Am. B*, Vol. 27, No.11, pp. 2384-2392.

Isaenko, L.; Yelisseyev, A.; Tkachuk, A.; Ivanova, S.; Vatnik, S.; Merkulov, A.; Payne, S.; Page, R. & Nostrand, M. (2001). New laser crystals based on KPb_2Cl_5 for IR region. *Mater. Sci. Eng. B*, Vol. 81, pp. 188-190.

Isaenko, L.; Yelisseyev, A.; Tkachuk, A. & Ivanova, S. (2008). New monocrystals with low phonon energy for mid-IR lasers. In *Mid-Infrared Coherent Sources and Applications;* Ebrahim-Zadeh, M., Sorokina, I. T., Eds.; Springer : New York, pp. 3-65.

Isaenko, L. I.; Merkulov, A. A.; Tarasova, A. Yu.; Pashkov, V. M. & Drebushchak, V. A. (2009a). Coefficients of thermal expansion of the potassium and rubidium halogenide plumbates. *J. Therm. Anal. Cal.*, Vol. 95, No.1, pp. 323-325.

Isaenko, L. I.; Mel'nikova, S. V.; Merkulov, A. A.; Pashkov, V. M. & Tarasova, A. Yu. (2009b). Investigation of the influence of gradual substitution K/Rb on the structure and phase transition in $K_xRb_{1-x}Pb_2Br_5$ solid solutions. *Physics of the Solid State*, Vol. 51, No.3, pp. 589-592.

Jenkins, N. W.; Bowman, S. R.; O'Connor, S.; Searles, S. K. & Ganem, J. (2003). Spectroscopic characterization of Er-doped KPb_2Cl_5 laser crystals. *Opt. Mater.*, Vol. 22, pp. 311-320.

John Peter, A. L. & Viraraghavan, T. (2005). Thallium : a review of public health and environmental concerns. *Environment International*, Vol. 31, pp. 493-501.

Kaminska, A.; Cybińska, J.; Zhydachevskii, Ya.; Sybilski, P.; Meyer, G. & Suchocki, A. (2011). Luminescent properties of ytterbium-doped ternary lanthanum chloride. *J. All. Comp.*, Vol. 509, pp. 7993-7997.

Lichkova, N. V.; Zagorodnev, V. N.; Butvina, L. N.; Okhrimchuk, A. G. & Shestakov, A. V. (2006). Preparation and optical properties of rare-earth-activated alkali metal lead chlorides. *Inorg. Mater.*, Vol. 42, No.1, pp. 81-88.

Malkin, B. Z.; Iskhakova, A. I.; Kamba, S.; Heber, J.; Altwein, M. & Schaak, G. (2001). Far-infrared spectroscopy investigation and lattice dynamics simulations in $CsCdBr_3$ and $CsCdBr_3{:}R^{3+}$ crystals. *Phys. Rev. B*, Vol. 63, pp. 075104-1/11.

Mel'nikova, S. V.; Isaenko, L. I.; Pashkov, V. M. & Pevnev, I. V. (2005). Phase transition in a KPb_2Br_5 crystal. *Phys. Sol. St.*, Vol. 47, No.2, pp. 332-336.

Mel'nikova, S. V.; Isaenko, L. I.; Pashkov, V. M. & Pevnev, I. V. (2006). Search for and study of phase transitions in some representatives of the APb_2X_5 family. *Phys. Sol. St.*, Vol. 48, No.11, pp. 2152-2156.

Merkulov, A. A.; Isaenko, L. I.; Pashkov, V. M.; Mazur, V. G.; Virovets, A. V. & Naumov, D. Yu. (2005). Crystal structure of KPb_2Cl_5 and KPb_2Br_5. *J. Struct. Chem.*, Vol. 46, No.1, pp. 103-108.

Neukum, J.; Bodenschatz, N. & Heber, J. (1994). Spectroscopy and upconversion of $CsCdBr_3{:}Pr^{3+}$. *Phys. Rev. B*, Vol. 50, No.6, pp. 3536-3546.

Nitsch, K.; Cihlář, A.; Malková, Z.; Rodová, M.; Vaněček, M. (1993). The purification and preparation of high-purity lead chloride and ternary alkali lead chloride single crystals. *J. Crystal Growth*, Vol. 131, pp. 612.

Nitsch, K.; Dušek, M.; Nikl, M.; Polák, K.; Rodová, M. (1995a). Ternary alkali lead chlorides : crystal growth, crystal structure, absorption and emission properties. *Prog. Cryst. Growth Charact.*, Vol. 30, pp. 1–22.

Nitsch, K.; Cihlář, A.; Nikl, M.; Rodová, M. (1995b). Growth of lead bromide and ternary alkalic lead bromide single crystals. *J. Electr. Eng.*, Vol. 46, No.8/s, pp. 82-84.

Nitsch, K. & Rodová, M. (1999). Influence of thermal treatment on supercooling of molten potassium lead chloride. *J. Elect. Eng.*, Vol. 50, No.2/s, pp. 35-37.

Nitsch, K.; Cihlář, A. & Rodová, M. (2004). Molten state and supercooling of lead halides. *J. Cryst. Growth*, Vol. 264, pp. 492–498.

Nostrand, M. C.; Page, R. H.; Payne, S. A.; Krupke, W. F.; Schunemann, P. G. & Isaenko, L. I. (1998). Spectroscopic data for infrared transitions in $CaGa_2S_4:Dy^{3+}$ and $KPb_2Cl_5:Dy^{3+}$. *Trends in Optics and Photonics;* Bosenberg, W. R. and Fejer, M. M., Eds.; Optical Society of America : Washington, DC; Vol. 19, pp. 524-528.

Nostrand, M. C.; Page, R. H.; Payne, S. A.; Krupke, W. F.; Schunemann, P. G. & Isaenko, L. I. (1999). Room temperature $CaGa_2S_4:Dy^{3+}$ laser action at 2.43 and 4.31 μm and $KPb_2Cl_5:Dy^{3+}$ laser action at 2.43 μm. *Trends in Optics and Photonics;* Fejer, M. M., Ingeyan, H., Keller, U., Eds.; Optical Society of America : Washington, DC; Vol. 26, pp. 441-449.

Nostrand, M. C.; Page, R. H.; Payne, S. A.; Isaenko, L. I. & Yelisseyev, A. P. (2001). Optical properties of Dy^{3+}- and Nd^{3+}-doped KPb_2Cl_5. *J. Opt. Soc. Am. B*, Vol. 18, No.3, pp. 264–276.

Okhrimchuk, A.; Butvina, L. & Dianov, E. (2002). Sensitization of MIR Tb^{3+} luminescence by Tm^{3+} ions in $CsCdBr_3$ and KPb_2Cl_5 crystals. *Advanced Solid-State Photonics (ASSP);* OSA Technical Digest Series; paper WB5-1, pp. 273-275.

Okhrimchuk, A. G.; Butvina, L. N.; Dianov, E. M.; Lichkova, N. V.; Zagorodnev, V. N. & Shestakov, A. V. (2006). New laser transition in a $Pr^{3+}:RbPb_2Cl_5$ crystal in the 2.3-2.5-μm range. *Quantum Electronics*, Vol. 36, No.1, pp. 41-44.

Okhrimchuk, A. G.; Butvina, L. N.; Dianov, E. M.; Shestakova, I. A.; Lichkova, N. V.; Zagorodnev, V. N. & Shestakov, A. V. (2007). Optical spectroscopy of the $RbPb_2Cl_5$:Dy3+ laser crystal and oscillation at 5.5 μm at room temperature. *J. Opt. Soc. Am. B*, Vol. 24, No.10, pp. 2690-2695.

Okhrimchuk, A. G. (2008). Dy^{3+} and Pr^{3+} doped crystals for mid-IR lasers. *Conference on Lasers and Electro-Optics & Quantum Electronic and Laser Science Conference*, Vol. 1-9, pp. 1531-1532.

Oyebola, O.; Hömmerich, U.; Brown, E.; Trivedi, S. B.; Bluiett, A. G. & Zavada, J. M. (2010). Growth and optical spectroscopy of Ho-doped KPb_2Cl_5 for infrared solid state lasers. *J. Cryst. Growth*, Vol. 312, No.8, pp. 1154-1156.

Popova, M. N.; Chukalina, E. P.; Malkin, B. Z.; Iskhakova, A. I.; Antic-Fidancev, E.; Porcher P. & Chaminade, J.-P. (2001). High-resolution infrared absorption spectra, crystal field levels, and relaxation processes in $CsCdBr_3:Pr^{3+}$. *Phys. Rev. B*, Vol. 63, pp. 075103-1/10.

Quagliano, J. R.; Cockroft, N. J.; Gunde, K. E. & Richardson, F. S. (1996). Optical characterization and electronic energy-level structure of Er^{3+}-doped $CsCdBr_3$. *J. Chem. Phys.*, Vol. 105, No.22, pp. 9812-9822.

Quimby, R. S.; Condon, N. J.; O'Connor, S. P.; Biswal, S. & Bowman, S. R. (2008). Upconversion and excited-state absorption in the lower levels of Er:KPb_2Cl_5. *Opt. Mater.*, Vol. 30, pp. 827-834.

Rademaker, K.; Krupke, W. F.; Page, R. H.; Payne, S. A.; Petermann, K.; Huber, G.; Yelisseyev, A. P.; Isaenko, L. I.; Roy, U. N.; Burger, A.; Mandal, K. C. & Nitsch, K. (2004). Optical properties of Nd^{3+}- and Tb^{3+}-doped KPb_2Br_5 and $RbPb_2Br_5$ with low nonradiative decay. *J. Opt. Soc. Am. B*, Vol. 21, No.12, pp. 2117-2129.

Rana, R. S. & Kaseta, F. W. (1983). Laser excited fluorescence and infrared absorption spectra of Pr^{3+}:$LaCl_3$. *J. Chem. Phys.*, Vol. 79, No.11, pp. 5280-5285.

Ren, P.; Qin, J. & Chen, C. (2003). A novel nonlinear optical crystal for the IR region : noncentrosymmetrically crystalline $CsCdBr_3$ and its properties. *Inorg. Chem.*, Vol. 42, No.1, pp. 8-10.

Riedener, T.; Egger, P.; Hulliger, J. & Güdel, H. U. (1997). Upconversion mechanisms in Er^{3+}-doped Ba_2YCl_7. *Phys. Rev. B*, Vol. 56, No.4, pp. 1800-1808.

Riedener, T. & Güdel, H. U. (1997). NIR-to-VIS upconversion of Er^{3+} in host materials with low-energy phonons. *Chimia*, Vol. 51, No.3, pp. 95-96.

Rodová, M.; Málková, Z. & Nitsch, K. (1995). Study of hygroscopicity and water-sensitivity of lead halides. *J. Electr. Eng.*, Vol. 46, No.8/s, pp. 79-81.

Roy, U. N.; Cui, Y.; Guo, M.; Groza, M.; Burger, A.; Wagner, G. J.; Carrig, T. J. & Payne, S. A. (2003). Growth and characterization of Er-doped KPb_2Cl_5 as laser host crystal. *J. Cryst. Growth*, Vol. 258, pp. 331–336.

Roy, U. N.; Hawrami, R. H.; Cui, Y.; Morgan, S.; Burger, A.; Mandal, K. C.; Noblitt, C. C.; Speakman, S. A.; Rademaker, K. & Payne, S. A. (2005). Tb^{3+}-doped KPb_2Br_5 : low-energy phonon mid-infrared laser crystal. *Appl. Phys. Lett.*, Vol. 86, pp. 151911/1-3.

Shakurov, G. S.; Malkin, B. Z.; Zakirov, A. R.; Okhrimchuk, A. G.; Butvina, L. N.; Lichkova, N. V. & Zagorodnev, V. N. (2004). High-frequency EPR of Tb^{3+}-doped KPb_2Cl_5 crystal. *Appl. Magn. Reson.*, Vol. 26, pp. 579-586.

Singh, N. B.; Suhre, D. R.; Green, K.; Fernelius, N. & Hopkins, F. K. (2005). Periodically poled materials for long wavelength infrared (LWIR) NLO applications. *J. Cryst. Growth*, Vol. 274, pp. 132-137.

Tarasova, A. Yu.; Isaenko, L. I.; Kesler, V. G.; Pashkov, V. M.; Yelisseyev, A. P.; Denysyuk, N. M. & Khyzhun, O. Yu. (2011). Electronic structure and fundamental absorption edges of KPb_2Br_5, $K_{0.5}Rb_{0.5}Pb_2Br_5$ and $RbPb_2Br_5$ single crystals. *J. Phys. Chem. Sol.*, submitted.

Tigréat, P.-Y.; Doualan, J.-L.; Moncorgé, R. & Ferrand, B. (2001). Spectroscopic investigation of a 1.55 μm emission band in Dy^{3+}-doped $CsCdBr_3$ and KPb_2Cl_5 single crystals. *J. Lumin.*, Vol. 94-95, pp. 107-111.

Tkachuk, A. M.; Ivanova, S. E.; Isaenko, L. I.; Yelisseyev, A. P.; Pustovarov, V. A.; Joubert, M.-F.; Guyot, Y. & Gapontsev, V. P. (2005). Emission peculiarities of TR^{3+}-doped KPb_2Cl_5 laser crystals under selective direct, upconversion and excitonic/host excitation of impurity centers. *Trends in Optics and Photonics*, Vol. 98, Issue Advanced Solid State Photonics, pp. 69-74.

Tkachuk, A. M.; Ivanova, S. E.; Joubert, M.-F.; Guyot, Y.; Isaenko, L. I. & Gapontsev, V. P. (2007). Upconversion processes in Er^{3+}:KPb_2Cl_5 laser crystals. *J. Lumin.*, Vol. 125, pp. 271-278.

Velázquez, M.; Ferrier, A.; Chaminade, J.-P.; Menaert, B. & Moncorgé, R. (2006a). Growth and thermodynamic characterization of pure and Er-doped KPb_2Cl_5. *J. Cryst. Growth*, Vol. 286, No.2, pp. 324–333. In the Figure 1b caption of this article, the Er^{3+} content was mistakenly reported to amount to 0.35 mol %. It is actually 0.35 wt % ≈ 1.32 mol % ≈ 6.10^{19} ions cm^{-3}, or x ≈ 0.013 in $K_{1-x}(V_K)_xPb_{2-x}Er_xCl_5$.

Velázquez, M.; Ferrier, A.; Pérez, O.; Péchev, S.; Gravereau, P.; Chaminade, J.-P. & Moncorgé, R. (2006b). A cationic order-disorder phase transition in KPb_2Cl_5. *Eur. J. Inorg. Chem.*, Vol. 20, pp. 4168–4178. During the typesetting and publishing processes of this article, a minus sign appeared in the "Structure of rare-earth defects" section, that must be systematically replaced by a plus sign.

Velázquez, M. & Pérez, O. (2007). Comments on "Growth and characterization of single crystal Er^{3+}:KPb_2Cl_5 as a mid-infrared laser material". *J. Cryst. Growth*, Vol. 307, pp. 500–501.

Velázquez, M.; Marucco, J.-F.; Mounaix, P.; Pérez, O.; Ferrier, A. & Moncorgé, R. (2009). Segregation and twinning in the rare-earth-doped KPb_2Cl_5 laser crystals. *Cryst. Growth & Design*, Vol. 9, No.4, pp. 1949-1955.

Vinogradova, N. N.; Galagan, B. I.; Dmitruk, L. N.; Moiseeva, L. V.; Osiko, V. V.; Sviridova, E. E.; Brekhovskikh, M. N. & Fedorov, V. A. (2005). Growth of rare-earth-doped K_2LaCl_5, K_2BaCl_4, and K_2SrCl_4 single crystals. *Inorganic Materials*, Vol. 41, No.6, pp. 654-657.

Virey, E.; Couchaud, M.; Faure, C.; Ferrand, B.; Wyon, C. & Borel, C. (1998). Room temperature fluorescence of $CsCdBr_3$:RE (RE=Pr, Nd, Dy, Ho, Er, Tm) in the 3-5 μm range. *J. All. Comp.*, Vol. 275-277, pp. 311-314.

Voda, M.; Al-Saleh, M.; Lobera, G.; Balda, R. & Fernández, J. (2004). Crystal growth of rare-earth-doped ternary potassium lead chloride single crystals by the Bridgman method. *Opt. Mater.*, Vol. 26, No.4, pp. 359–363.

Vtyurin, A. N.; Isaenko, L. I.; Krylova, S. N.; Yelisseyev, A.; Shebanin, A. P.; Turchin, P. P.; Zamkova, N. G. & Zinenko, V. I. (2004). Raman spectra and elastic properties of KPb_2Cl_5 crystals. *Phys. Stat. Sol. (c)*, Vol. 1, No.11, pp. 3142-3145.

Vtyurin, A. N.; Isaenko, L. I.; Krylova, S. N.; Yelisseyev, A.; Shebanin, A. P. & Zamkova, N. G. (2006). Vibrational spectra of KPb_2Cl_5 and KPb_2Br_5 crystals. *Comp. Mat. Sci.*, Vol. 36, pp. 212-216.

Wang, Y.; Li, J.; Tu, C.; You, Z.; Zhu, Z. & Wu, B. (2007). Crystal growth and spectral analysis of Dy3+ and Er3+ doped KPb_2Cl_5 as a mid-infrared laser crystal. *Cryst. Res. Technol.*, Vol. 42, No.11, pp. 1063-1067.

Zhou, G.; Han, J.; Zhang, S.; Cheng, Z. & Chen, H. (2000). Synthesis, growth and characterization of a new laser upconversion crystal Ba_2ErCl_7. *Prog. Cryst. Growth and Charact. Mater.*, Vol. 40, pp. 195-2000.

Part 3

Nonlinearity in Solid-State Lasers

5

Chirped-Pulse Oscillators: Route to the Energy-Scalable Femtosecond Pulses

Vladimir L. Kalashnikov
Institut für Photonik, Technische Universität Wien
Austria

1. Introduction

In the last decade, femtosecond pulse technology has evolved extremely rapidly and allowed achieving a few-optical-cycle pulse generation directly from a solid-state oscillator (Brabec & Krausz (2000); Steinmeyer et al. (1999)). Applications of such pulses range from medicine and micro-machining to fundamental physics of light-matter interaction at unprecedented intensity level and time scale (Agostini & DiMauro (2004); Gattass & Mazur (2008); Hannaford (2005); Krausz & Ivanov (2009); Martin & Hynes (2003); Mourou et al. (2006); Pfeifer et al. (2006)). In particular, high-energy solid-state oscillators nowadays allow high-intensity experiments such as direct gas ionization (Liu et al. (2008)), where the level of intensity must be of the order of 10^{14} W/cm^2. Such an intensity level enables pump-probe diffraction experiments with electrons, direct high-harmonic generation in gases, production of nm-scale structures at a surface of transparent materials, etc. The required pulse energies have to exceed one and even tens of micro-joules at the fundamental MHz repetition rate of an oscillator (Südmeyer et al. (2008)).

Such energy frontiers have become achievable due to the chirped pulse amplification outside an oscillator (Diels & Rudolph (2006); Koechner (2006); Mourou et al. (2006)). However, the amplifier technology is i) complex, ii) expensive, iii) noise amplification is unavoidable, and iv) accessible pulse repetition rates lie within the kHz range. The last is especially important because the signal rates in, for example, electron experiments are usually low and an improvement factor of 10^3–10^4 due to the higher repetition rate of the pulses significantly enhances the signal-to-noise ratio. Therefore, it is desirable to find a road to the direct over-microjoule femtosecond pulse generation at the MHz pulse repetition rates without an external amplification.

In principle, a cavity dumping allows increasing the pulse energy from a solid-state oscillator (Huber et al. (2003); Zhavoronkov et al. (2005)) but it makes the system more complex. There are the few alternative ways of increasing the oscillator pulse energy E, which is a product of the average power P_{av} and the repetition period T_{rep}: by increasing the cavity length and/or increasing the power (Apolonski et al. (2000); Cho et al. (1999); der Au et al. (2000); Südmeyer et al. (2008)). The impediment is that a high-energy pulse with $E = P_{av}T_{rep}$ suffers from instabilities owing to nonlinear effects caused by the high pulse peak power $P_0 \propto E/T$ (T is the pulse width). The leverage is to stretch a pulse, i.e. to increase its width T and thereby to decrease its peak power below the instability threshold.

One may grasp an implementation of this strategy on basis of the so-called solitonic conception of ultrashort pulse. If the pulse parameters change slowly during one cavity round-trip (that is a reasonable approximation for a typical solid-state oscillator), the ultrashort pulse can be treated as a dissipative soliton governed by both linear and nonlinear factors of an oscillator (Akhmediev & Ankiewicz (2005); Haus (1975a); Kärtner et al. (2004)). From a solitonic standpoint, the solid-state oscillators can be subdivided into i) those operating in the anomalous dispersion regime and ii) those operating in the normal dispersion regime.

In the anomalous dispersion regime, there are the simple relations between the pulse parameters E, T, and P_0 as well as the oscillator parameters β (that is the net-group-delay-dispersion coefficient) and γ (that is the self-phase modulation coefficient of a nonlinear medium, e.g. active crystal, air, optical plate, etc.) (Agrawal (2006))

$$T = \sqrt{|\beta|/\gamma P_0}, \quad E = 2\,|\beta|/\gamma T. \tag{1}$$

Since the peak power P_0 has to be kept lower than some threshold value P_{th} in order to avoid the pulse destabilization, the energy scaling requires a pulse stretching which can be provided by only the substantial dispersion growth (see Eq. (1)):

$$E = 2\sqrt{P_{th}\,|\beta|/\gamma}. \tag{2}$$

In the normal dispersion regime, a pulse is stretched and its peak power is reduced due to appearance of the so-called chirp ψ (Haus et al. (1991)). The chirp means that the instantaneous frequency varies with time and, as a result, the pulse becomes stretched by ψ times in the time domain in comparison with a chirp-free pulse (Kharenko et al. (2011)). Since a pulse is chirped, an oscillator operating in such a regime is called as "*chirped-pulse oscillator*" (Fernández González (2008)). As a result of chirping, an oscillator becomes extremely stable at a comparatively low level of dispersion. The spectrally broad pulse from such an oscillator can be substantially compressed (the compression factor is $\approx \psi$) with the proportional growth of its peak power P_0. An implementation of this regime promises a substantial enhancement of energy scalability in comparison with the law (2) and this strategy is the objective of the recent review.

The chapter is structured in the following way. In the first part, the physical principles of operation of a chirped-pulse oscillator are considered. The decisive contribution of dissipative effects such as the spectral filtering and the self-amplitude modulation into formation of a chirped-pulse is emphasized. Then, the concept of the chirped pulse as the chirped dissipative soliton is formulated. The underlying model is based on the so-called complex nonlinear Ginzburg-Landau equation (Akhmediev & Ankiewicz (1997); Aranson & Kramer (2002)). The main theoretical results concerning the chirped dissipative solitons of this equation are reviewed. They are based on both analytical and numerical integration techniques. The former can be divided into exact and approximated approaches. The last is most powerful and allows reducing the analysis of a chirped-pulse oscillator to the construction of two-dimensional "master diagram" comprising main properties of the chirped dissipative soliton (Kalashnikov et al. (2006)). The concept of the soliton energy scalability is analyzed and the different approaches to this concept are compared. In the course of analysis, a parallel between the chirped-pulse solid-state oscillators and the all-normal-dispersion fiber lasers (Rühl (2008); Wise et al. (2008)) is drawn.

In the last part of review some experimental achievements in the field of solid-state chirped-pulse oscillators are surveyed. Both broadband bulk and Yb:doped thin-disk oscillators are considered and the prospects for a further energy-scaling are estimated.

2. Operational principles and theory of chirped-pulse oscillators

In this section, the theory of a chirped-pulse oscillator will be outlined. As will be shown, the dissipative factors of a laser play a decisive role in the formation of an ultrashort pulse which can be treated as a chirped dissipative soliton. The underlying model is based on the complex nonlinear Ginzburg-Landau equation. Formulation and integration of this equation as well as interpretation of the obtained results are the important steps required for comprehension of the operational principles of chirped-pulse oscillators.

Firstly, we shall characterize the principal factors governing the oscillator dynamics and formulate the muster equation (i.e. the complex nonlinear Ginzburg-Landau equation) modeling this dynamics. Then, the physical principles of the chirped pulse formation will be considered. It will be shown, that the chirped pulse exists due to a phase balance supported by a self-amplitude modulation. As a result, a chirped dissipative soliton emerges. Only two exact shapes for such a soliton are known. They are the partial solutions of the complex nonlinear Ginzburg-Landau equation and have a limited range of applications. Therefore, the approximated approaches to the muster equation are of interest and they will be surveyed in a nutshell.

The first approach is based on the regularized adiabatic approximation (Ablowitz & Horikis (2009); Kalashnikov (2010); Podivilov & Kalashnikov (2005)) and the second class of approaches exploits the so-called Galerkin truncation (Blanchard & Brüning (1992); Malomed (2002)). The main advantage of the approximated methods is that they project the initial problem with infinitely many degrees of freedom to the finite-dimensional one that, thereby, makes the chirped dissipative soliton parameters to be easily traceable. We shall give emphasis to two practically important outputs of the approximated models of a chirped dissipative soliton viz. to the concepts of i) "master diagram" (Kalashnikov & Apolonski (2010); Kalashnikov et al. (2006)) and ii) "dissipative soliton resonance" (Chang et al. (2008b)). These concepts allow truncating the soliton parametrical space and provide the thorough grasp of the oscillator energy scalability. A possible application of the theory to the all-normal-dispersion fiber lasers will be implied in the course of consideration. In the course of this section, we shall touch upon some issues of a chirped-pulse oscillator stability.

2.1 Operational principles of a chirped-pulse oscillator

Let's begin with the consideration of principal factors governing the ultrashort pulse formation in an oscillator. Their schematic representation is shown in Fig. 1. In the slowly varying envelope approximation (Oughstun (2009)), the laser field envelope A evolves under influence of the operators corresponding to each of the factors represented in the diagram. The slowly varying envelope approximation is valid until $T \gg 1/\omega_0$ and $\Delta \ll \omega_0$ (ω_0 is the carrier frequency, Δ is the spectral half-width of ultrashort pulse and T is its width) and has proved its usefulness for a theory of ultrashort pulse propagation even in the limit of $T \to 1/\omega_0$ (Brabec & Krausz (1997)).

We shall consider below a $(1+1)$-dimensional field envelope $A(z, \tau)$, where $z \in [0, NL_{cav}]$ is the propagation distance (i.e. the distance taken along the arrows in Fig. 1) and $\tau \equiv [t -$

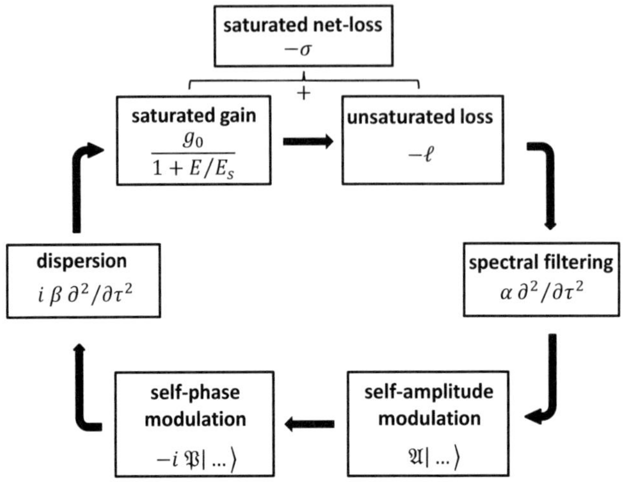

Fig. 1. Schematic representation of principal factors contributing to the ultrashort pulse formation and dynamics.

$z\,(dk/d\omega)|_{\omega=\omega_0}] \in [0, T_{rep}]$ is the local or "reduced" time ($t \in [0, NT_{rep}]$ is the time, ω is the frequency deviation measured from ω_0, $k(\omega)$ is the wave number). An oscillator is a naturally periodic system with the repetition time $T_{rep} \equiv L_{cav}/c$ (c is the light speed, L_{cav} is the oscillator cavity length for a circular scheme of Fig. 1 or the double cavity length for a linear oscillator) and the repetition ("cavity round-trip") number N.

2.1.1 Saturable gain and linear loss

The dissipative factors, which are generic for all types of oscillators, are the saturable gain and the linear (i.e. power-independent) loss. The latter includes both intracavity and output losses. In the simplest case, their common contribution into laser dynamics can be described as

$$\frac{\partial A(z,\tau)}{\partial z} = -\sigma A(z,\tau) = \frac{g_0 A(z,\tau)}{1+E/E_s} - \ell A(z,\tau), \tag{3}$$

where g_0 is the unsaturated gain defined by a pump (i.e. the gain coefficient for a small signal), $E \equiv \int_0^{T_{rep}} |A|^2 d\tau$ is the intracavity field energy ($|A|^2$ has a dimension of power), $E_s \equiv \hbar\omega_0 S/\sigma_g$ is the gain saturation energy (σ_g is the gain cross-section and S is the laser beam area). Multiple propagation of the pulse through an active medium during one cavity round-trip, as it takes a place in a thin-disk oscillator (see Sec. 3) has to be taken into account by a corresponding multiplier before E in (3). The power-independent ("linear") loss coefficient is ℓ.

Since an oscillator in a steady-state regime operates in the vicinity of lasing threshold (where σ =0 by definition), one may expand σ (Kalashnikov et al. (2006)):

$$\sigma(E) \approx \delta\left(\frac{E}{E^*} - 1\right), \tag{4}$$

where E^* is the energy of continuous-wave operation corresponding to σ =0, and

$\delta \equiv (d\sigma/dE)|_{E=E^*}$.

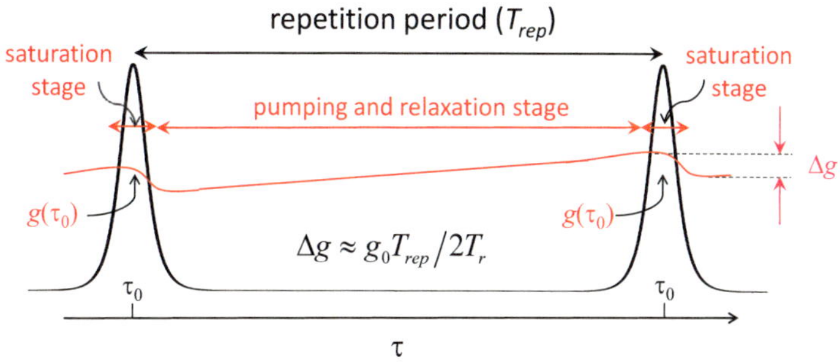

Fig. 2. Gain dynamics (red curve corresponds to $g(\tau)$) affected by pumping, gain relaxation, and gain saturation in the steady-state (i.e. z-independent) pulsing regime (black curve corresponds to $|A(\tau)|^2$).

Eqs. (3,4) do not take into account the time-dependence of σ due to the so-called dynamic gain saturation (see Fig. 2) (Kalashnikov (2008)). The dynamic gain saturation emerges when $E/E_s \to 1$ and the gain is saturated by rather "fluent" energy $\int_0^\tau |A|^2 d\tau'$. For instance, a Ti:sapphire oscillator with S =130 μm^2 and E =200 nJ has $E/E_s \approx 0.2$. That is the dynamic gain saturation begins to play a role in the dynamics. In this case, Eq. (3) has to be supplemented with the rate equation for a four-level active medium (Herrmann & Wilhelmi (1987)) that results in

$$\begin{aligned} \frac{\partial A(z,\tau)}{\partial z} &= \left(g(z,\tau) - \ell\right) A(z,\tau), \\ \frac{\partial g(z,\tau)}{\partial \tau} &= \frac{P_p}{S_p}\frac{\sigma_a}{\hbar\omega_a}\left(g_{\max} - g(z,\tau)\right) - \frac{|A(z,\tau)|^2}{SE_s} g(z,\tau) - \frac{g(z,\tau)}{T_r} \end{aligned} \tag{5}$$

Here the maximum gain g_{max}, the absorption cross-section σ_a, the absorption frequency ω_a, the absorbed pump power P_a, the pump beam area S_a, and the relaxation time T_r. For a pulse with $T \ll T_{rep}$ (this requirement is trivial), one may use the expansion in E/E_s-series (Haus (1975b); Kalashnikov, Kalosha, Mikhailov, Poloyko, Demchuk, Koltchanov & Eichler (1995))

$$\sigma(z,\tau) = \ell - g(z,\tau_0)\exp\left(-\frac{\int_{\tau_0}^{T_{rep}} |A|^2 d\tau'}{E_s}\right) \approx \ell - g(z,\tau_0)(z)\left(1 - \frac{\int_{\tau_0}^{T_{rep}} |A|^2 d\tau'}{E_s} + \ldots\right). \tag{6}$$

Here $g(\tau_0)$ is the gain at the pulse peak, which appears at some repetitive instant $\tau \equiv \tau_0$ (Fig. 2). $g(\tau_0)$ can be expressed iteratively by integration of the second equation of (5) (Jasapara et al. (2000)):

$$\begin{aligned} g(z,\tau_0) &= g_0\left(z - cT_{rep}, \tau_0\right)\exp\left(-\frac{E}{2E_s} - \frac{T_{rep}}{T_r} - \frac{P_p}{S_p}\frac{\sigma_a T_{rep}}{\hbar\omega_a}\right) + \\ &+ \frac{g_{\max} P_p \sigma_a T_{rep}/S_p\hbar\omega_a}{E/2E_s + T_{rep}/T_r + P_p\sigma_a T_{rep}/S_p\hbar\omega_a}\left[1 - \exp\left(-\frac{T_{rep}}{T_r} - \frac{P_p}{S_p}\frac{\sigma_a T_{rep}}{\hbar\omega_a}\right)\right]. \end{aligned} \tag{7}$$

On the one hand, the contribution of $\sigma(z,\tau)$ (Eqs. (3) or (5)) to the oscillator dynamics is important because the dissipative soliton emerges spontaneously from a destabilized continuous-wave regime of an oscillator (Soto-Crespo et al. (2002)). The continuous-wave solution of Eq. (3) corresponds to the condition $\partial A/\partial z = -\sigma A = 0$, which results in

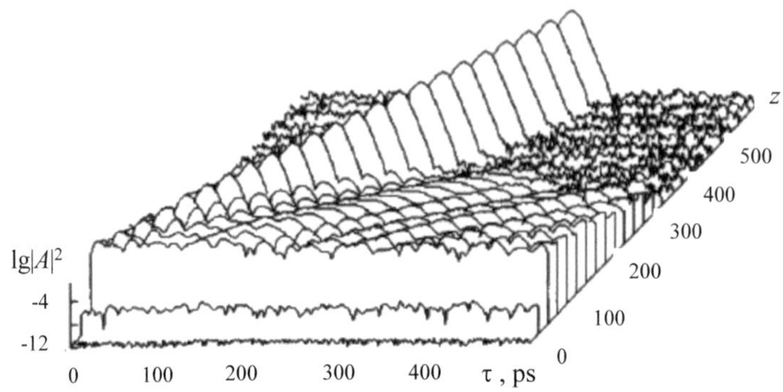

Fig. 3. Evolution of the normalized power for an oscillator studied in (Kalashnikov, Kalosha, Mikhailov & Poloyko (1995)). Figure illustrates the stages of ultrashort pulse formation with the initial noise amplification (vacuum instability), the continuous-wave formation, its destabilization, and the further noise reduction to an initial ("vacuum") level at the stage, when a dissipative soliton becomes "mature". The propagation distance z is normalized to cT_{rep}, i.e. it is the cavity round-trip number. Adapted from (Kalashnikov, Kalosha, Mikhailov & Poloyko (1995))

P_{av} =0 (so-called "vacuum"), and $P_{av} = E_s (g_0 - \ell)/\ell T_{rep}$ (actually, continuous-wave). The continuous-wave develops from the vacuum instability (the stage from z =0 to $z \approx$100 in Fig. 3).

On the other hand, the stability condition for the mature soliton is the vacuum stability that is $\sigma >$0. This condition will play a crucial role in the sequel. It means physically that there exists no an amplification outside the dissipative soliton (see the stage after $z \approx$300 in Fig. 3, where the vacuum becomes stable). If the gain dynamics contributes (Fig. 2), such a criterion becomes even stronger (Kalashnikov et al. (2006)):

$$\sigma > \Delta g \approx g_0 T_{rep}/2T_r. \tag{8}$$

Another important aspect of gain dynamics results from the energy-dependence of σ which causes a negative-feedback: the destabilizing growth (reduction) of soliton energy increases (decreases) σ >0 and, thereby, leads to the soliton attenuation (amplification) that tends in turn to the soliton stabilization.

2.1.2 Spectral filtering

As will be shown in Sec. 2.2, the spectral filtering is the key factor for a chirped-pulse oscillator. Such a filtering results from spectrally-dependent gain and/or loss and begins to contribute already at an initial stage of pulse formation. Fig. 3 demonstrates smoothing of an initial noise structure since the generation starts. In the spectral domain, a spectral filter provides a minimum loss at some frequency and, thereby, "selects" it (so-called "longitudinal mode selection"). The initially broad noise spectrum, corresponding to the short time spikes shrinks that corresponds to a smoothed time structure. The latter tends asymptotically to a continuous-wave. But a continuum-wave can be unstable in the presence of nonlinear factors (see next Sec. 2.1.3) so that an ultrashort pulse develops (Fig. 3).

Let us assume that a spectral filter has a profile $\Phi(\omega)$. Then, Eq. (3) has to be supplemented with

$$\frac{\partial A(z,\omega)}{\partial z} = \Phi(\omega) A(z,\omega), \quad (9)$$

where $A(z,\omega)$ is the Fourier-image of $A(z,\tau)$.

Usually, the Lorenzian shape for $\Phi(\omega)$ is assumed (Haus (1975a)):

$$\Phi(\omega) = \frac{g(\omega_0)}{1 + i\omega/\Omega}, \quad (10)$$

where $g(\omega_0)$ is the gain coefficient or the maximum transmission coefficient (e.g., for an output mirror). It is assumed that spectral filtering is centered at ω_0. Ω is the filter bandwidth.

If $2\Delta \ll \Omega$, one may expand (10) and proceed in the time-domain

$$\frac{\partial A(z,\tau)}{\partial z} \approx g(\omega_0) \left[1 - \frac{1}{\Omega}\frac{\partial}{\partial \tau} + \frac{1}{\Omega^2}\frac{\partial^2}{\partial \tau^2} - h.o.t. \right] A(z,\tau), \quad (11)$$

where *h.o.t.* means the higher in $\partial/\partial\tau$-order terms, which are negligible as a rule (for some important exceptions see, e.g. (Akhmediev & Ankiewicz (2005); Kalashnikov et al. (2011))).

2.1.3 Self-amplitude modulation

A self-amplitude modulation provides a pulse-power discrimination so that the net-gain increases with $|A|^2$ (at least up to some power level). This factor forms and stabilizes a pulse (Fig. 3). The mechanisms of self-amplitude modulation are various (Paschotta (2008); Weiner (2009)) but there are two ones, which are widespread in the solid-state oscillators: i) Kerr-lens mode-locking (KLM) (Spence et al. (1991)), and ii) mode-locking due to a semiconductor saturable absorber mirror (SESAM) (Keller et al. (1996)).

The KLM uses a self-focusing of laser beam inside some nonlinear element (e.g., active medium) that changes diffractional loss (so-called hard aperture mode-locking) or overlapping between the lasing and pumping beams (so-called soft aperture mode-locking) (Paschotta (2008)). It is important, that i) response of this mechanism to a laser field is practically instantaneous (i.e. it is power-dependent), and ii) the mechanism is strongly interrelated with the self-phase modulation (see next Sec. 2.1.4) because both phenomena are caused by the same nonlinear process.

A detailed modeling of spatial variations of laser beam is cumbersome and unpractical. Therefore, some reduction of dimensionality is required. For instance, one may consider an evolution of only zero-order Gaussian beams under action of factors presented in Fig. 1 (Kalashnikov, Kalosha, Mikhailov & Poloyko (1995); Kalosha et al. (1998)). However, such an approach remains to be cumbersome because it needs considering the detailed geometrical structure of an oscillator. Hence, it is usable to reduce the KLM to an action of some effective fast saturable absorber with a response function $\mathfrak{U}|A(z,\tau)\rangle \equiv \mathfrak{F}(|A(z,\tau)|^2)A(z,\tau)$ so that Eqs. (3,9) have to be supplemented with

$$\frac{\partial A(z,\tau)}{\partial z} = \mathfrak{F}(|A(z,\tau)|^2)A(z,\tau). \quad (12)$$

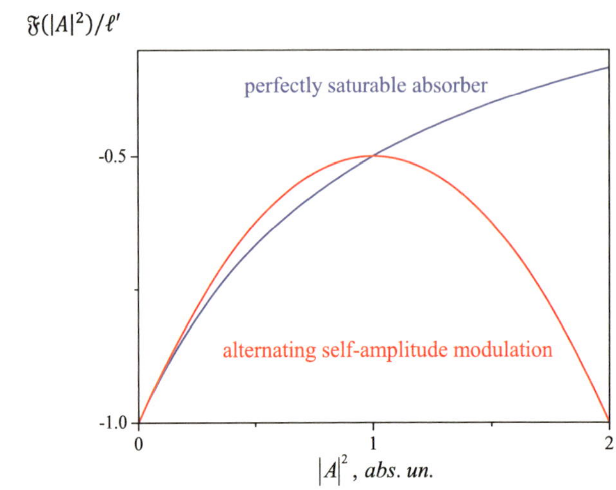

Fig. 4. Idealized self-amplitude modulation functions $\mathfrak{F}\left(|A(z,\tau)|^2\right)$.

From this point of view, one may consider the soft and hard aperture mode-locking as (see Fig. 4) i) "alternating" self-amplitude modulation for the former with

$$\mathfrak{F}\left(|A(z,\tau)|^2\right) \approx -\ell' + \kappa\left(1 - \zeta|A(z,\tau)|^2\right)|A(z,\tau)|^2, \tag{13}$$

where a loss saturation changes into a loss enhancement with the power growth; and ii) perfectly saturable absorber for the latter with

$$\mathfrak{F}\left(|A(z,\tau)|^2\right) \approx -\frac{\ell'}{1+\kappa|A(z,\tau)|^2}, \tag{14}$$

where the loss ℓ' vanishes asymptotically with the power growth.

Here ℓ' is the low-power saturable loss (i.e. unsaturated saturable loss) and it has not to be confused with ℓ (i.e. unsaturable loss). The ℓ'-coefficient can be absorbed by σ (Eq. 3) (for such "absorption", the right-hand side of Eq. 14 has to be rewritten as

$-\ell' + \ell'\kappa|A|^2 / \left(1+\kappa|A|^2\right)$). The κ-parameter is the self-amplitude modulation coefficient (i.e. the inverse power of loss saturation), ζ is the coefficient of self-amplitude alteration.

Physical sense of the curves presented in Fig. 4 can be explained in the following way. Self-focusing inside a nonlinear element is provided by the cubic nonlinearity therefore the leading term in (12) is proportional to $|A|^2A$. If a hard aperture is used, the beam squeezing due to self-focusing leads to the monotonic decrease of the diffraction loss (Eq. (14) and the blue curve in Fig. 4). If a soft aperture is used, the overlapping between the lasing and pumping beams and, thereby, the gain increases initially with power. Then the overlapping becomes complete and the gain reaches its maximum. Further growth of power worsens the overlapping and the gain decreases with power (Eq. (13) and the red curve in Fig. 4).

Now let's consider a SESAM as the self-amplitude modulator. As a rule, the nonlinear properties of a semiconductor structure are complex. Despite this complexity, one may model the nonlinear response of a SESAM by Eq. (12) if $T \gg T_r^E$ (T_r^E is the decomposition time of excitons which is lesser than 100 fs as a rule) and $T_r^E \ll T_r^F$ (T_r^F is the recombination time of free carriers which exceeds tens of picoseconds) (Haus & Silberberg (1985)). Then, ℓ' is the modulation depth, and $\kappa \equiv 2\sigma^A T_r^E / \hbar\omega_0 S_A$ is the inverse saturation power of absorber (S_A is the beam area on an absorber, σ^A is the absorption cross-section). Since T equals to a few of picoseconds for an uncompressed chirped pulse, the approximation $T \gg T_r^E$ is quite precise for a chirped-pulse oscillator.

2.1.4 Self-phase modulation

Active crystal, air, optical plates, etc. have a third-order nonlinearity so that their refractivity indexes n are power-dependent: $n = n_0 + \gamma |A|^2$ (n_0 is the refractive index for a small signal, γ is the self-phase modulation coefficient for a medium) (Agrawal (2006)). The coefficient γ can be estimated as: $\gamma = 8\pi^2 n_0 n_2 / \lambda^2$ for the case of strong beam focusing inside a nonlinear element (e.g. inside an active crystal or optical plate) or $\gamma = 4\pi n_2 L / \lambda S$ for the case of weak focusing (e.g. inside an active thin-disk or air) (Weiner (2009)). Here λ is the central wavelength of an oscillator, L is the length of a nonlinear medium, n_2 is its nonlinear refractive index, and S is the beam area.

Since the γ-coefficient depends on beam area in the last formula, the self-focusing can affect the self-phase modulation through the S-change. As a result, the refractive index acquires the corrections with higher-order powers of $|A|$. Then, the contribution of self-phase modulation can be modeled as

$$\frac{\partial A(z,\tau)}{\partial z} = -i\mathfrak{P}\,|A(z,\tau)\rangle = -i\left[\gamma + \chi |A(z,\tau)|^2\right] |A(z,\tau)|^2 A(z,\tau), \tag{15}$$

where χ describes the higher-order correction to the self-phase modulation coefficient.

Despite the factors considered above, the self-phase modulation is non-dissipative effect and does not change the pulse energy.

2.1.5 Group-delay dispersion

The last effect required for the ultrashort pulse generation is the group-delay dispersion. This effect causes frequency dependence of the wave propagation constant k: $k(\omega) = k(\omega_0) + k_1\omega + k_2\omega^2 + h.o.t.$ Since an ultrashort pulse has a broad spectrum, such a dependence cannot be neglected. The coefficients k and k_1 can be included in the definitions of the soliton wave-number and the reduced time τ, respectively. Then, the contribution of the group-delay dispersion is defined as

$$\frac{\partial A(z,\tau)}{\partial z} = i\beta \frac{\partial^2 A(z,\tau)}{\partial \tau^2} + h.o.t. \tag{16}$$

Here $\beta \equiv L_{cav}\left(dk/d\omega\right)\big|_{\omega=\omega_0}$ is the group-delay dispersion coefficient and, in our notations, it is positive for a normal dispersion and negative for an anomalous dispersion.

Both self-phase modulation and group-delay dispersion are non-dissipative effects and contribute to the pulse phase profile. In the solitonic regime, these contributions have to be balanced therefore both γ and β are the key control parameters of an oscillator.

Higher-order terms in (16) are called as the higher-order dispersions and become especially important for broadband oscillators (such as Ti:sapphire, Cr:YAG, Cr:ZnSe, etc.). Some aspects of their contribution to the characteristics of a chirped-pulse oscillator will be considered in Sec. 2.3.1, as well.

2.1.6 Haus master equation

The pulse changes during one cavity round-trip are, as a rule, small for a solid-state oscillator. That allows joining Eqs. (3,11,12,15,16):

$$\frac{\partial A(z,\tau)}{\partial z} = \left[-\sigma(E) + \frac{g(\omega_0)}{\Omega^2}\frac{\partial^2}{\partial\tau^2} + \mathfrak{F}\left(|A|^2\right)\right] A(z,\tau) + i\left\{\beta\frac{\partial^2}{\partial\tau^2} - \gamma|A|^2 - \chi|A|^4\right\} A(z,\tau). \tag{17}$$

Eq. (17) is the modified Haus master equation introduced in (Haus et al. (1991)). Formally, it can be rated as a complex nonlinear Ginzburg-Landau equation (Akhmediev & Ankiewicz (1997); Aranson & Kramer (2002)). The last has an extremely wide horizon of applications covering quantum optics, modeling of Bose-Einstein condensation, condensate-matter physics, study of non-equilibrium phenomena, and nonlinear dynamics, quantum mechanics of self-organizing dissipative systems, and quantum field theory. In laser physics, this equation provides an adequate description of ultrashort pulses in both solid-state (Haus et al. (1991)) and fiber (Ding & Kutz (2009); Komarov et al. (2005)) lasers, as well as, pulse propagation in nonlinear fibers. As will be shown in Sec. 2.3, the theory of chirped-pulse oscillators is based mainly on study of soliton-like solutions of (17). Hereinafter, we shall attend to two main modifications of (17): i) cubic-quintic nonlinear Ginzburg-Landau equation (CQNGLE) with $\mathfrak{F}\left(|A(z,\tau)|^2\right)$ defined by (13), and ii) generalized nonlinear Ginzburg-Landau equation (GNGLE) with $\mathfrak{F}\left(|A(z,\tau)|^2\right)$ defined by (14).

The exact analytical solutions of the complex nonlinear Ginzburg-Landau equation are known only for a few of cases, when they represent the dissipative solitons and some algebraic relations on the parameters of the equation are imposed (Akhmediev & Afanasjev (1996)). More general solutions can be revealed on basis of the algebraic nonperturbative techniques (Conte (1999)), which, nevertheless, are not developed sufficiently still. The perturbative methods allow obtaining the dissipative soliton solutions (Malomed & Nepomnyashchy (1990)) in the vicinity of the Schrödinger solitonic sector (Agrawal (2006)) of CQNGLE. Another approximate methods utilize the reduction of infinite-dimensional space of CQNGLE to the finite-dimensional one on basis of the method of moments (Tsoy & Akhmediev (2005)) or the variational method (Ankiewicz et al. (2007); Bale & Kutz (2008)). At last, there is the direct approximate integration technique for the complex Ginzburg-Landau equation with an arbitrary $\mathfrak{F}\left(|A(z,\tau)|^2\right)$ (Podivilov & Kalashnikov (2005)). Applications of these methods to the theory of chirped-pulse oscillator will be reviewed in Sec. 2.3.

2.2 Physical principles of chirped dissipative soliton formation

A dissipative soliton is named as "chirped" one if its phase $\phi(\tau)$ is time-dependent. The chirp results from a joint action of normal dispersion ($\beta>0$) and self-phase modulation. Eq. (17) suggests that a round-trip contribution of the time-dependent phase to the pulse phase change is $-\beta|A(\tau)|\left[d\phi(\tau)/d\tau\right]^2 - \gamma|A(\tau)|^3$ (blue curve in left Fig. 5). Simultaneously, a phase contribution of the pulse envelope is $\beta\partial^2|A(\tau)|/\partial\tau^2$ (red curve in left Fig. 5). Hence, the phase balance is *possible* for a chirped pulse propagating in the normal dispersion regime (Kalashnikov et al. (2008); Renninger et al. (2011)). It should be noted, that the Schrödinger

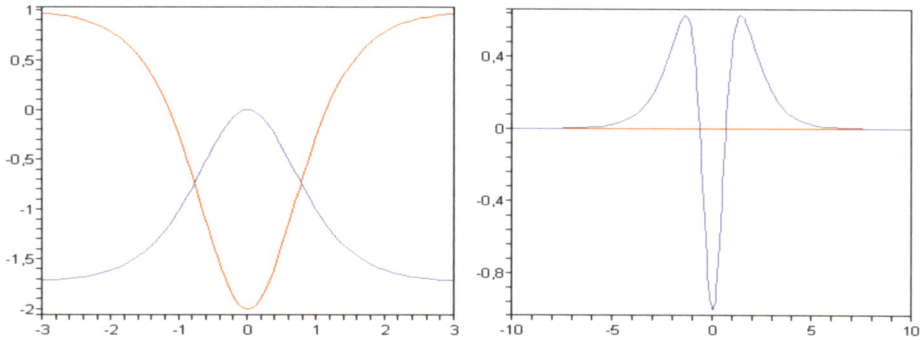

Fig. 5. One cavity round-trip contribution to the time-dependent phase $\phi(\tau)$ (left) and the amplitude $|A(\tau)|$ (right) for a chirped pulse. Left: the blue curve corresponds to contribution from the time-dependent phase, the red curve corresponds to contribution from the pulse envelope. The horizontal axes correspond to τ in arb. un.

soliton (Agrawal (2006)) developing in the anomalous dispersion regime (β <0) exists owing to a phase balance, which is possible only in the absence of chirp.

The simple condition for a phase balance can be expressed as $\psi^2 = 2 + \gamma P_0 T^2/\beta$ if $A(\tau) = \sqrt{P_0}\mathrm{sech}(\tau/T)^{1+i\psi}$ (ψ is the chirp parameter, i.e. $\phi(\tau) = \psi \ln[\mathrm{sech}(\tau/T)]$), $g(\omega_0)/\Omega^2\beta \ll 1$, and $\chi = 0$.

However unlike the Schrödinger soliton, the only phase balance is not sufficient for a chirped dissipative soliton existence. There is the amplitude perturbation (right Fig. 5) causing the soliton spreading. The source of this perturbation is the time-dependent phase, which contributes through a dispersion: $-\beta\left[2\left(dA(\tau)/d\tau\right)\left(d\phi(\tau)/d\tau\right) + A(\tau)\left(d^2\phi(\tau)/d\tau^2\right)\right]$. In addition, there exists a dissipation owing to spectral filtering $\left(g(\omega_0)/\Omega^2\right) d^2\left|A(\tau)\right|/d\tau^2$ and a net-loss with the coefficient σ >0 (see Eq. (17)). Hence, some self-amplitude modulation is required for the soliton stabilization.

The obvious mechanism, which acts against the pulse spreading and dissipation is the nonlinear gain provided by $\mathfrak{F}\left(|A(z,t)|^2\right)$. But for a chirped pulse, there exists an additional squeezing mechanism resulted from the spectral filtering and defined by the term

$-\left(g(\omega_0)/\Omega^2\right)|A(\tau)|\left(d\phi(\tau)/d\tau\right)^2$ in Eq. (17).

Physically, the action of the last mechanism can be explained in the following way (see (Haus et al. (1991); Proctor et al. (1993))). The chirp causes the substantial spectral broadening so that $T\Delta \gg 1$ (see left Fig. 6, where the Wigner function (Diels & Rudolph (2006)) for a chirped pulse is shown). As a result of such broadening, the spectral filtering becomes conspicuous. It cuts off the spectral components located at the pulse wings (right Fig. 6) that results in the pulse shortening. Thus, the balance of both dissipative and non-dissipative factors can support the self-sustained ultrashort pulse in a chirped-pulse oscillator. The theory of such self-sustained pulse, or "chirped dissipative soliton", will be considered below.

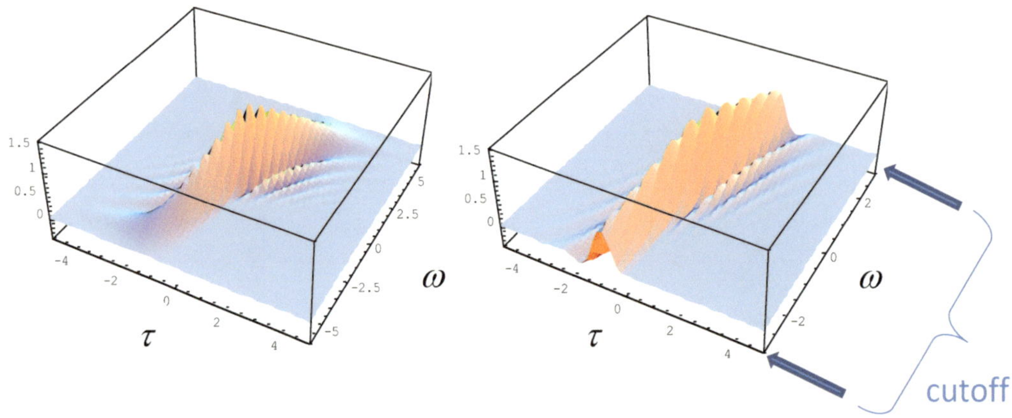

Fig. 6. Left: the Wigner function of a chirped pulse (arb. un.). Right: spectral cut-off resulting in the pulse shortening.

2.3 Theory of chirped dissipative soliton

It is convenient to systemize the theory of chirped-pulse oscillators from the point of view of approaches based on the concept of a chirped dissipative soliton. Such approaches can be divided into i) numerical, ii) exact analytical, and iii) approximated analytical ones. Of course, this division is conventional and a realistic analysis combines different approaches, as a rule.

2.3.1 Numerical modeling of chirped dissipative soliton

The master equation (17) is multiparametrical and the direct numerical simulations face the challenge of coverage of a multidimensional parametrical space. That requires the extensive calculations and the obtained results have to be carefully sorted. Moreover, the real parametrical space of a chirped dissipative soliton can differ from a separate set of the independent parameters of (17). In this case, the numerical identification of a relevant parametrical space is intricate problem. Nevertheless, a numerical approach provides most complete and adequate description of a real-world oscillator. In particular, such important characteristics of a mode-locked oscillator as its stability and self-starting ability, allocation of laser elements and noise properties can be analyzed mainly numerically. In addition, the numerical simulations are the testbed for any analytical model.

Extensive numerical simulations of the CQNGLE in the normal dispersion regime, that is relevant for the theory of chirped-pulse oscillator, have been carried out by N.N. Akhmediev with coauthors (Akhmediev et al. (2008); Soto-Crespo et al. (1997)). The simulations have allowed finding the pulse stability regions for some two-dimensional projections of the CQNGLE parametrical space. Unfortunately, the problem is that these regions cannot be compared directly with a parametrical space of concrete oscillator generally owing to dropping of the energy-dependence of net-loss parameter σ (i.e. δ-parameter in the notations of (Akhmediev et al. (2008); Soto-Crespo et al. (1997))). Additionally, the energy belongs rather to a set of control parameters of an oscillator than it is some derivative parameter of a solitonic solution.

Nevertheless, the numerical results of (Akhmediev et al. (2008); Soto-Crespo et al. (1997)) provide the qualitative grasp of the dissipative soliton properties. Fig. 7 demonstrates

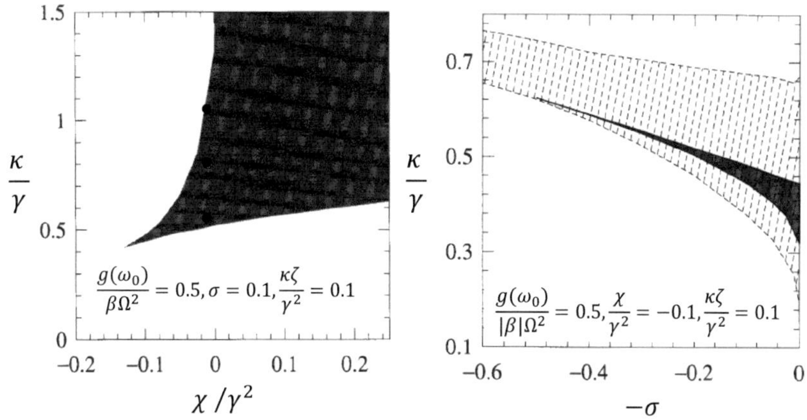

Fig. 7. Regions where the stable dissipative solitons exist in the normal dispersion regime (filled). For comparison, the hatched region in the right figure corresponds to the soliton developing in the anomalous dispersion regime. The parameters correspond to Eqs. (13,17). Adapted from (Soto-Crespo et al. (1997)).

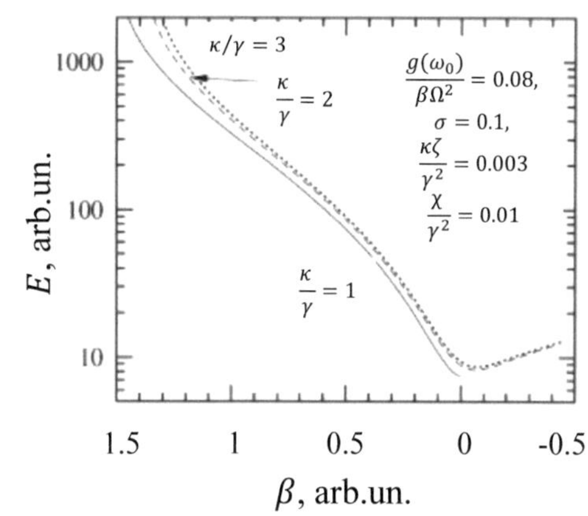

Fig. 8. Soliton energy versus dispersion β. Adapted from (Akhmediev et al. (2008)).

the numerically calculated regions where the stable dissipative solitons of Eqs. (13,17) exist (Soto-Crespo et al. (1997)). One may see (left figure) that the self-limited self-phase modulation (χ <0) tends to destabilize the dissipative soliton (the stability region shrinks). The right Fig. 7 demonstrates another interesting property: the *isogain* region (i.e. the region of pulse existence where σ =const) is strongly confined in the κ-dimension. The dissipative solitons with such properties will be called the *positive-branch* solitons (the sense of this term will be clear in sequel).

The physically important property of the positive-branch dissipative soliton is the swift growth of its energy along an isogain with the dispersion β. It means that such a soliton is *energy scalable* (see Fig. 8). Simultaneously, the energy dependence on the self-amplitude parameter κ is weaker. As will be shown in Secs. 2.3.2.2 and 2.3.2.3, such scaling properties of chirped-pulse oscillator can be explained on the basis of analytical approaches.

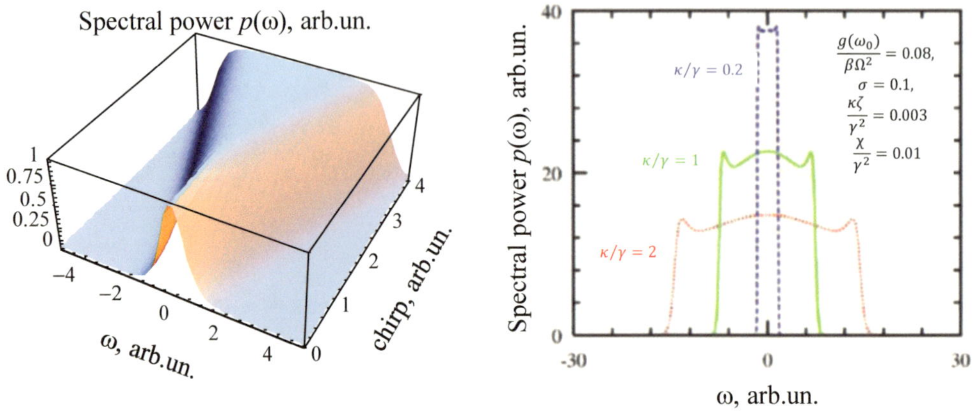

Fig. 9. Chirped dissipative soliton spectra. Left: spectrum vs. chirp parameter ψ for a chirped soliton $\sqrt{P_0}\text{sech}(\tau/T)^{1+i\psi}$. Right: spectra of the CQNGLE solitons adapted from (Akhmediev et al. (2008)).

The remarkable feature of a chirped-pulse oscillator is that its spectrum, which broadens with a chirp, has truncated edges (left Fig. 9) and flat, concave, convex and concave-convex tops (left and right Figs. 9, see also (Chong et al. (2008a); Kalashnikov et al. (2006; 2005))). The last spectral type has a modification with sharp spikes at the spectrum edges (Chong et al. (2008a); Kalashnikov & Chernykh (2007)).

The truncated spectral shape results from a rapid chirp growth at the spectrum edges (Kalashnikov et al. (2005)). Simultaneously, a chirp growing with dispersion causes the pulse stretching and, thereby, reduces its peak power that prevents the soliton destabilization (Chong et al. (2008a); Kalashnikov et al. (2006; 2005); Soto-Crespo et al. (1997)). Such a stretched pulse can be dechirped and compressed outside an oscillator cavity by approximately factor of ψ, where ψ is the dimensionless chirp parameter: $\psi \approx T\Delta/4 \approx (1/4)d^2\phi(\omega)/d\omega^2$ (T is the soliton width before compression, Δ is the soliton spectral half-width, $\phi(\omega)$ is the spectral phase) (Kalashnikov et al. (2005)). The ψ-value can be estimated approximately as

$$\psi \approx \frac{3}{\left(g(\omega_0)/\Omega^2\beta\right) + \left(\kappa/\gamma\right)}, \tag{18}$$

which explains the chirp growth with dispersion (β), self-phase modulation (γ), and broadening of a spectral filter (Ω).

Since the spectral chirp increases abruptly at spectrum edges, the satellites appear at the soliton wings after dechirping. Such satellites have uncompensated chirp, large frequency deviation and contain a part ($\approx$10%) of full energy (Kalashnikov et al. (2005)).

Numerical simulations demonstrate the spectral broadening with the dispersion decrease (Chong et al. (2008a); Kalashnikov et al. (2006; 2005)) although some spectral width reduction can appear for the small normal dispersions (Soto-Crespo et al. (1997)). Also, a spectrum broadens with a growth of the nonlinear phase shift and the spectral filtering (Chong et al. (2008a)). The pulse energy increases with the nonlinear phase shift (Chong et al. (2008a)) and dispersion (Kalashnikov et al. (2006; 2005); Soto-Crespo et al. (1997)).

Variety of the numerical results allows identifying the control parameters of a chirped-pulse oscillator with β (dispersion), Ω (spectral width of filter or gainband), γ (self-phase modulation coefficient), and κ (coefficient of self-amplitude modulation) (see Eqs. (13,17)) and conjecturing that their dimensionless combination in the form of (Kalashnikov et al. (2006))

$$C \equiv \frac{g(\omega_0)}{\beta\Omega^2}\frac{\gamma}{\kappa} = \frac{\alpha}{\beta}\frac{\gamma}{\kappa} \tag{19}$$

is relevant to the description of the dissipative soliton properties (the abbreviation $\alpha \equiv g(\omega_0)/\Omega^2$ is introduced). Such a conjecture is suggested by the analytical theory as well (see Secs. 2.3.2.2 and 2.3.2.3) and gives a deep insight into the properties of chirped-pulse oscillators. The most important understanding is that the properties of chirped dissipative soliton are defined primarily by not absolute values of parameters but their relations: spectral dissipation to dispersion and self-phase modulation to self-amplitude one. This observation allows a unified standpoint at the very different lasers operating in the normal dispersion regime. For instance, the fiber lasers operate on comparatively high levels of dispersion ($\approx 0.01 \div 0.1$ ps^2) but the self-phase modulation is high, as well. As a result, the C-parameter does not differ substantially from that for a solid-state oscillator. This allows including the fiber lasers into consideration (e.g. see works of *F. Wise* with coauthors).

The detailed analysis of this topic will be presented in Secs. 2.3.2.2 and 2.3.2.3 but it should be noted here that the relevance of combined parameters like C suggests the existence of low-dimensional hyper-surfaces in a multidimensional parametrical space of dissipative soliton which characterizes both solid-state and fiber lasers. The important examples of such low-dimensional parametrical representations of a dissipative soliton are based on the closely related concepts of the "*master diagram*" (Kalashnikov et al. (2006)) (Fig. 10) and the "*dissipative soliton resonance*" (Chang et al. (2008b)) (Fig. 11). It has been demonstrated, that both master diagram and dissipative soliton resonance are sufficiently robust structures and remain in the oscillators with a parameter management (Chang et al. (2008a); Kalashnikov et al. (2006)).

Significance and structure of the master diagram will be described in detail in Secs. 2.3.2.2 and 2.3.2.3 but one has to note here that the isogains (i.e. curves with $\sigma = const$) play a special role in the structure of representations of Figs. 10,11 (e.g., see (Bélanger (2007))). In particular, the singular curve $\sigma = 0$ denotes the soliton stability threshold. It was found numerically that the net-loss parameter σ increases almost linearly with κ/γ (Kalashnikov et al. (2005)). Also, this parameter increases initially with the dispersion lowering but then it decreases and approaches zero (Fig. 12). Really, the soliton loses its stability at some $\sigma = \Delta g > 0$ (Eq. (8)) due to dynamic gain saturation. The loss of stability means that the noise out of the pulse begins to amplify.

This observation testifies that the saturable gain is the decisive factor in a pulse stabilization. As has been found (Kalashnikov & Chernykh (2007); Kalashnikov et al. (2006)), the energy-dependence of σ is required to stabilize a chirped dissipative soliton so that some types of chirped dissipative solitons lose a stability in the absence of gain saturation (Kharenko et al. (2011)). Simultaneously, the gain dynamics during one cavity round-trip (so-called dynamic gain saturation, see Eq. (5)) can cause a soliton destabilization. Numerical simulations of (Kalashnikov (2008)) suggest that the stability region becomes confined on the (*Pump–Dispersion*)-plane in the presence of dynamic gain saturation. This means that there are some minimum and maximum dispersions as well as minimum and maximum pump

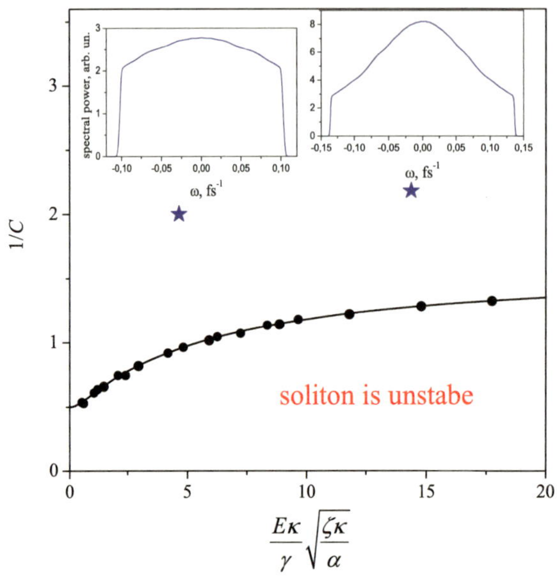

Fig. 10. The master diagram adapted from (Kalashnikov et al. (2006)). The circles obtained from simulations of a Ti:sapphire oscillator correspond to the isogain σ =0. The numerical spectra (inserts) are placed in vicinity of the corresponding parametrical points (blue stars).

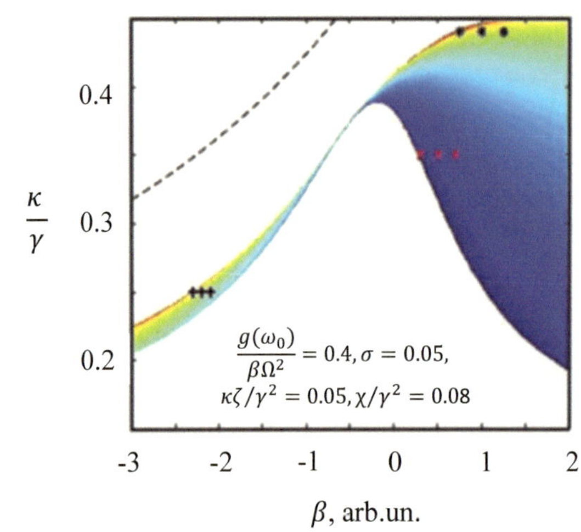

Fig. 11. Contour plot of the energy of stable soliton solution of the CQNGLE in the two-dimensional parameter space of dispersion β and nonlinear gain κ. Color scales the pulse energy from low (blue) to high (red) levels. The dashed line is an analytical approximation to the resonance curve (Chang et al. (2008b)). Adapted from (Grelu et al. (2010)).

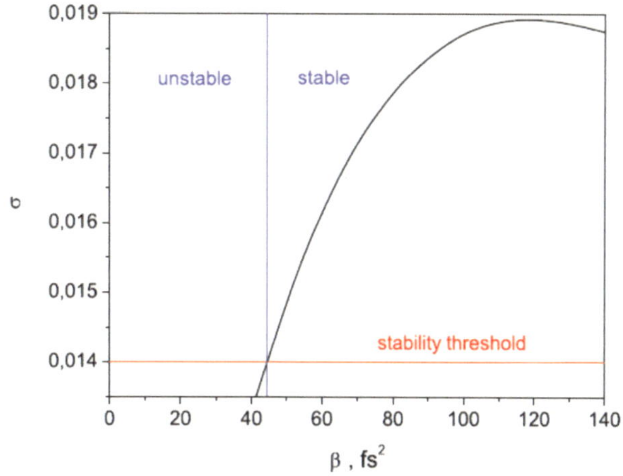

Fig. 12. Dependence of the net-loss parameter σ on dispersion. The stability threshold (σ_{th}) is determined by the dynamic gain saturation. Adapted from (Kalashnikov et al. (2005)).

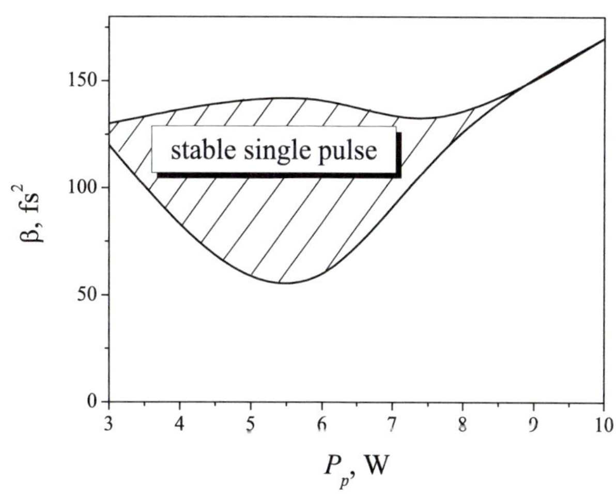

Fig. 13. Region of the chirped dissipative soliton stability (shaded area) in the presence of dynamic gain saturation (Eqs. (5,13,17)) with $\zeta = 0.6\gamma$, $\kappa = 0.04\gamma$, g_{max} =0.33, ℓ =0.22, S_p =100 μm^2, S =130 μm^2, T_r =3.5 μs. Double transit through a Ti:sapphire crystal per one round-trip (T_{rep} =21 ns) was considered. Adapted from (Kalashnikov (2008)).

powers providing a stable pulse (Fig. 13). The destabilizing mechanism was identified with an appearance of satellite in front of the pulse that results from the dynamic gain saturation and causes the energy transfer from a pulse to a satellite.

A special problem, which can be explored chiefly numerically is the stability of a chirped-pulse oscillator against the higher-order dispersions, i.e. in the presence of a frequency-dependence of the dispersion coefficient $\beta(\omega)$. Such a dependence is substantial in broadband chirped-pulse oscillators like Ti:sapphire (Kalashnikov et al. (2005)), Cr:YAG (Sorokin et al. (2008)) and Cr:ZnSe (Kalashnikov et al. (2011)). As a result, the master equation (17) has to

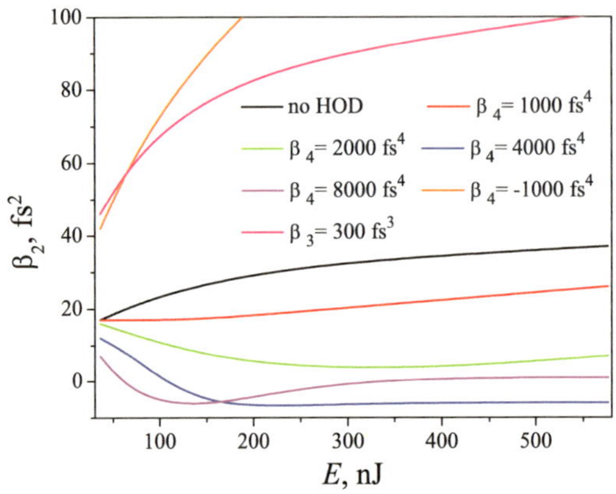

Fig. 14. Dependence of the stability threshold on the energy and a higher-order dispersion (HOD). Adapted from (Kalashnikov et al. (2008)).

be modified so that (e.g., see (Kalashnikov et al. (2008)))

$$\frac{\partial A(z,\tau)}{\partial z} = \left[-\sigma(E) + \frac{g(\omega_0)}{\Omega^2}\frac{\partial^2}{\partial\tau^2} + \mathfrak{F}\left(|A|^2\right)\right] A(z,\tau) + i\left\{-\sum_{n=2}^{N} i^n \beta_n \frac{\partial^n}{\partial t^n} - \gamma|A|^2 - \chi|A|^4\right\} A(z,\tau), \tag{20}$$

where β_n are the dispersion coefficients of nth−order.

The results of simulations for a Ti:sapphire oscillator with $N \leq 4$ are shown in Fig. 14. In fact, that is the representation of the master diagram presented in Fig. 10. The pulse is stable above the corresponding curves. One may conclude that, in general case, odd higher-order dispersions destabilize a pulse so that a large positive β_2 is required for pulse stabilization ($\beta_2 \equiv \beta(\omega_0)$).

It was conjectured that a source of destabilization in this case is an excitation of dispersive waves which is caused by resonance with a continuous-wave perturbation (Kalashnikov (2011); Kalashnikov et al. (2008)). Such an excitation appears if the resonance condition is satisfied: $k(\omega) \equiv \beta_2\omega^2 + \sum_{n=3} \beta_n \omega^n = q$, where $k(\omega)$ is the wavenumber of linear wave (i.e. perturbation) and q is the wavenumber of soliton. If there is some frequency which satisfies this condition, the generation of a dispersive wave begins at the expense of a soliton that perturbs the latter.

The effect of even higher-order dispersions β_{2n} ($n > 1$) is more complicated. If the sign of β_{2n} is such that the net-dispersion decreases toward the edges of soliton spectrum, the pulse can be stabilized by only larger dispersion β_2. For the opposite sign of β_{2n}, the pulse becomes stable within a wider range of dispersions and the stability border can penetrate even into anomalous dispersion region (Fig. 14).

The chirped dissipative soliton destabilized by higher-order dispersions behaves chaotically: its parameters, in particular the peak power and the central frequency, shake

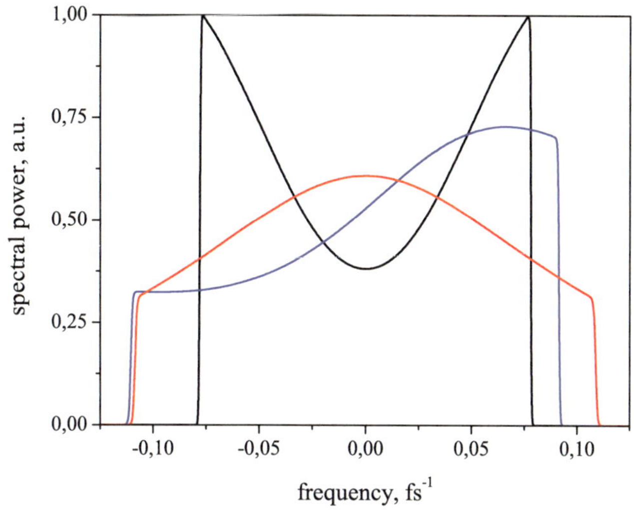

Fig. 15. Calculated spectral profiles of a Ti:sapphire oscillator for $\beta_2 = 110$ fs^2, β_4=0 fs^4 (red), 2000 (blue), 10000 (black), β_3=0 fs^3 (black, red), 300 (blue). Adapted from (Kalashnikov et al. (2005)).

(Kalashnikov et al. (2008); Sorokin et al. (2008)). This regime is called as "*chaotic mode-locking*" (Deissler & Brand (1994)) and, as a rule, separates the regimes of chirped and chirp-free pulse generation. As the extreme case, the noisy mode-locking (Horowitz et al. (1997)) can develop: pulse envelope and spectrum become completely fragmentized on a small scale but preserve their localization on a large scale so that the averaged envelope and spectrum are comparatively smooth (Kalashnikov (2011); Kalashnikov et al. (2008)).

Besides an effect on the pulse stability, higher-order dispersions distort the pulse spectrum (Fig. 15). As a rule, the maximum spectral amplitude is located within a range, where the normal dispersion is larger (Kalashnikov et al. (2005)).

Also, numerical simulations have reviled that a chirped-pulse oscillator can suffer from the long-periodic pulsations (both regular and chaotic) of the pulse peak power even in the absence of higher-order dispersions (Kalashnikov & Chernykh (2007)). Such an instability growing with the dispersion was attributed to an excitation of internal perturbation modes of dissipative soliton. An important consequence of this instability is the modulation of pulse spectrum, especially, the growth of spikes at the spectrum edges (Fig. 16).

One may conclude that the numerical modeling has provided with a rich information concerning the properties of a chirped dissipative soliton: i) dependencies of main pulse parameters on the parameters of master equations have been obtained, ii) spectral profiles of chirped dissipative soliton have been classified, iii) it has been conjectured that a true parametric space of chirped-pulse oscillator has a reduced dimension, iv) pulse energy scalability has been demonstrated and concepts of "master-diagram" and "dissipative soliton resonance" have been formulated, and, at last, v) pulse stability region have been explored extensively and decisive contributions of the gain saturation and the higher-order dispersions to the pulse instability have been shown.

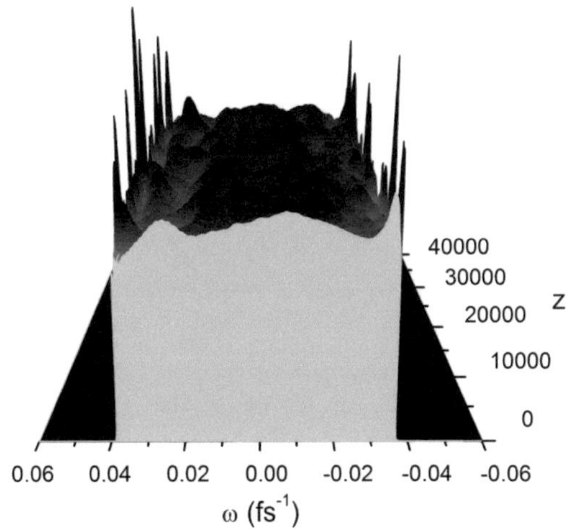

Fig. 16. Evolution of distorted spectrum profile of a Ti:sapphire oscillator, β =150 fs^2. Adapted from (Kalashnikov & Chernykh (2007)).

Nevertheless, a systematic understanding of the numerical results is hardly possible without an exploration of adequate analytical models of chirped-pulse oscillators.

2.3.2 Analytical models of chirped dissipative soliton

Below, basic analytical models of chirped dissipative solitons will be surveyed. For convenience, we divide these models into three groups which are based on i) exact soliton solution of the master equation, ii) solutions obtained from the adiabatic approximation, and iii) approximated solutions obtained from the variational method and the method of moments.

2.3.2.1 Exact soliton solution of the CQNGLE

Eq. (17) is nonintegrable and its exact soliton solutions are known for only a few of cases, when some algebraic relations on the parameters of the equation are imposed (Akhmediev & Ankiewicz (1997)). More general solutions can be revealed on basis of the algebraic nonperturbative techniques (Conte (1999)) which, nevertheless, are not developed sufficiently still. However, a few of known exact solutions can provide with some insight into properties of a chirped-pulse oscillator.

For the CQNGLE (Egs. (13,17)), the sole soliton solution is known (Soto-Crespo et al. (1997); van Saarloosa & Hohenberg (1992)). It can be expressed in the following form (Renninger et al. (2008))

$$A(z,\tau) = \sqrt{\frac{\mathrm{A}}{\mathrm{B} + \cosh\left(\tau/T\right)}} \exp\left[-\frac{i\psi}{2}\ln\left(\mathrm{B} + \cosh\left(\tau/T\right)\right) + iqz\right] \qquad (21)$$

where A, B, T, ψ, and q are the real constants characterizing pulse amplitude, shape, width, chirp, and wavenumber, respectively. It is important to emphasize, that it is a partial solution and exists for only certain algebraic relations imposed on the parameters of Eqs. (13,17).

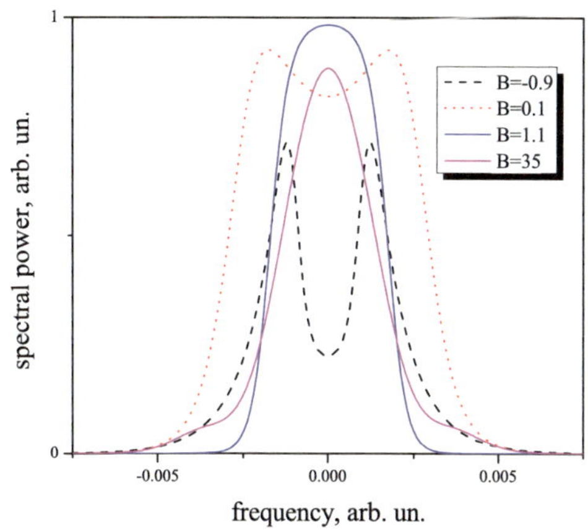

Fig. 17. Spectral profiles of a dissipative soliton (21), β/α =250, $\zeta = -0.5\gamma$ for B <1 and 0.5γ for B >1. The scales are arbitrary.

In spite of the fact that Eq. (21) represents only partial solution existing on some fixed hypersurface of parametrical space, it reproduces main tendencies of chirped-pulse oscillators: the energy, the chirp and the pulse duration increase with the dispersion. Also, the ratio of the self-amplitude modulation to the self-phase modulation (i.e. κ/γ) required for (21) decreases with the growth of the ratio of the dispersion to the spectral filtering (i.e. β/α) (Renninger et al. (2008); Soto-Crespo et al. (1997)). The most interesting property of (21) is that it reproduces a variety of experimental and numerical spectra classified above (e.g., see Fig. 9). Examples of analytical spectra are shown in Fig. 17 in dependence on the shape parameter B (the expressions for the parameters of (21) are taken from (Renninger et al. (2008))). A steepness of spectrum edges depends on the chirp and increases with it so that the spectrum becomes truncated for $\psi \gg 1$.

The interesting property of solution (21) is that, even in the absence of fourth-order dispersion (Fig. 15) and perturbations (Fig. 16) it reproduces the concave spectra (black dashed curve in Fig. 17) which are often observable in an experiment. The corresponding condition is B <1 but the self-amplitude modulation is not saturable, i.e. ζ <0. The last condition means that the corresponding solution is unstable against a pulse collapse. It was suggested that three factors can stabilize the dissipative soliton in this case: i) gain saturation dynamics, ii) contribution of higher order terms in the self-amplitude modulation law of (13), and iii) breather-like behavior of a pulse in a real-world oscillator (Renninger et al. (2008)).

The soliton solution (21) provides with some important insights but there is some crucial disadvantage in the approach considered: in both isogain (Soto-Crespo et al. (1997)) and non-isogain (Renninger et al. (2008)) representations the exact solution imposes the strict relations on the parameters of (17). As a result, the chirped dissipative soliton can not be traced within a broad multidimensional parametrical range. Hence, the obtained picture is rather sporadic and is of interest only in the close relation with the numerical results (see above).

An additional information can be obtained on the basis of the perturbative method (Malomed & Nepomnyashchy (1990)). The important property of the corresponding solution is that it is continuously extendible to the "Schrödinger" one: $A(z,t) \propto \cosh(t/T)^{-1+i\psi}$

when $\zeta \rightarrow 0$ (i.e. in the absence of saturation of self-amplitude modulation). We shall call this chirped dissipative soliton as the "negative branch" one. This perturbative technique provides with a quite accurate approximation for a low-energy chirped dissipative soliton when $\sigma\zeta/\kappa \ll 1$ (Kalashnikov (2009a)) and the corresponding parametrical range of pulse existence is $C \leq 2 - 4\sqrt{6\sigma\zeta/5\kappa}$ ($\sigma\zeta/\kappa \leq 1/4$).

As a further step in the development of analytical techniques required for the dissipative soliton exploration, two approximate methods will be considered below.

2.3.2.2 Adiabatic theory of chirped dissipative soliton

The adiabatic method considered in this section is developed in three main steps: i) the condition of $T \gg \sqrt{\beta}$ allows the adiabatic approximation for Eq. (17), ii) regularization procedure is applied to the expression for the soliton frequency deviation that excludes nonphysical solutions, and iii) the method of stationary phase for the Fourier image of soliton complex envelope is applied that gives the expression for soliton spectrum and energy. The last step requires $\psi \gg 1$, i.e. $\beta \gg \alpha$ and $\gamma \gg \kappa$ (Eq. (18)).

Let's begin with the traveling wave reduction of (17):

$$A(z,\tau) = \sqrt{P(\tau)} \exp\left[i\phi(\tau) - iqz\right], \tag{22}$$

where $P(\tau)$ is the instant power, $\phi(\tau)$ is the phase, and q is the soliton wave-number. In the adiabatic limit $T \gg \sqrt{\beta}$, the substitution of (22) in the CQNGLE (13,17) results in

$$\begin{aligned} &\beta\Omega(\tau)^2 = q - \gamma P(\tau) - \chi P(\tau)^2, \\ &\beta\left(\frac{\Omega(\tau)}{P(\tau)}\frac{dP(\tau)}{d\tau} + \frac{d\Omega(\tau)}{d\tau}\right) = \kappa P(\tau)\left(1 - \zeta P(\tau)\right) - \sigma - \alpha\Omega(\tau)^2, \end{aligned} \tag{23}$$

where $\Omega \equiv d\phi(t)/dt$ is the instant frequency.

Since $P(\tau) \geq 0$ by definition, there is the maximum frequency deviation Δ from the carrier frequency: $\Delta^2 = q/\beta$. Thus, Eqs. (23) lead, after some algebra (Kalashnikov (2009a)), to

$$\begin{aligned} &\frac{d\Omega}{d\tau} = \frac{\sigma + \alpha\Omega^2 - \frac{\kappa(\Upsilon-\gamma)}{4\chi^2}\left(2\chi + \zeta\gamma - \zeta\Upsilon\right)}{\beta\left[4\chi\beta\Omega^2 - (\Upsilon-\gamma)\Upsilon\right]}(\Upsilon-\gamma)\Upsilon, \\ &\Upsilon = \sqrt{\gamma^2 + 4\beta\chi\left(\Delta^2 - \Omega^2\right)}. \end{aligned} \tag{24}$$

The singularity points of Eq. (24) (the regularity condition is $|d\Omega/d\tau| < \infty$) impose the restrictions on the Δ value:

$$\Delta^2 = \frac{\gamma}{16\zeta\beta\left(\frac{C}{b}+1\right)} \times \\ \times\left[\frac{2\left(3+\frac{C}{b}+\frac{4}{b}\right)\left(2+\frac{C}{2}+\frac{3b}{2} \pm \sqrt{(C-2)^2 - 16a\left(1+\frac{C}{b}\right)}\right)}{1+\frac{C}{b}} - 12 - 3c - 9b - \frac{32a}{b}\right], \tag{25}$$

where three control parameters are $a \equiv \sigma\zeta/\kappa$, $b \equiv \zeta\gamma/\chi$, and $C \equiv \alpha\gamma/\beta\kappa$. These three parameters define the parametric dimensionality of a chirped dissipative soliton.

In the limit of $\chi \to 0$, one has (Kalashnikov et al. (2006))

$$\gamma P_0 = \beta\Delta^2 = \frac{3\gamma}{4\zeta}\left(1 - \frac{C}{2} \pm \sqrt{\left(1-\frac{C}{2}\right)^2 - 4a}\right),$$
$$\frac{d\Omega}{d\tau} = \frac{\beta\zeta\kappa}{3\gamma^2}\left(\Delta^2 - \Omega^2\right)\left(\Omega^2 + \Xi^2\right),\ \beta\Xi^2 = \frac{\gamma}{\zeta}(1+C) - \frac{5}{3}\gamma P_0, \tag{26}$$

where the equation for the peak power P_0 (or, equally, for Δ) results from the regularity condition.

Eqs. (25,26) demonstrate two branches of soliton solutions corresponding to opposite signs before a square root. In correspondence with these two signs, the solutions will be called as "positive-branch" and "negative-branch" chirped dissipative solitons.

In the cubic nonlinear limit of Eqs. (13,20) ($\chi \to 0$, $\zeta \to 0$), which admits an exact chirped dissipative soliton solution (Haus et al. (1991)), one has

$$A(z,\tau) = \sqrt{P_0}\mathrm{sech}\left(\frac{\tau}{T}\right)^{1-i\psi} e^{-iqz},$$
$$\alpha\Delta^2 = \frac{3\sigma C}{2-C},\ \gamma P_0 = \beta\Delta^2,\ \psi = \frac{3\gamma}{\kappa(1+C)},\ T = \frac{3\gamma}{\kappa\Delta(1+C)}. \tag{27}$$

Next assumption allows a further simplification. Since the phase $\phi(\tau)$ is a rapidly varying function due to a large chirp (the necessary requirements are $\beta \gg \alpha$ and $\gamma \gg \kappa$, see Eq. (18)), one may apply the method of stationary phase to the Fourier image $e(\omega)$ of $A(\tau)$ (Podivilov & Kalashnikov (2005)). Then, the spectral profile corresponding to (23,24) can be written as

$$p(\omega) \equiv |e(\omega)|^2 \approx \frac{\pi(\Upsilon-1)\left((\Upsilon-1)Cb + 4\left(2\omega^2-\Delta^2\right)\right)\mathrm{H}\left(\Delta^2-\omega^2\right)}{C\Upsilon\left((\Upsilon-1)\left(C\left(a+b+b^2+\omega^2\right)+b\left(\Delta^2-\omega^2\right)\right)-2(b+1)\left(\Delta^2-\omega^2\right)\right)}, \tag{28}$$

The expression for $p(\omega)$ allows obtaining the soliton energy by integration: $E = \int_{-\Delta}^{\Delta}\frac{d\omega}{2\pi}p(\omega)$. The soliton parametrical space is given by the following normalizations: $\tau' = \tau(\kappa/\zeta)\sqrt{\kappa/\alpha\zeta}$, $\Delta'^2 = \Delta^2\alpha\zeta/\kappa$, $\Omega'^2 = \Omega^2\alpha\zeta/\kappa$, $P' = \zeta P$, and $E' = E(\kappa/\gamma)\sqrt{\kappa\zeta/\alpha}$ (primes will be omitted thereafter). Hence $\Upsilon \equiv \sqrt{1+4\left(\Delta^2-\Omega^2\right)/bC}$ and H is the Heaviside's function in Eq. (28).

As it has been demonstrated in (Kalashnikov (2009a)), the truncated at $\pm\Delta$ spectra (28) have convex, concave, and concave-convex vertexes (Fig. 18, left). When $\chi \to 0$, only convex truncated spectra remain (Podivilov & Kalashnikov (2005)):

$$p(\omega) \approx \frac{6\pi\gamma}{\zeta\kappa}\frac{\mathrm{H}\left(\Delta^2-\omega^2\right)}{\Xi^2+\omega^2},\ E = \frac{6\gamma}{\kappa\zeta\Xi}\arctan\left(\frac{\Lambda}{\Xi}\right). \tag{29}$$

At last, for the cubic nonlinear limit of Eq. (13,20) ($\chi \to 0$, $\zeta \to 0$) one may obtain (Kalashnikov (2009b))

$$p(\omega) \approx \frac{6\pi\beta}{\kappa(1+C)}\mathrm{H}\left(\Delta^2-\omega^2\right),\ E = \frac{6\beta\Delta}{(1+C)\kappa}. \tag{30}$$

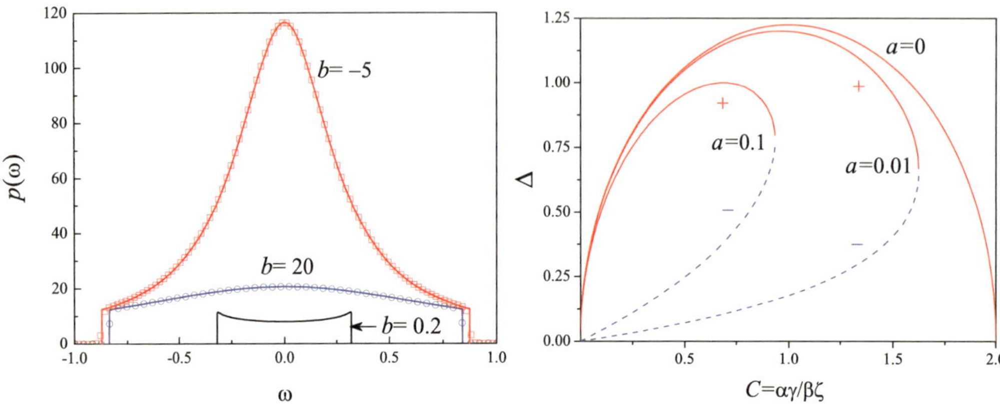

Fig. 18. Left: Spectra corresponding to the positive-branch (i.e. + sign in Eqs. (25,26)) of chirped dissipative soliton for $C=1$, $a=0.01$ and different values of b. $b>0$ corresponds to an enhancement of the self-phase modulation with the power growth and $|b|\to\infty$ corresponds to an absence of the quintic self-phase modulation, i.e. $|\chi|\to 0$. Squares and circles represent the numerical spectral profiles. Adapted from (Kalashnikov (2009a)). Right: Normalized spectral half-widths in dependence on C for the different values of a ($\chi=0$). Positive branch - red solid curves, negative branch - dashed blue curves. Adapted from (Kalashnikov (2009b)).

That is the truncated flat-top spectrum, which approximates the spectrum of exact solution of the cubic nonlinear complex Ginzburg-Landau equation.

The important property of the integration procedure considered is that it is applicable to various types of self-amplitude modulation. For instance, one has for Eqs. (14,20) (Kalashnikov (2009b); Kalashnikov & Apolonski (2009)):

$$\begin{gathered}\kappa P_0 = \frac{\alpha}{\ell' C}\Delta^2 = \frac{3}{4C}\left[2\left(1-\varsigma\right) - C \pm \sqrt{\Upsilon}\right],\\ \frac{d\Omega}{d\tau} = \frac{\alpha}{3\beta}\frac{\left(\Delta^2-\Omega^2\right)\left(\Xi^2-\Omega^2\right)}{\Delta^2-\Omega^2+\gamma/\kappa\beta},\quad \frac{\alpha}{\ell'}\Xi^2 = \frac{2\alpha}{3\ell'}\Delta^2+1-\varsigma+C,\\ p\left(\omega\right) \approx \frac{6\pi\beta^2}{\alpha\gamma}\frac{\Delta^2-\omega^2+\gamma/\kappa\beta}{\Xi^2-\omega^2}\mathrm{H}\left(\Delta^2-\omega^2\right),\end{gathered} \tag{31}$$

where $\varsigma \equiv \sigma/\ell'$, and $\Upsilon \equiv (2-C)^2 - 4\varsigma\left(2-\varsigma+C\right)$. For the above introduced normalizations with the replacement of ζ by κ, the dimensionless energy is

$$E = \frac{6\Delta}{C^2}\left[1 - \frac{\left(\Xi^2-\Delta^2-C\right)\operatorname{arctanh}\left(\frac{\Delta}{\Xi}\right)}{\Delta\Xi}\right]. \tag{32}$$

It should be noted, that the presented technique based on the adiabatic approximation and the regularization of $d\Omega/dt$ is analogous to that of (Ablowitz & Horikis (2009)). However, an approximate integration in the spectral domain allows us the further reduction of parametric space dimension and the construction of physically meaningful master diagrams that makes the soliton properties to be easily traceable.

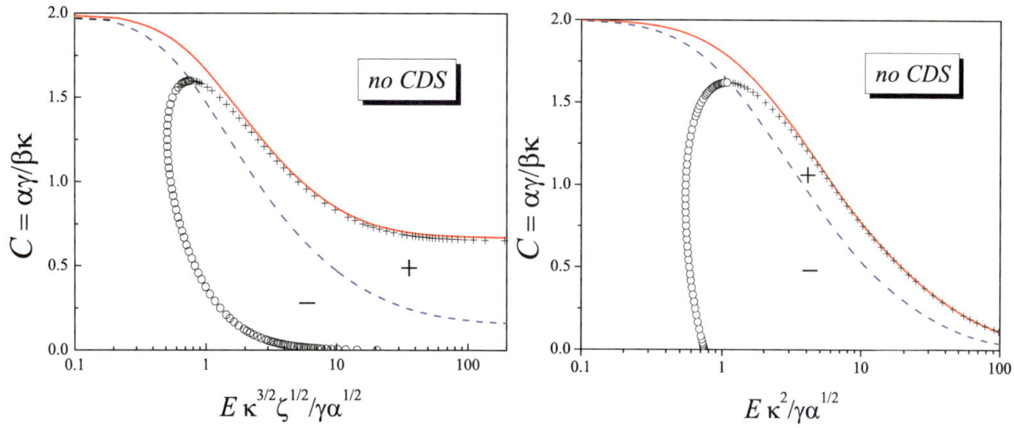

Fig. 19. Left: master diagram for the chirped dissipative soliton (CDS) of Eqs. (13,17), χ =0. There exists no soliton above the red curve (therein σ <0, i.e. the vacuum of (17) is unstable). a =0 (zero-level isogain) along the red curve. Blue dashed curve divides the regions, where the positive (+) and negative (−) branches of (25,26) exist. Crosses (circles) correspond to the +(−) branch for a=0.01. The normalization for energy is restored. Adapted from (Kalashnikov (2009a)). Right: master diagram for the chirped dissipative solitons corresponding to Eqs. (31,32). Red curve is the stability threshold (ς =0), blue curve divides the regions of positive and negative branches which are shown for ς =0.01 by crosses and circles, respectively. Adapted from (Kalashnikov (2009b)).

A *master diagram* connects the normalized energy with the parameter C along an isogain. The example of master diagram for the CQNGLE with vanishing quintic self-phase modulation (i.e. $\chi \to 0$ in (15)) is presented in Fig. 19 (left). The chirped dissipative soliton has a two-dimensional parametric space in this case. The solid red curve a =0 shows the border of the soliton existence. Above this border, the vacuum of Eq. (20) is unstable. One may see that the stability threshold on C decreases with E. Physically, that means, for instance, the dispersion growth and the self-phase modulation reduction required for obtaining the more energetic pulses. The blue dashed curve divides the existence regions for the positive and negative branches of soliton (the corresponding signs in Eq. (25,26)). The branches merge along this curve. Crosses (circles) represent the curve along which there exists the positive (negative) branch for some fixed value of a (*isogain* curve).

The master diagram reveals four significant differences between the positive and negative branches of chirped dissipative soliton. The first one is that the negative branch has lower energy than the positive one for a fixed C. The second difference is that the positive branch has a finite limit on C for $E \to \infty$. For the zero-level isogain such a limit is C $\approx$0.66 in the case considered. This value decreases with χ >0 and increases with χ <0. The existence of such "resonant" C allows concluding that the positive branch is energy scalable. This means that the soliton energy growth does not require a substantial change of C, i.e. the change of relation between spectral dissipation and self-phase modulation from one side and dispersion and self-amplitude modulation from another side. Such a property corresponds to the notion of the dissipative soliton resonance of (Grelu et al. (2010)). In the terms of (Kalashnikov et al. (2006)), the resonant dissipative soliton is an overdeveloped soliton with the finger-like spectrum ($\Delta \gg \Xi$ in (29)) that is the pulse with the almost Lorentz spectrum and the truncated flat-top temporal profile.

The negative branch is not energy scalable, that is the energy growth along this branch needs a substantial decrease of C (e.g., owing to a dispersion growth or a self-phase modulation reduction). The third difference is that the positive branch verges on a=0 within a whole range of E. The fourth difference is that the negative branch has a Schrödinger soliton limit (27) for $\zeta, \chi \to 0$ (Eq. (13,17)). The detailed description of master diagrams, pulse characteristics, spectra and temporal profiles of chirped dissipative solitons in the presence of quintic self-phase modulation $\chi \neq 0$ can be found in (Kalashnikov (2009a)).

It is of interest to compare the master diagrams for the chirped dissipative solitons of the CQNGLE (13,17) and the complex nonlinear Ginzburg-Landau equation with a perfectly saturable absorber (14,17) (see (31,32)). Master diagram for the latter is shown in Fig. 19 (right). Its structure is similar to that in Fig. 19 (left) but there exist two important differences. First, the C-parameter tends asymptotically to zero with the energy growth. This means that there is no dissipative soliton resonance in this case. Secondly, the energy on an isogain for the negative soliton branch remains almost constant with $C \to 0$. One may call this phenomena as the dissipative soliton antiresonance. Nevertheless, in spite of absence of perfect energy scalability (i.e. absence of the dissipative soliton resonance), the energy of positive soliton branch can be scaled with C in agreement with the following approximate asymptotical law:

$$E \approx \frac{18\ell' \beta^2}{\gamma \alpha^{3/2}}. \tag{33}$$

This scaling law assumes $C \ll 1$ and gives the energy at the zero-level isogain $\varsigma = 0$, i.e. at the stability threshold. The quadratic dependence of energy on dispersion is very promising because comparatively moderate dispersions can provide high energies. Also, it is very important that the energy is proportional to the cube of a gain bandwidth that offers advantages to the broadband active media. In agreement with Eq. (33), approximately linear energy growth with the modulation depth as well as faster than linear energy growth with the spectral filter bandwidth have been demonstrated experimentally for a dissipative soliton fiber laser (Lecaplain et al. (2011)).

The spectral width of the dissipative soliton is of interest because it defines the pulse width after extra-resonator compression: $T \approx 2/\Delta$. For the negative branch having narrower spectrum than the positive branch, the spectral width decreases with the C-decrease (Fig. 18, right) owing to the dispersion growth ($C \propto 1/\beta$) which stretches the pulse when the energy remains almost constant (compare with (Chong et al. (2008b))). When the energy E changes weakly along an isogain corresponding to the negative branch, the spectrum broadens with the α-increase ($C \propto \alpha$), i.e. the gainband narrowing (Fig. 18, right). The explanation is that the growth of spectral filtering enhances a cut-off of spectral components located on the pulse edges (Fig. 6, right). As a result, the pulse shortens and, for a fixed energy, the peak power P_0 increases. Since $\Delta^2 \propto P_0$ (Eqs. (26,27,31)), the spectrum broadens. For the positive branch, the spectrum initially broadens with the C-decrease (Fig. 18, right) because the energy E and, consequently, the self-phase modulation grow along an isogain (Fig. 19). However, further decrease of C narrows the spectrum like the negative branch in agreement with the numerical results of (Siegel et al. (2008)). It is important to note that the dependence of σ on the energy for a fixed C is inverse for the negative and positive branches: σ increases with E for the former and decreases for the latter. Hence, Fig. 18 (right) allows concluding that the spectrum broadens with energy (up to some $\Delta = const$ for the positive branch). Such a result has been corroborated by the numerical simulations (Siegel et al. (2008)).

Another important characteristic of chirped dissipative soliton is its chirp, which has to be $\gg$1 to provide the energy scalability. The chirp at the pulse center (where $\tau = 0$ by definition) normalized to $\kappa/\zeta\beta$ is (Kalashnikov (2009a))

$$\psi|_{\tau=0} \equiv \left.\frac{d\Omega}{d\tau}\right|_{\Omega=0} = -a - \frac{b^2}{4}\left(1 - \Upsilon|_{\Omega=0}\right)\left(1 + \frac{2}{b} - \Upsilon|_{\Omega=0}\right), \tag{34}$$

where $\Upsilon = \sqrt{1 + 4\left(\Delta^2 - \Omega^2\right)/cb}$ is the normalized version of the corresponding expression from (24). For a soliton solution (i.e. an isolated solution with the appropriate asymptotic $\lim_{\tau \to \pm\infty} P(\tau) = 0$), the chirp has to be positive that agrees in the limits of $\beta/\alpha \gg 1$ and $\gamma/\kappa \gg 1$ with the analytical theory presented in (Soto-Crespo et al. (1997)) where $0 < \psi \approx 3\gamma/(1 + C)\kappa < \beta/\alpha$ (here ψ is defined as the parameter in the phase profile ansatz $\phi(\tau) \equiv \psi \ln\sqrt{P(\tau)}$). It is interesting that the positive (i.e. energy scalable) soliton branch disappears when the self-phase modulation saturation is strong: $0 > b \geq -2$. In this case, the "antisoliton" appears which has a truncated (i.e. $\lim_{\tau \to T} P(\tau) = 0$) parabolic-top temporal profile with $\lim_{\tau \to T} \Omega(\tau) = \pm\infty$ (Kalashnikov (2009a)). Formally, such a soliton has the temporal and spectral shapes exchanged in comparison with those for an ordinary chirped dissipative soliton. Such an "antisoliton" has been observed in (Liu (2010)) and possesses an enhanced energy scalability.

One may conclude that the adiabatic theory of chirped dissipative solitons provides with a deep insight into physics of chirped-pulse oscillators. The pulse characteristics become easily traceable and finding of the true parametrical space of soliton allows looking at an extremely broad range of oscillators from a unified point of view. Nevertheless, the underlying approximations of: i) strong domination of the nondissipative effects (dispersion and self-phase modulation) over the dissipative ones (spectral filtering and self-amplitude modulation), ii) negligible contribution of the higher-order dispersions, and iii) distributed character of a laser system impose some restrictions. Second approximation is sound for the thin-disk oscillators based on comparatively narrow-band active media like Yb:YAG. The last approximation is well-grounded for the high-energy solid-state oscillators in general. Moreover, the simple analytical expressions for the complex spectral amplitude of soliton allow developing the perturbation theory in spectral domain (e.g., see (Kalashnikov (2010; 2011); Kalashnikov et al. (2011))). Nevertheless, another analytical approaches can shed light on some properties of chirped-pulse oscillators which are beyond the scope of the adiabatic theory.

2.3.2.3 Truncation of phase space: variational approximation and method of moments

Both variational approximation and method of moments allow truncating the space of (unknown) solutions of (17) to a sub-space of soliton-like ones by some appropriate ansatz (for an overview see (Anderson et al. (2001); Ankiewicz et al. (2007); Malomed (2002); Perez-Garcia et al. (2007))). The ansatz, as a rule, is the known analytical solution (21) or its reduced representation. As a result, the complex dynamics of (17) becomes to be reduced to the comparatively simple one described by a set of ordinary differential equations governing an evolution of the ansatz parameters (pulse amplitude, width, chirp, etc.). As a result, the problem becomes semi-analytical (Tsoy et al. (2006); Usechak & Agrawal (2005)).

The variational approximation can be sketched in the following way. The factors governing the pulse dynamics are divided into two parts: dissipative and nondissipative ones. The last underlying the nonlinear Schrödinger equation (Eqs. (15,16)) can be described by the Lagrangian. In absence of higher-order dispersions and quintic self-phase modulation, the corresponding Lagrangian density is

$$\mathscr{L}=\frac{i}{2}\left[A^*(z,\tau)\frac{\partial A(z,\tau)}{\partial\tau}-A(z,\tau)\frac{\partial A^*(z,\tau)}{\partial\tau}\right]-\frac{\beta}{2}\frac{\partial A(z,\tau)}{\partial\tau}\frac{\partial A^*(z,\tau)}{\partial\tau}+\frac{\gamma}{2}|A(z,\tau)|^2. \quad (35)$$

Eq. (35) defines the nonlinear Schrödinger equation through the Lagrange-Euler equation. The dissipative part of the CQNGLE (see Eqs. (13,20)) is described by the driving force:

$$\mathscr{Q}=-i\sigma A(z,\tau)+\frac{ig_0}{1+E_s^{-1}\int_{-\infty}^{\infty}|A|^2d\tau'}\left[A(z,\tau)+\tau\frac{\partial^2}{\partial\tau^2}A(z,\tau)\right]+ \\ +i\kappa\left[|A(z,\tau)|^2-\zeta|A(z,\tau)|^4\right]A(z,\tau). \quad (36)$$

The force-driven Lagrange-Euler equations

$$\frac{\partial\int_{-\infty}^{\infty}\mathscr{L}d\tau}{\partial\mathbf{f}}-\frac{\partial}{\partial z}\frac{\partial\int_{-\infty}^{\infty}\mathscr{L}d\tau}{\partial\mathbf{f}}=2\Re\int_{-\infty}^{\infty}\mathscr{Q}\frac{\partial A^*}{\partial\mathbf{f}}d\tau \quad (37)$$

allow obtaining a set of the ordinary first-order differential equations for a set $\mathbf{f}$ of the soliton parameters if one assumes the soliton shape in the form of some trial function $A(z,\tau)\approx\mathscr{F}[\mathbf{f}(z,\tau)]$. One may chose (Kalashnikov & Apolonski (2010))

$$\mathscr{F}=A_0(z)\,\mathrm{sech}\left(\frac{\tau}{T(z)}\right)\exp\left[i\left(\phi(z)+\psi(z)\ln\left(\mathrm{sech}\left(\frac{\tau}{T(z)}\right)\right)\right)\right], \quad (38)$$

with $\mathbf{f}=\{A_0(z),T(z),\phi(z),\psi(z)\}$ describing amplitude, width, phase, and chirp of dissipative soliton, respectively.

Substitution of (38) into (37) results in four equations for the soliton parameters. These equations are completely solvable for a steady-state propagation (i.e. when $\partial_z A_0=\partial_z T=\partial_z\psi=0$, but $\partial_z\phi\neq 0$).

As it taken a place in the adiabatic theory (see Fig. 19, left), the dissipative soliton is completely characterized by two-dimensional master diagram if $C \ll 1$ (Fig. 20, left). The red curve in Fig. 20 corresponds to the stability threshold obtained from the adiabatic theory (see Eq. (29)): the soliton is unstable on the right of this curve. The positive branch solution with convex spectrum, which is predicted by the adiabatic theory, cannot be obtained from the Schrödinger-like ansatz (38) so that the solution for the latter is situated on the left of the curves corresponding to the different values of γ/κ shown in Fig. (20). These curves are the zero-level isogains for pulses (38). One has note, that the requirement of $\gamma\gg\kappa$ is not essential for the variational approximation. Nevertheless, one may see that all solutions have a single asymptotic (dashed curve) for $C \ll 1$ so that the master diagram is two-dimensional

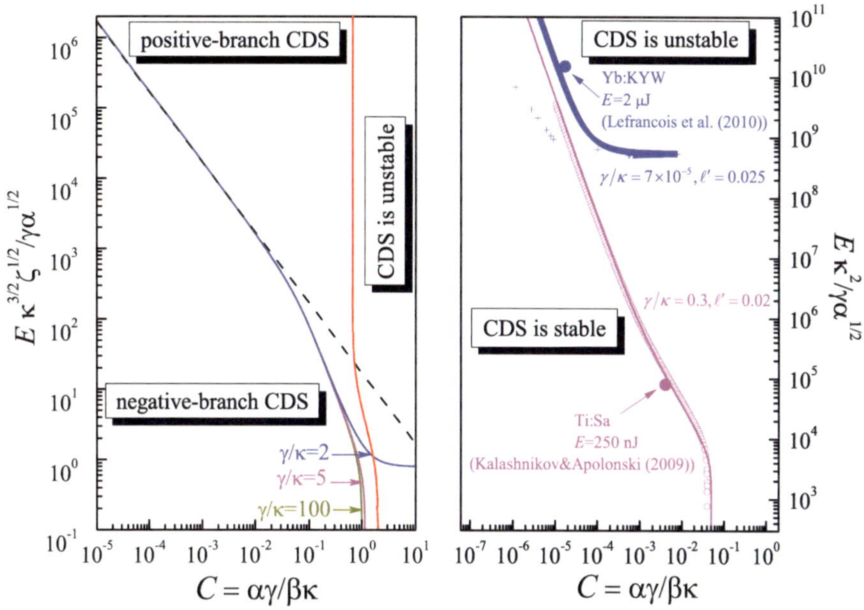

Fig. 20. Left: master diagram of a chirped dissipative soliton (CDS) for the self-amplitude modulation (13) (CQNGLE). The red solid curve corresponds to the stability threshold obtained from the adiabatic approximation (29). Another colored curves correspond to the thresholds obtained from the variational approximations with the ansatz (38) for the different values of γ/κ. The dashed line corresponds to the asymptotic $C \ll 1$. Right: analogous master diagram for the self-amplitude modulation (14). Crosses and circles show the stability thresholds for a chirped dissipative soliton (CDS) obtained from the variational approximation. The solid line corresponds to the adiabatic approximation with $\ell' = 0.02$. The parameters are shown in Figure. Points correspond to the experimental operational points of two solid-state oscillators. Right figure is adapted from (Kalashnikov & Apolonski (2010)).

(i.e. it does not depend on the γ/κ-value) in this limit. The asymptotical values of the pulse parameters along the dashed curve are:

$$E \approx \frac{17\beta}{\sqrt{\alpha\kappa\zeta}}, \quad T \approx \frac{8}{C}\frac{\gamma}{\kappa}\sqrt{\frac{\alpha\zeta}{\kappa}}. \tag{39}$$

The importance of these scaling rules (compare with Eq. (33)) is that they correspond to the flat-top spectrum relevant to (38), which has the spectral chirp with a weak frequency-dependence within the range of $[-\Delta, \Delta]$ (Podivilov & Kalashnikov (2005)). As a result, such a pulse is almost perfectly compressible that is the energy loss due to satellite generation in the process of compression is minimal.

In the case of self-amplitude modulation defined by (14), the master diagram obtained with the help of the variational approximation and the ansatz (38) is shown in Fig. 20, right. Crosses and circles correspond to the stability thresholds (zero-level isogain) dictated by the variational approximation and correspond to the parameters shown in Figure. The solid curve is based on the adiabatic theory (see Eqs. (31,32)). The points correspond to the operational points of two different oscillators (bulk and thin-disk solid-state ones). The asymptotical energy scaling law is matched perfectly by Eq. (33).

Two observations are of interest. Firstly, the asymptotical scaling law is valid for the systems with the extremely different ratio γ/κ (e.g., for a thin-disk oscillator with suppressed self-phase modulation and a Ti:sapphire oscillator with a comparatively large nonlinear phase shift). This means that a true parametrical space of chirped dissipative soliton remains two-dimensional. Secondly, the ansatz (38) covers the space of adiabatic solutions better than it takes a place in the case of CQNGLE (compare Figs. 20 left and right). The last results from the difference of spectral profiles of (31) and (29). The former corresponding to the self-amplitude modulation (14) has not finger-like profiles and is nearer to a flat-top spectrum of (38).

More general ansatz (21) allows describing a variety of spectra appearing in the normal dispersion regime (Bale & Kutz (2008)) and, that is especially important, one may simulate an unsteady pulse dynamics in both distributed (Ankiewicz et al. (2007)) and undistributed (Bale & Kutz (2008)) laser systems. In comparison with the full-sized simulations based on Eq. (20), the truncation of phase space reduces the problem of pulse dynamics to the solution of a set of nonlinear ordinary first-order differential equations for the pulse parameters ($\partial_z \mathbf{f} \neq 0$ in Eq. (37) in this case). As a result, the task of optimization of undistributed oscillator can be solved and, for instance, the optimal placement of laser elements and the optimal dispersion map parameter can be defined (Bale et al. (2010; 2009; 2008b)). Also, an important advantage of the approach considered is that it allows investigating the dynamic stability of soliton solutions like (21,38) (e.g., see (Bale et al. (2008a))). However, it is important to remember that the results obtained on the basis of simulations in a truncated phase space require verification by full-sized simulations of Eqs. (17) or (20).

The method of moments (Maimistov (1993); Perez-Garcia et al. (2007)) is akin to the variational one. It considers a set of moments of the nonlinear Schrödinger equation (which consists of nondissipative terms of Eq. (17) and omits higher-order dispersions, quintic self-phase modulation, etc.) with the subsequent study of their evolution under action of the dissipative terms, higher-order phase nonlinearity, higher-order dispersions, etc. For instance, the first five moments are (Chang et al. (2008b); Maimistov (1993))

$$\begin{aligned} &E = \int_{-\infty}^{\infty} |A|^2 d\tau, \;\; P = \int_{-\infty}^{\infty} \left(A\frac{\partial A^*}{\partial \tau} - A^*\frac{\partial A}{\partial \tau} \right) d\tau, \;\; I_1 = \int_{-\infty}^{\infty} \tau |A|^2 d\tau, \\ &I_2 = \int_{-\infty}^{\infty} \tau^2 |A|^2 d\tau, \;\; I_3 = \int_{-\infty}^{\infty} \tau \left(A\frac{\partial A^*}{\partial \tau} - A^*\frac{\partial A}{\partial \tau} \right) d\tau, \;\; etc. \end{aligned} \tag{40}$$

Here first two moments are the energy E and the momentum P. If one denotes all dissipative terms as well as higher-order dispersions and phase nonlinearities as $R[A]$, the evolution of moments can be described in the following way:

$$\begin{aligned} &\frac{dE}{dz} = i\int_{-\infty}^{\infty} (AR^* - A^*R)\, d\tau, \;\; \frac{dP}{dz} = -i\int_{-\infty}^{\infty} \left(\frac{\partial A}{\partial \tau}R^* - \frac{\partial A^*}{\partial \tau}R \right) d\tau, \\ &\frac{dI_1}{dz} = -2i\beta P + i\int_{-\infty}^{\infty} \tau\, (AR^* - A^*R)\, d\tau, \;\; \frac{dI_2}{dz} = 2i\beta I_3 + i\int_{-\infty}^{\infty} \tau^2\, (AR^* - A^*R)\, d\tau, \\ &\frac{dI_3}{dz} = -i\int_{-\infty}^{\infty} \left(4\beta \left| \frac{\partial A}{\partial \tau} \right|^2 + \gamma |A|^4 \right) d\tau + 2i\int_{-\infty}^{\infty} \tau \left(\frac{\partial A}{\partial \tau}R^* - \frac{\partial A^*}{\partial \tau}R \right) d\tau - i\int_{-\infty}^{\infty} (AR^* - A^*R)\, d\tau. \end{aligned} \tag{41}$$

The system (41) is exact and the next step is analogous to that for the variational approximation: we truncate the phase space by substitution of a trial function in place of A. As a result, one has a set of nonlinear ordinary first-order differential equations for evolvable pulse parameters which allows exploring the pulse dynamics and stability like the variational approach (e.g., see (Tsoy & Akhmediev (2005); Tsoy et al. (2006); Usechak & Agrawal (2005))).

The most impressive result of the method of moments concerning the theory of chirped-pulse oscillators is conception of the so-called dissipative soliton resonance which is a representation of master diagram asymptotic $E \to \infty$ (see Sec. 2.3.2.2). The corresponding approximate condition is (Chang et al. (2008b)):

$$\left(6.333 + \frac{3.8}{b}\right) C = 2, \tag{42}$$

where the normalizations of (25) are used.

Eq. (42) gives approximately two-fold deviation from the results of the numerical simulations (see dashed curve in Fig. 11 and (Chang et al. (2009))) and the adiabatic theory (Kalashnikov (2009a)) but, nevertheless, it provides with the correct representation of soliton parametric space and the tendencies of its asymptotic behavior. In particular, the constant spectral width is predicted in the limit of $E \to \infty$ when $C \to const$ in agreement with the adiabatic theory (see Fig. 18 (right), positive branch with $C = const$ and decreasing a; and Figs. 19). Since $\Delta \to const$, the peak power P_0 remains constant, as well (Eq. (26)). Hence, the energy scaling is provided by scaling of the pulse width.

One may conclude that the phase space truncation methods have allowed obtaining the correct asymptotic representation of the parametric space of chirped dissipative soliton and revealing the structure of this space in agreement with the results of both numerical simulations and adiabatic theory. As a result, a unified point of view on a variety of chirped-pulse oscillators becames possible. Moreover, the advantages of the truncation methods are that they allow exploring the unsteady pulse evolution as well as the undistributed laser systems. Another important advantage of these methods is that they permit to widen the dimensionality of Eqs. (17,20) by including the transverse spatial dimensions and to consider a space-time dynamics of an oscillator. Some partial reductions of the space-time dynamics have been used in the numerical simulations (e.g., (Kalosha et al. (1998))) and semi-analytical matrix formalism has been developed (Jirauschek & Kärtner (2006); Kalashnikov (2003)). Nevertheless, the space-time theory of chirped pulse oscillators is not developed to date.

3. Experimental realizations of solid-state chirped-pulse oscillators

In this part, some experimental realizations of chirped-pulse oscillators will be considered. Although the theory of chirped dissipative soliton is suitable for both solid-state and fiber lasers, we confine ourself exclusively to the former. Also, the energy scalable solid-state oscillators operating in the anomalous (i.e. negative in our terms) dispersion regime (Steinmeyer et al. (1999); Südmeyer et al. (2008)) are beyond the scope of this review. Taking into account these limitations, one may divide the oscillators considered into two classes different by an active medium thickness, diode-pump ability and average power scalability: i) bulk and ii) thin-disk oscillators. The last class possesses excellent average power scaling properties and diode-pump ability (Giessen & Speiser (2007)) but the available active media have comparatively narrow gainbands (i.e. the large α-parameter in Eq. (19)). The bulk crystalline media, vice-versa, have limitations of power scaling but, as a rule, are broadband (Sorokina (2003)). Of course, such a division is not fundamental and the thin-disk laser systems evolve in a direction of broader available gainbands (Südmeyer et al. (2009)).

The femtosecond pulse breakthrough for a Ti:sapphire oscillator has been closely connected with the dispersion compensation technique (Spence & Sibbett (1991)). As a result, the first normal (i.e. positive) dispersion regime in this broadband mode-locked oscillator has been observed and characterized immediately (Proctor et al. (1993)). The regime obtained has demonstrated the truncated spectra of chirped picosecond pulses. It was observed that this regime is disconnected from the anomalous dispersion one due to instability in the vicinity of zero dispersion. The connection with the theoretical model of (Haus et al. (1991)), where the mechanism of chirped-pulse formation has been suggested, has been declared. The next step was an observation that a chirped-pulse oscillator allows the pulse energy scaling (Apolonski et al. (2000); Fernandez et al. (2004)). The result was a swift rise of the pulse energy from a Ti:sapphire chirped-pulse oscillator (Fernández et al. (2007); Naumov et al. (2005); Siegel et al. (2009)). The decisive steps were the cavity period T_{rep} growth and the positive dispersion control within a broad spectral range (Pervak et al. (2008)). The results are shown in Fig. 21. The near-1-μJ pulses with a compressibility down to sub-40 fs have been obtained. The preferable mode-locking technique for a highest-energy pulse generation in the case of a reduced oscillator repetition-rate is based on a semiconductor saturable absorber (Kalashnikov et al. (2006); compare Eqs. (33,39) where the scaling law is more promising for a perfectly saturable absorber). A further pulse energy growth by the means of a growth of cavity period is problematical because the gain relaxation time of 3 μs for this medium is about of an ultimate T_{rep} and the gain dynamics begins to destabilize the pulse (Kalashnikov et al. (2006)).

As a development of the chirped-pulse oscillator technique, a further shift into the infrared generation range taken a place (Cankaya et al. (2011); Lin (2010); Sorokin et al. (2008); Tan et al. (2011)). The characteristics of a Cr:forsterite broadband oscillator approach the Ti:sapphire ones (Cankaya et al. (2011)). The problem is that the relative contribution of higher-order dispersions increases in the mid-infrared range. As a result, the pulse spectra becomes distorted and the chaotic mode-locking becomes the principal destabilizing mechanism (Kalashnikov et al. (2011); Sorokin et al. (2008)). Nevertheless, the potential of such media as Cr-doped chalcogenides allows reaching around-1-μJ sub-100 fs pulses at 2.5 μm wavelength.

The technique of averaged power scaling based on using the thin-disk diode-pumped media (Giessen & Speiser (2007)) has allowed reaching a highest sub-picosecond pulse energies in the anomalous (negative) dispersion regime (Südmeyer et al. (2008)). In such a regime the scaling law obtained from the variational approximation for a perfectly saturable absorber is

$$E \propto \frac{|\beta|}{\gamma\sqrt{\alpha}} \tag{43}$$

and the comparison with Eq. (33) shows that the dispersion required for a pulse stabilization in a chirped-pulse oscillator can be substantially lower. Such a conclusion is supported by data of (Palmer et al. (2008; 2010)) where the high-energy pulses of approximately 400 fs width are realized by using a moderate positive dispersion (750 fs^2 and ranging from 1500 to 4700 fs^2 for different configurations) for the pulse stabilization. Another advantage of

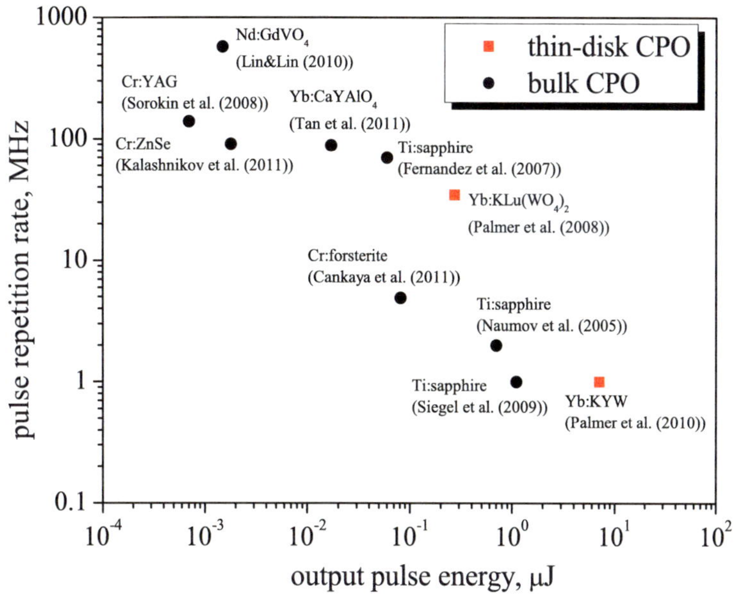

Fig. 21. Experimental realizations of chirped-pulse solid-state oscillators.

the positive dispersion regime over the negative one is the cubic energy scaling with a gain bandwidth versus the linear one for the latter (Eqs. (33,43)). Therefore, the use of more bradband materials like $Yb:Lu_2O_3$, $Yb:LuScO_3$, and Yb:CALGO (Südmeyer et al. (2009)) is very promising for a further energy scaling in the positive dispersion regime.

One may conclude that the experimental realization of chirped-pulse oscillators is at the beginning of the development. Although the scaling potential of Ti:sapphire, it seems, is close to exhaustion, the mid-infrared broadband media like Cr:ZnSe and Cr:ZnS are extremely promising for a further development which can be achieved by enhancement of pumping sources and chirped mirrors. On the other hand, a swift growth of pulse energy of thin-disk oscillators is expected. As the numerical simulations have demonstrated (Kalashnikov & Apolonski (2010)), the pulses with 100 μJ energies are completely accessible in a thin-disk chirped-pulse oscillator with perfectly saturable absorber (see scaling rule (33)). A further progress will be based on an optimization of oscillator, SESAM and pump designs; on a progress in the chirped mirror technology; on using the more broadband active media and, probably, on a transition to the Kerr-lens mode-locking technique.

4. Conclusion

A rapid progress in the field of energy scaling of femtosecond pulse oscillators has yielded three main fruits: i) thin-disk solid-state oscillators operating in the anomalous-dispersion regime, ii) both bulk and thin-disk oscillators operating in the normal dispersion regime, and iii) fiber lasers operating in the net-normal dispersion regime. In this review, the second type of oscillators has been under consideration. It was affirmed, that the source of energy scalability is a formation of chirped dissipative soliton therefore these oscillators are named "chirped-pulse oscillators". The main factors governing the chirped dissipative soliton have been considered and the decisive contribution of dissipative effects such as saturable gain, spectral filtering and self-amplitude modulation has been analyzed.

The theory of such solitons demonstrates that the relevant control parameters for a chirped dissipative soliton are some specific dimensionless combinations of oscillator parameters but not their absolute values. As a result, a chirped dissipative soliton "lives" in a low-dimensional parametric space and a variety of chirped-pulse oscillators (including the fiber ones) can be considered from a unified point of view. The insight into nature of dissipative soliton parametric space underlies the interrelated concepts of "master diagram" and "dissipative soliton resonance". The achievement of these concepts is the formulation of extremely simple scaling rules which relate the pulse parameters with those of an oscillator. As a result, the properties of chirped-pulse oscillator becomes to be easily traceable within an extremely broad parametric range. Also, these rules allow comparing between different mode-locking mechanisms and demonstrate the advantage of the normal dispersion regime over the anomalous dispersion one in further energy scaling.

The stability of chirped-pulse oscillators has been considered with respect to gain saturation, higher-order dispersions and internal spectral perturbations. The numerical simulations demonstrate a variety of destabilization scenarios including the chaotic mode-locking induced by the odd-order dispersions and by the dispersion decrease towards the pulse spectrum edges. It was conjectured that the underlying mechanism is a resonance with dispersive waves but the corresponding analytical theory remains to be undeveloped, to date. The noise properties and the self-starting ability of mode-locking in the positive dispersion regime remain to be controversial, as well.

The experimental realizations of chirped-pulse solid-state oscillators have been surveyed. For convenience, the oscillators have been divided into bulk and thin-disk ones. At this moment, the bulk active media are more broadband than the thin-disk ones. There are another differences (gain relaxation times, thermal properties, diode-pump abilities, etc.), as well. But it is clear at this moment, that the chirped-pulse oscillators promise (or have fulfilled, as for a Ti:sapphire oscillator) an outstanding scalability due to: i) reduced dispersion, ii) potentially broader spectra, and iii) rapid energy growth with the gain bandwidth in comparison with oscillators operating in the anomalous dispersion regime. One may hope, that a chirped-pulse oscillator is superior to an anomalous dispersion one in stability, starting-ability and noise properties and that sub- and over-100 μJ femtosecond pulse generation will be achievable directly from an oscillator as a result of development of this technique.

5. Acknowledgements

The research was funded by the Austrian Science Fund (FWF): P20293-N16.

6. References

Ablowitz, M. J. & Horikis, T. P. (2009). Solitons in normally dispersive mode-locked lasers, *Phys. Rev. A* Vol. 79(No. 6): 063845.

Agostini, P. & DiMauro, L. F. (2004). The physics of attosecond light pulses, *Rep. Prog. Phys.* Vol. 67(No. 6): 813–855.

Agrawal, G. (2006). *Nonlinear Fiber Optics*, Academic, San Diego.

Akhmediev, N. & Ankiewicz, A. (1997). *Solitons: Nonlinear Pulses and Beams*, Chapman & Hall, London.

Akhmediev, N. & Ankiewicz, A. (eds) (2005). *Dissipative Solitons*, Springer, Berlin.

Akhmediev, N. N. & Afanasjev, V. V. (1996). Singularities and special solutions of the cubic-quintic complex Ginzburg-Landau equation, *Phys. Rev. A* Vol. 53(No. 1): 1190–1201.

Akhmediev, N., Soto-Crespo, J. M. & Grelu, P. (2008). Roadmap to ultra-short record high-energy pulses out of laser oscillator, *Physics Letters A* Vol. 372(No. 17): 3124–3128.

Anderson, D., Lisak, M. & Berntson, A. (2001). A variational approach of nonlinear dissipative pulse propagation, *Pramana J. Phys.* Vol. 57(No. 5–6): 917–936.

Ankiewicz, A., Akhmediev, N. & Devine, N. (2007). Dissipative solitons with a Lagrangian approach, *Optical Fiber Technology* Vol. 13(No. 2): 91–97.

Apolonski, A., Poppe, A., Tempea, G., Spielmann, C., Udem, T., R.Holzwarth, Hänsch, T. & Krausz, F. (2000). Strong-field double ionization of ar below the recollision threshold, *Phys. Rev. Lett.* Vol. 85(No. 4): 740–743.

Aranson, I. S. & Kramer, L. (2002). The world of the complex Ginzburg-Landau equation, *Rev. Mod. Phys.* Vol. 74(No. 1): 99–143.

Bale, B. G., Boscolo, S., Kutz, J. N. & Turitsyn, S. K. (2010). Intracavity dynamics in high-power mode-locked fiber lasers, *Phys. Rev. A* Vol. 81(No. 3): 033828.

Bale, B. G., Boscolo, S. & Turitsyn, S. K. (2009). Diddipstive dispersion-managed solitons in mode-locked lasers, *Opt. Lett.* Vol. 34(No. 21): 3286–3288.

Bale, B. G. & Kutz, J. N. (2008). Variational method for mode-locked lasers, *J. Opt. Soc. Am. B* Vol. 25(No. 7): 1193–1202.

Bale, B. G., Kutz, J. N., Chong, A., Renninger, W. H. & Wise, F. W. (2008a). Spectral filtering for high-energy mode-locking in normal dispersion fiber lasers, *J. Opt. Soc. Am. B* Vol. 25(No. 10): 1763–1770.

Bale, B. G., Kutz, J. N., Chong, A., Renninger, W. H. & Wise, F. W. (2008b). Spectral filtering for mode-locking in the normal dispersive regime, *Opt. Lett.* Vol. 33(No. 9): 941–943.

Bélanger, P.-A. (2007). Stable operation of mode-locked fiber lasers: similariton regime, *Optics Express* Vol. 15(No. 17): 11033–11041.

Blanchard, P. & Brüning, E. (1992). *Variational Methods in Mathematical Physics: A Unified Approach*, Springer, Berlin.

Brabec, T. & Krausz, F. (1997). Nonlinear optical pulse propagation in the single-cycle regime, *Phys. Rev. Lett.* Vol. 78(No. 17): 3282–3285.

Brabec, T. & Krausz, F. (2000). Intense few-cycle laser fields: Frontiers of nonlinear optics, *Rev. Mod. Phys.* Vol. 72(No. 2): 545–591.

Cankaya, H., Akturk, S. & Sennaroglu, A. (2011). Direct generation of 81 nJ pulses and external compression to a subpicosecond regime with a 4.9 MHz chirped-pulse multipass-cavity Cr^{4+}:forsterite oscillator, *Opt. Lett.* Vol. 36(No. 9): 1572–1574.

Chang, W., Ankiewicz, A., Soto-Crespo, J. M. & Akhmediev, N. (2008a). Dissipative soliton resonance in laser models with parameter management, *J. Opt. Soc. Am. B* Vol. 25(No. 12): 1972–1977.

Chang, W., Ankiewicz, A., Soto-Crespo, J. M. & Akhmediev, N. (2008b). Dissipative soliton resonances, *Phys. Rev. A* Vol. 78(No. 2): 023830.

Chang, W., Soto-Crespo, J. M., Ankiewicz, A. & Akhmediev, N. (2009). Dissipative soliton resonances in the anomalous dispersion regime, *Phys. Rev. A* Vol. 79(No. 3): 033840.

Cho, S. H., Bouma, B. E., Ippen, E. P. & Fujimoto, J. G. (1999). Low-repetition-rate high-peak-power Kerr-lens mode-locked Ti:Al_2O_3 laser with a multiple-pass cavity, *Opt. Lett.* Vol. 24(No. 6): 417–419.

Chong, A., Renninger, W. H. & Wise, F. W. (2008a). Properties of normal-dispersion femtosecond fiber lasers, *J. Opt. Soc. Am. B* Vol. 25(No. 2): 140–148.

Chong, A., Renninger, W. H. & Wise, F. W. (2008b). Route to the minimum pulse duration in normal-dispersion fiber lasers, *Opt. Lett.* Vol. 33(No. 22): 2638–2640.

Conte, R. (ed.) (1999). *The Painlevè Property. One Century Later*, Springer, New York.

Deissler, R. J. & Brand, H. R. (1994). Periodic, quasiperiodic, and chaotic localized solutions of the quintic complex Ginzburg-Landau equation, *Phys. Rev. Lett.* Vol. 72(No. 4): 478–481.

der Au, J. A., Spühler, G. J., Südmeyer, T., Paschotta, R., Hövel, R., Moser, M., Erhard, S., Karzewski, M., Gissen, A. & Keller, U. (2000). 16.2-W average power from a diode-pumped femtosecond Yb:YAG thin disk laser, *Opt. Lett.* Vol. 25(No. 11): 859–861.

Diels, J.-C. & Rudolph, W. (2006). *Ultrashort Laser Pulse Phenomena: Fundamentals, Techniques, and Applications on a Femtosecond Time Scale*, Elsevier, Amsterdam.

Ding, E. & Kutz, J. N. (2009). Operating regimes, split-step modeling, and the Haus master mode-locking model, *J. Opt. Soc. Am. B* Vol. 26(No. 12): 2290–2300.

Fernandez, A., Fuji, T., Poppe, A., Fürbach, A., Krausz, F. & Apolonski, A. (2004). Chirped-pulse oscillators: a route to high-power femtosecond pulses without external amplification, *Opt. Lett.* Vol. 29(No. 12): 1366–1368.

Fernández, A., Verhoef, A., Pervak, V., Lermann, G., Krausz, F. & Apolonski, A. (2007). Generation of 60-nJ sub-40-fs pulses at 70 MHz repetition rate from a Ti:sapphire chirped pulse-oscillator, *Appl. Phys. B* Vol. 87(No. 3): 395–398.

Fernández González, A. (2008). *Chirped-Pulse Oscillators: Generating Microjoule Femtosecond Pulses at Megahertz Repetition Rate*, VDM Verlag.

Gattass, R. R. & Mazur, E. (2008). Femtosecond laser micromachining in transparent materials, *Nature Photonics* Vol. 2(No. 4): 219–225.

Giessen, A. & Speiser, J. (2007). Fifteen years of work on thin-disk lasers: results and scaling laws, *IEEE J. Selected Top. in Quantum. Electron.* Vol. 13(No. 3): 598–609.

Grelu, P., Chang, W., Ankiewicz, A., Soto-Crespo, J. M. & Akhmediev, N. (2010). Dissipative soliton resonance as a guideline for high-energy pulse laser oscillators, *J. Opt. Soc. Am. B* Vol. 27(No. 11): 2336–2341.

Hannaford, P. (ed.) (2005). *Femtosecond Laser Spectroscopy*, Springer, Boston.

Haus, H. A. (1975a). Theory of mode locking with a fast saturable absorber, *J. Applied Phys.* Vol. 46(No. 7): 3049–3058.

Haus, H. A. (1975b). Theory of mode locking with a slow saturable absorber, *IEEE J. Quantum Electron.* Vol. 11(No. 9): 736–746.

Haus, H. A., Fujimoto, J. G. & Ippen, E. P. (1991). Structures for additive pulse mode locking, *J. Opt. Soc. Am. B* Vol. 8(No. 10): 2068–2076.

Haus, H. A. & Silberberg, Y. (1985). Theory of mode locking of a laser diode with a multiple-quantum-well structure, *J. Opt. Soc. Am. B* Vol. 2(No. 7): 1237–1243.

Herrmann, J. & Wilhelmi, B. (1987). *Lasers for ultrashort light pulses*, Elsevier, Amsterdam.

Horowitz, M., Barad, Y. & Silberberg, Y. (1997). Noiselike pulses with a broadband spectrum generated from a erbium-doped fiber laser, *Opt. Lett.* Vol. 22(No. 11): 799–801.

Huber, R., Adler, F., Leitenstorfer, A., Beutter, M., Baum, P. & Riedle, E. (2003). 12-fs pulses from a continuous-wave-pumped 200-nJ Ti:sapphire amplifier at a variable repetition rate as high as 4 MHz, *Opt. Lett.* Vol. 28(No. 21): 2118–2120.

Jasapara, J., Kalashnikov, V. L., Krimer, D. O., Poloyko, I. G., Lenzner, M. & Rudolph, W. (2000). Automodulations in cw Kerr-lens mode-locked solid-state lasers, *J. Opt. Soc. Am. B* Vol. 17(No. 2): 319–326.

Jirauschek, C. & Kärtner, F. X. (2006). Gaussian pulse dynamics in gain media with Kerr nonlinearity, *J. Opt. Soc. Am. B* Vol. 23(No. 9): 1776–1784.

Kalashnikov, V. & Apolonski, A. (2010). Energy scalability of mode-locked oscillators: a completely analytical approach to analysis, *Optics Express* Vol. 18(No. 24): 25757–25770.

Kalashnikov, V. L. (2003). Propagation of the gaussian ultrashort pulse in a nonlinear laser medium. Commented Maple computer algebra worksheet, available at `http://info.tuwien.ac.at/kalashnikov/selff.html`.

Kalashnikov, V. L. (2008). Chirped-pulse oscillators: an impact of the dynamic gain saturation. Preprint arXiv:0807.1050v1 [physics.optics], available at `http://lanl.arxiv.org/abs/0807.1050`.

Kalashnikov, V. L. (2009a). Chirped dissipative solitons of the complex cubic-quintic nonlinear Ginzburg-Landau equation, *Phys. Rev. E* Vol. 80(No. 4): 046606.

Kalashnikov, V. L. (2009b). The unified theory of chirped-pulse oscillators, *in* M. Bertolotti (ed.), *Proc. SPIE: Nonlinear Optics and Applications III*, Vol. 7354, SPIE, p. 73540T.

Kalashnikov, V. L. (2010). Chirped dissipative solitons, *in* L. F. Babichev & V. I. Kuvshinov (eds), *Nonlinear Dynamics and Applications*, Vol. 16, Republic Institute of Higher School, Minsk, pp. 199–206.

Kalashnikov, V. L. (2011). Dissipative solitons: Perturbations and chaos formation, *in* C. H. Skiadas, I. Dimotikalis & C. SkiadasBabichev (eds), *Chaos Theory. Modeling, Simulation and Applications*, World Scientific Publishing Company, London, pp. 58–67.

Kalashnikov, V. L. & Apolonski, A. (2009). Chirped-pulse oscillators: A unified standpoint, *Phys. Rev. A* Vol. 79(No. 4): 043829.

Kalashnikov, V. L. & Chernykh, A. (2007). Spectral anomalies and stability of chirped-pulse oscillators, *Phys. Rev. A* Vol. 75(No. 3): 033820.

Kalashnikov, V. L., Fernández, A. & Apolonski, A. (2008). High order dispersion in chirped-pulse oscillators, *Optics Express* Vol. 16(No. 6): 4206–4216.

Kalashnikov, V. L., Kalosha, V. P., Mikhailov, V. P. & Poloyko, I. G. (1995). Self-mode locking of four-mirror-cavity solid-state lasers by Kerr self-focusing, *J. Opt. Soc. Am. B* Vol. 12(No. 3): 462–467.

Kalashnikov, V. L., Kalosha, V. P., Mikhailov, V. P., Poloyko, I. G., Demchuk, M. I., Koltchanov, I. G. & Eichler, H. J. (1995). Frequency-shift mode locking of continuous-wave solid-state lasers, *J. Opt. Soc. Am. B* Vol. 12(No. 11): 2078–2082.

Kalashnikov, V. L., Podivilov, E., Chernykh, A. & Apolonski, A. (2006). Chirped-pulse oscillators: theory and experiment, *Applied Physics B* Vol. 83(No. 4): 503–510.

Kalashnikov, V. L., Podivilov, E., Chernykh, A., Naumov, S., Fernandez, A., Graf, R. & Apolonski, A. (2005). Approaching the microjoule frontier with femtosecond laser oscillators: theory and comparison with experiment, *New Journal of Physics* Vol. 7(No. 1): 217.

Kalashnikov, V. L., Sorokin, E. & Sorokina, I. T. (2011). Chirped dissipative soliton absorption spectroscopy, *Optics Express* Vol. 19(No. 18): 17480–17492.

Kalosha, V. P., Müller, M., Herrmann, J. & Gatz, S. (1998). Spatiotemporal model of femtosecond pulse generation in Kerr-lens mode-locked solid-state lasers, *J. Opt. Soc. Am. B* Vol. 15(No. 2): 535–550.

Kärtner, F. X., Morgner, U., Schibli, T., Ell, R., Haus, H. A., Fujimoto, J. G. & Ippen, E. P. (2004). Few-cycle pulses directly from a laser, *in* F. X. Kärtner (ed.), *Few-cycle laser pulse generation and its applications*, Springer, Berlin, pp. 73–136.

Keller, U., Weingarten, K. J., Kärtner, F. X., Kopf, D., Braun, B., Jung, I. D., Fluck, R., Hönninger, C., Matuschek, N. & , J. A. d. A. (1996). Semiconductor saturable absorber mirrors (SESAMs) for femtosecond to nanosecond pulse generation in solid-state lasers, *IEEE J. Sel. Top. Quantum Electron.* Vol. 2(No. 3): 435–453.

Kharenko, D. S., Shtyrina, O. V., Yarutkina, I. A., Podivilov, E. V., Fedoruk, M. P. & Babin, S. A. (2011). Highly chirped dissipative solitons as a one-parameter family of stable solutions of the cubic-quintic Ginzburg-Landau equation, *J. Opt. Soc. Am. B* Vol. 28(No. 10): 2314–2319.

Koechner, W. (2006). *Solid-State Laser Ingeneering*, Springer, New York.

Komarov, A., Leblond, H. & Sanchez, F. (2005). Quantic complex Ginzburg-Landau model for ring fiber laser, *Phys. Rev. E* Vol. 72(No. 2): 025604(R).

Krausz, F. & Ivanov, M. (2009). Attosecond physics, *Rev. Mod. Phys.* Vol. 81(No. 1): 163–234.

Lecaplain, C., Baumgartl, M., Schreiber, T. & Hideur, A. (2011). On the mode-locking mechanism of a dissipative-soliton fiber oscillator, *Optics Express* Vol. 19(No. 27): 26742–26751.

Lin, J-H. & Lin, K-H. (2010). Multiple pulsing and harmonic mode-locking in an all-normal-dispersion Nd:GdO_4 laser using a nonlinear mirror, *J. Phys. B.* Vol. 43(No. 6): 065402.

Liu, X. (2010). Pulse evolution without wave breaking in a strongly dissipative dispersive laser system, *Phys. Rev. A* Vol. 81(No. 5): 053819.

Liu, Y., Tschuch, S., Rudenko, A., Dürr, M., Siegel, M., Morgner, U., Moshammer, R. & Ullrich, J. (2008). Strong-field double ionization of air below the recollision threshold, *Phys. Rev. Lett.* Vol. 101(No. 5).

Maimistov, A. I. (1993). Evolution of solitary waves which are approximately solitons of a nonlinear Schrödinger equation, *J. Exp. Theor. Phys.* Vol. 77(No. 5): 727–731.

Malomed, B. A. (2002). Variational methods in nonlinear fiber optics and related fields, *in* E. Wolf (ed.), *Variational methods in nonlinear fiber optics and related fields*, Vol. 43, Elsevier, Amsterdam, pp. 71–193.

Malomed, B. A. & Nepomnyashchy, A. A. (1990). Kinks and solitons in the generalized Ginzburg-Landau equation, *Phys. Rev. A* Vol. 42(No. 10): 6009–6014.

Martin, M. M. & Hynes, J. T. (eds) (2003). *Femtochemistry and femtobiology: ultrafast events in molecular science*, Elsevier, Amsterdam.

Mourou, G. A., Tajima, T. & Bulanov, S. V. (2006). Optics in the relativistic regime, *Rev. Mod. Phys.* Vol. 78(No. 2): 309–692.

Naumov, S., Fernandez, A., Graf, R., Dombi, P. & Apolonski, F. K. A. (2005). Approaching the microjoule frontier with femtosecond laser oscillators, *New Journal of Physics* Vol. 7(No. 1): 216.

Oughstun, K. E. (2009). *Electromagnetic and Optical Pulse Propagation 2: Temporal Pulse Dynamics in Dispersive, Attenuative Media*, Springer.

Palmer, G., Schultze, M., Siegel, M., Emons, M., Bünding, U. & Morgner, U. (2008). Passively mode-locked Yb:KLu$(WO_4)_2$ thin-disk oscillator operated in the positive and negative dispersion regime, *Opt. Lett* Vol. 33(No. 14): 1608–1610.

Palmer, G., Schultze, M., Siegel, M., Emons, M., Lindemann, A. L., Pospiech, M., Steingrube, D., Lederer, M. & Morgner, U. (2010). 12 MW peak power from a two-crystal Yb:KYW chirped-pulse oscillator with cavity-dumping, *Optics Express* Vol. 18(No. 18): 19095–19100.

Paschotta, R. (2008). *Encyclopedia of Laser Physics and Technology*, Wiley.

Perez-Garcia, V. M., Torres, P. & Montesinos, G. D. (2007). The method of moments for nonlinear Schrödinger equations: theory and applications, *SIAM J. Appl. Math.* Vol. 67(No. 4): 990–1015.

Pervak, V., Teisett, C., Siguta, A., Naumov, S., Krausz, F. & Apolonski, A. (2008). High-dispersive mirrors for femtosecond lasers, *Optiacs Express* Vol. 16(No. 14): 10220–10233.

Pfeifer, T., Spielmann, C. & Gerber, G. (2006). Femtosecond X-ray science, *Rep. Prog. Phys.* Vol. 69(No. 2): 443–505.

Podivilov, E. & Kalashnikov, V. L. (2005). Heavily-chirped solitary pulses in the normal dispersion region: new solutions of the cubic-quintic complex Ginzburg-Landau equation, *JETP Letters* Vol. 82(No. 8): 467–471.

Proctor, B., Westwig, E. & Wise, F. (1993). Characterization of a Kerr-lens mode-locked Ti:sapphire laser with positive group-velocity dispersion, *Opt. Lett.* Vol. 18(No. 19): 1654–1656.

Renninger, W. H., Chong, A. & Wise, F. W. (2008). Dissipative solitons in normal-dispersion fiber lasers, *Phys. Rev. A* Vol. 77(No. 2): 023814.

Renninger, W. H., Chong, A. & Wise, F. W. (2011). Pulse shaping and evolution in normal-dispersion mode-locked fiber lasers, *IEEE J. Sel. Top. Quantum Electron.* Vol. PP(No. 99): DOI:10.1109/JSTQE.2011.2157462.

Rühl, A. (2008). *The normal dispersion regime in passively mode-locked fiber oscillators*, Cuvillier Verlag, Göttingen.

Siegel, M., Palmer, G., Emons, M., Schultze, M., Ruchl, A. & Morgner, U. (2008). Pulsing dynamics in Ytterbium based chirped-pulse oscillator, *Optics Express* Vol. 16(No. 19): 14314–14320.

Siegel, M., Pfullmann, N., Palmer, G., Rausch, S., Binhammer, T., Kovacev, M. & Morgner, U. (2009). Microjoule pulse energy from a chirped-pulse Ti:sapphire oscillator with cavity dumping, *Opt. Lett.* Vol. 34(No. 6): 740–742.

Sorokin, E., Kalashnikov, V. L., Mandom, J., Guelachvili, G., Picquè, N. & Sorokina, I. T. (2008). Cr^{4+}:YAG chirped-pulse oscillator, *New Journal of Physics* Vol. 10(No. 8): 083022.

Sorokina, I. T. (2003). Crystalline mid-infrared lasers, *in* I. T. Sorokina & V. K. L. (eds), *Solid-state mid-infrared laser sources*, Springer-Verlag, Berlin, pp. 255–349.

Soto-Crespo, J. M., Akhmediev, N. N., Afanasjev, V. V. & Wabnitz, S. (1997). Pulse solutions of the cubic-quintic complex Ginzburg-Landau equation in the case of normal dispersion, *Phys. Rev. E* Vol. 55(No. 4): 4783–4796.

Soto-Crespo, J. M., Akhmediev, N. & Town, G. (2002). Continuous-wave versus pulse regime in a passively mode-locked laser with a fast saturable absorber, *J. Opt. Soc. Am. B* Vol. 19(No. 2): 234–242.

Spence, D. E., Kean, P. N. & Sibbett, W. (1991). 60-fsec pulse generation from a self-mode-locked Ti:sapphire laser, *Opt. Lett.* Vol. 16(No. 1): 42–44.

Spence, D. E. & Sibbett, W. (1991). Femtosecond pulse generation by a dispersion-compensated, coupled-cavity, mode-locked Ti:sapphire laser, *J. Opt. Soc. Am. B* Vol. 18(No. 19): 2053–2060.

Steinmeyer, G., Sutter, D. S., Gallmann, L., Matuschek, N. & Keller, U. (1999). Frontiers in ultrashort pulse generation: pushing the limits in linear and nonlinear optics, *Science* Vol. 286(No. 19): 1507–1512.

Südmeyer, T., Kränkel, C., Baer, C. R. E., Heckl, O. H., Saraceno, C. J., Golling, M., Peters, R., Petermann, K., Huber, G. & Keller, U. (2009). High-power ultrafast thin disk laser oscillators and their potential for sub-100-femtosecond pulse generation, *Appl. Phys. B* Vol. 97(No. 2): 281–295.

Südmeyer, T., Marchese, S. V., Hashimoto, S., Baer, C. R. E., Gingras, G., Witzel, B. & Keller, U. (2008). Femtosecond laser oscillators for high-field science, *Nature Photonics* Vol. 2(No. 10): 599–604.

Tan, W. D., Tang, D. Y., Xu, C. W., Zhang, J., Xu, X. D., Li, D. Z. & Xu, J. (2011). Evidence of dissipative solitons in Yb^{3+}:$CaYAlO_4$, *Optics Express* Vol. 36(No. 19): 18495–18500.

Tsoy, E. N. & Akhmediev, N. N. (2005). Bifurcations from stationary to pulsating solitons in the cubic-quintic complex Ginzburg-Landau equation, *Physics Letters A* Vol. 343(No. 6): 417–422.

Tsoy, E. N., Ankiewicz, A. & Akhmediev, N. (2006). Dynamical models for dissipative localized waves of the complex Ginzburg-Landau equation, *Phys. Rev. E* Vol. 73(No. 3): 036621.

Usechak, N. G. & Agrawal, G. P. (2005). Semi-analytic technique for analyzing mode-locked lasers, *Optics Express* Vol. 13(No. 6): 2075–2081.

van Saarloosa, W. & Hohenberg, P. C. (1992). Fronts, pulses, sources and sinks in generalized complex Ginzburg-Landau equations, *Physica D: Nonlinear Phenomena* Vol. 56(No. 4): 303–367.

Weiner, A. M. (2009). *Ultrafast Optics*, Wiley.

Wise, F. W., Chong, A. & Renninger, W. H. (2008). High-energy femtosecond fiber lasers based on pulse propagation at normal dispersion, *Laser & Photon. Rev.* Vol. 2(No. 1–2): 58–73.

Zhavoronkov, N., Maekawa, H., Okuno, H. & Tominaga, K. (2005). All-solid-state femtosecond multi-kilohertz laser system based on a new cavity-dumped oscillator design, *J. Opt. Soc. Am. B* Vol. 22(No. 3): 567–571.

6

Intra-Cavity Nonlinear Frequency Conversion with Cr^{3+}-Colquiriite Solid-State Lasers

H. Maestre, A. J. Torregrosa and J. Capmany
Photonic Systems Group, Universidad Miguel Hernández
Elche, Alicante,
Spain

1. Introduction

The availability of laser radiation in new wavelengths is a research topic of permanent interest. Nonlinear optical frequency conversion of a primary laser radiation is one of the widely exploited resources for this purpose. Since the birth of nonlinear optics with the first demonstration of second harmonic generation of laser radiation, new ways and new materials are seek. Due to the inherent low conversion efficiency of nonlinear optical mixing related to the decreasing efficiency of higher order terms in the nonlinear Taylor expansion of the optical polarization with the electric field, second-order nonlinear optics referred to as three-wave mixing processes is in general preferred. It is well stated that nonlinear optical frequency conversion efficiency improves with increasing power density –and thus amplitude- of the primary optical field, material nonlinear response and interaction length as well as with providing a suitable phase relation among the interacting electromagnetic waves (Armstrong et al., 1962). It was soon realized that conversion of pulsed lasers that emit high peak power pulses could result efficient in a simple manner, by just single-passing the interacting waves through a suitable nonlinear medium, generally an anisotropic crystal.

The case of continuous-wave is more challenging due to the lower peak power available in comparison with pulsed lasers. A skilled technique –intracavity conversion-, where the nonlinear crystal is placed inside the cavity of the primary laser, was developed as a way to boost conversion efficiency in the continuous-wave case, particularly in case of solid-state lasers. Because the circulating intracavity power is typically around two orders of magnitude higher than in the output of a solid-state laser where typical optimal output coupling is achieved with a few percent transmission, a rough estimation of the power circulating inside the cavity is given by $P_{intra} \approx \frac{P_{out}}{T}$ with $T \approx 0.02 - 0.06$. As a result, a small fraction of intracavity conversion can provide an amount of useful converted power that equals the primary laser output in the absence of any conversion. The key for efficient intracavity conversion resides in confining as much as possible the primary laser within the cavity and to provide its output coupling through conversion of energy to another wavelength rather than by an outcoupling mirror (Smith, 1970).

Early intracavity nonlinear conversion focused on realizing a single conversion process like SHG or SFM in the primary laser, based on perfect phase matching in birefringent nonlinear crystals. A very useful alternative to phase matching is quasiphase matching based on domain-engineered nonlinear ferroelectric crystal. With quasiphase matching more than one nonlinear interaction (conversion) can be realized simultaneously in the same physical nonlinear crystal. A high degree of flexibility for the generation of new wavelengths can be achieved when the primary laser can oscillate simultaneously in different wavelengths and the nonlinear crystal can realize several different nonlinear conversion processes among those wavelengths. For this reason, gain materials with a broad gain bandwidth combined with a domain-engineered ferroelectric crystal become particularly useful.

Cr^{3+}-doped colquiriites can provide tunable lasing in a wide spectral range around 750-950 nm much like the well-established Ti^{3+}-Sapphire laser. However, Cr^{3+}-doped colquiriites have the practical advantage of diode laser pumping with red diodes around 660-690 nm. It is worth mentioning that the use of these lasers in intracavity nonlinear frequency generation has been little explored so far. Throughout this chapter it is intended to show the potential of these systems in the 1500-1700 nm spectral band. For this aim, the characteristics of chromium-doped colquiriite crystals are described firstly. Next, a description of the basic physics involving second-order nonlinear interactions is presented, where the allowed processes for three-wave mixing are detailed. Due to the dispersion of nonlinear media, a review of techniques for controlling the phase-matching among interacting waves is made, paying special attention to the flexibility and versatility offered by quasi-phase matching (QPM) in periodically poled ferroelectric crystals. Coherent radiation by intracavity wavelength conversion is described from the inclusion of such nonlinear crystals inside the resonator of a Cr-doped colquiriite lasers. Before conclusions, some examples of wavelength converting systems based on optical parametric oscillation and difference frequency generation are reported.

2. Characteristics of Cr^{3+}-doped colquiriite crystals

Cr^{3+} is one of the most popular ions used in inorganic solid-state lasers. It was also employed as the optical gain material in the first experimental demonstration of lasing action conducted by T. H. Maiman in 1960 (ruby laser) (Maiman, 1960). It is usually included in the group of transition metal lasers.

2.1 Chromium doped laser crystals

Crystals for solid-state lasers based on such ion are made of host materials containing aluminium. During growth of the crystal, some amount of Al^{3+} ions is replaced with Cr^{3+} ions and, therefore, a Cr^{3+}-doped crystal is obtained. Depending on the host, the intensity of the crystal electric field varies and, thus, Cr^{3+} ions are affected in a different manner by different hosts. The crystal electric field perturbs, among other properties, optical properties of the laser ion (energy levels, absorption, stimulated cross section, radiative and non-radiative lifetimes, transition linewidth, etc…). A high crystal field intensity induces narrow emission linewidths (as in ruby lasers), a moderate crystal field intensity allows both narrow and broad emission linewidths (as in alexandrite laser), whereas in a low intensity crystal electric field a broad emission linewidth is obtained, which is desirable for tunable lasers and characteristic of colquiriite (fluoride) lasers. Typical and well-known colquiriite host materials are $LiCaAlF_6$ (LiCAF) (Payne et al., 1988), $LiSrAlF_6$ (LiSAF) (Payne et al., 1989) and $LiSrGaAlF_6$ (LiSGAF) (Smith et al., 1992).

2.2 Energy levels, absorption and stimulated emission in Cr^{3+}-doped colquiriites

Cr^{3+}-doped colquiriites belong to the family of lasers called vibronic lasers. In such laser gain media exist a strong interaction of the electronic states with lattice vibrations (phonons). This vibrational–electronic interaction leads to a pronounced homogeneous broadening and thus to a broad optical gain bandwidth. There are four main energy levels in Cr^{3+} ions ($^4T_1$, $^4T_2$, ^2E and $^4A_2$) as it is shown in Fig. 1.

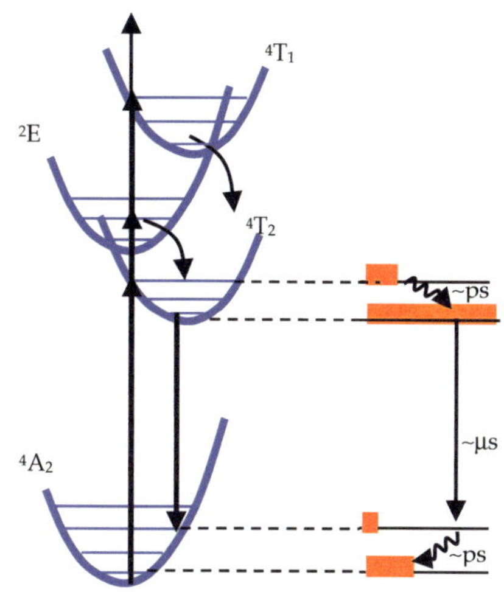

Fig. 1. Cr^{3+} ion energy levels in colquiriites.

Due to the electric forces acting on Cr^{3+} ion and in contrast to other Cr^{3+}-doped media, in Cr^{3+}-colquiriites the ^2E energy level lays above the $^4T_2$ and below the $^4T_1$ levels respectively. In these crystals the absorption and thus the channels allowing direct laser pumping are due to transitions between the $^4A_2$ ground state and ^2E or $^4T_1$ upper states ($^4A_2 \rightarrow {}^2E$ or $^4A_2 \rightarrow {}^4T_1$). At room temperature the $^4A_2 \rightarrow {}^2E$ transition can be pumped by means of commercially available red light diode lasers. The electrons excited to the ^2E state decay rapidly in a non-radiative fashion to the $^4T_2$ state, becoming the latter the most populated level and hence the upper laser level. Radiative rate from $^4T_2 \rightarrow {}^4A_2$ (much greater than that from $^2E \rightarrow {}^4A_2$) gives rise to stimulated emission in a four-level like laser system. Depending on the colquiriite host crystal, different stimulated emission peaks and tuning ranges can be obtained. As examples and for a higher detail on absorption and emission bands, Cr:LiSAF and Cr:LiCAF are shown in figures 2.a and 2.b respectively. Since colquiriites are uniaxial crystals, Cr^{3+} emission and absorption in such materials is strongly polarization dependant (parallel to c-axis (π-polarization) and perpendicular to c-axis (σ-polarization)).

As it can be seen Cr:LiSAF and Cr:LiCAF have quite similar absorption spectra but main differences arise from the emission spectra. Absorption peaks and the relative strengths of the peaks occur at nearly the same wavelengths and have similar values but, nevertheless, absolute absorption strengths for Cr:LiSAF are roughly twice as strong as those of Cr:LiCAF. Strongest absorption peaks for the π-polarization are approximately at 0.425 and 0.630 μm. Strongest absorption peaks for the σ-polarization are approximately at 0.423 and 0.622 μm. The absorption peak for wavelengths above 500 nm is stronger for the π-

polarization whereas the peaks for wavelengths below 500 nm are stronger for the σ-polarization. Emission from Cr:LiSAF , has a maximum emission peak at about 840 nm and a linewidth of about 197 nm for both polarizations and the π-polarized emission spectrum is approximately 3 times more intense than the σ-polarized emission. On the other hand, emission from Cr:LiCaF has a maximum peak at about 775 nm and a linewidth of about 132 nm for both polarizations and the π-polarized emission is approximately 1.5 times more intense than the σ-polarized.

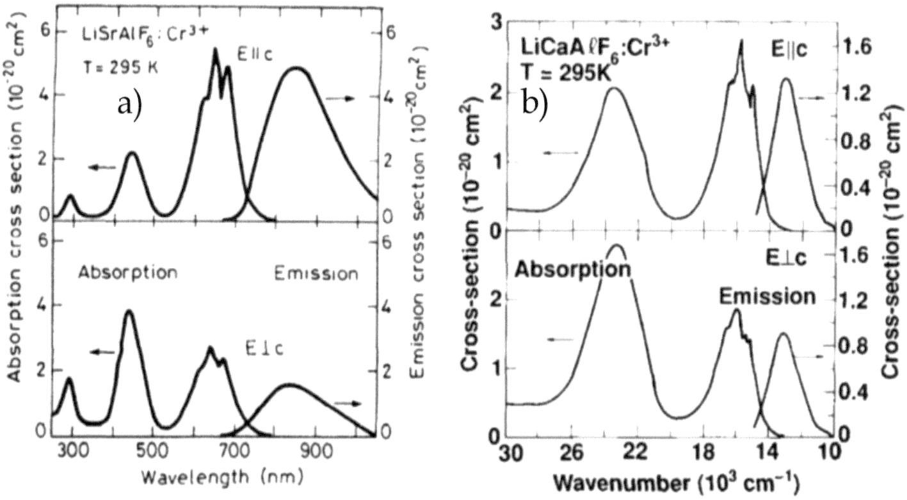

Fig. 2. Cr^{3+} absorption and emission cross-section spectra in a) LiSAF (Reprinted by permission from (Payne et al., 1989), American Institute of Physics) and b) LiCAF hosts (Reprinted by permission from (Payne et al., 1988), Institute of Electrical and Electronics Engineers (IEEE)).

2.3 Ti:Sapphire vs. Cr^{3+}-doped colquiriites

Cr^{3+}-doped colquiriites can be viewed as promising alternatives to titanium sapphire (Ti:Sapphire) lasers. Cr^{3+} doped fluorides (colquiriites) such as Cr^{3+}:LiSAF, Cr^{3+}:LiSGaF, and Cr^{3+}:LiCAF, belonging to solid-state vibronic lasers, can provide broadly tunable operation, high intrinsic slope efficiencies (>50%) and enable efficient laser operation with electrical-to-optical conversion efficiencies exceeding 10% (Demirbas et al., 2009). The main laser parameters are summarized in Table 1. Their peak emission cross section and their absorption bands are placed around 800 nm and 650 nm respectively, allowing direct diode pumping with commercially available AlGaInAs diode lasers around 670 nm, significantly reducing the total cost of the laser system.

In contrast to this, Ti:Sapphire lasers are typically pumped by argon gas lasers or frequency-doubled diode pumped neodymium lasers, since high power direct diode pumping is not currently available, which are bulky and thus making the overall system cost high and limiting wide-spread use. In addition, their anisotropic character readily allows polarized laser oscillation without additional polarizing elements or Brewster angle operation.

Gain material	$Ti^{3+}:Al_2O_3$	$Cr^{3+}:LiSrAlF_6$	$Cr^{3+}:LiCaAlF_6$	$Cr^{3+}:LiSrGaAlF_6$
Tuning range (nm)	660-1180	782-1042	754-871	777-977
Cross-section (σ_{em}) [x10^{-20} cm^2]	41	4.8	1.3	3.3
Room-temperature fluorescence lifetime (τ_f) [μs]	3.2	67	175	88
$\sigma_{em} \cdot \tau_f$ [10^{-20}xcm^2x μs]	131	322	228	290
Efficiency [%]	64	54	69	60
Thermal conductivity [W/k·m]	28	3.1	5.1	3.6
$T_{1/2}$, $\tau_f(T_{1/2})=0.5\cdot\tau_{rad}$	100	69	255	88
Refractive index	1.76	1.4	1.39	1.4

Table 1. Comparison of important laser parameters of the Ti:Sapphire, Cr:LiSAF, Cr:LiSGaF, and Cr:LiCAF gain media (Payne et al., 1988; Payne et al., 1989; Smith et al., 1992; Demirbas et al., 2009).

Among Cr^{3+}-doped colquiriites and despite the lower emission cross section of Cr:LiCAF as compared to other colquiriites, it offers the longest fluorescence lifetime and a high thermal conductivity, which reduces thermal lensing effects caused by high power pumping and preserves a good-quality mode when scaling power. Moreover, Cr:LiCAF suffers lower quantum defect due its blue shifted emission spectrum, as well as lower excited-state absorption (ESA) and upconversion rate (Eichenholz and Richardson, 1998).

3. Nonlinear optics and materials

In conventional (linear) optics, the interaction between an electro-magnetic wave and a dielectric material results in an induced polarization of the material which depends linearly upon the incident electric field. In non linear optical materials, the relation between the induced polarization and the incident electric field is no longer linear and can be described by means of a power series expansion of the electric susceptibility (equation (1)).

$$P_i(t) = \varepsilon_0 \chi_{ij}^{(1)} E_j(t) + \varepsilon_0 \chi_{ijk}^{(2)} E_j(t) E_k(t) + \varepsilon_0 \chi_{ijkl}^{(3)} E_j(t) E_k(t) E_l(t) + \ldots \quad (1)$$

The first-order term of the expansion gives rise to linear optics effects (refractive index, reflection, refraction, dispersion…) whereas higher order terms of the expansion (such as $\chi_{ijk}^{(2)}$ or $\chi_{ijkl}^{(3)}$) give rise to nonlinear effects which are responsible of the generation of new frequencies other than that of the incident wave. Such terms are tensors relating electric fields of same or different frequency and polarization to the optically induced nonlinear polarization. If the incident wave intensity is strong enough, the terms $\chi_{ijk}^{(2)}$ and $\chi_{ijkl}^{(3)}$ lead to second and third-order nonlinear effects respectively. In the following sections we will focus on second-order nonlinear processes.

3.1 Second-order nonlinear processes

Optical nonlinear second-order processes are determined by the second term ($\chi^{(2)}$) of the optical susceptibility series expansion or the more extensively used nonlinear coefficient $d = \frac{1}{2}\chi_{ijk}^{(2)}$ Such processes may be regarded as the interaction of one photon of frequency ω_1 and wavenumber k_1 combining with other photon of frequency ω_2 and wavenumber k_2 to yield a third photon of frequency ω_3 and wavenumber k_3. Allowed processes for three-wave mixing of continuous-wave (CW) signals are represented in Fig. 3: second-harmonic generation (SHG), sum-frequency generation (SFG), difference frequency generation (DFG), optical parametric generation (OPG), optical parametric amplification (OPA) and optical parametric oscillation (OPO).

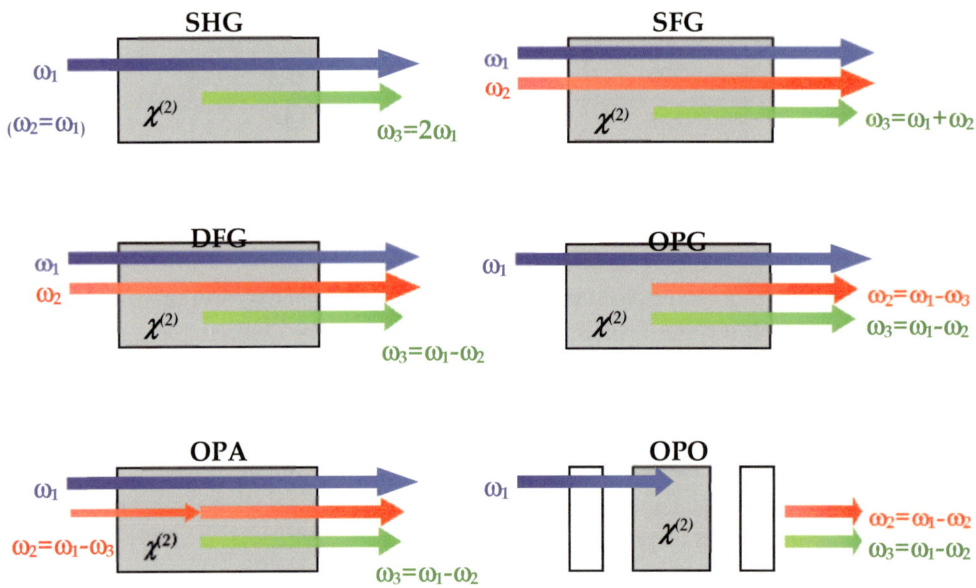

Fig. 3. Interacting waves and different second-order nonlinear optical processes.

In order to select a nonlinear process and to obtain efficient non linear interactions, it is necessary to fulfill both the photon energy, equation (2.a), and momentum conservation along the non linear crystal length. The momentum conservation condition is also known as the phase-matching condition. The phase-matching condition is represented in equation (2.b) where the *l*, *m* and *n* subscripts stand for the higher and lower frequency interacting waves respectively.

$$\omega_l = \omega_m + \omega_n \tag{2a}$$

$$k_l = k_m + k_n \rightarrow \frac{\omega_l}{c} n(\omega_l) = \frac{\omega_m}{c} n(\omega_m) + \frac{\omega_n}{c} n(\omega_n) \tag{2b}$$

The conversion efficiency, i.e. the amount of incoming power converted into a different frequency, is proportional to the square of effective nonlinear coefficient d_{eff} (the nonlinear coefficient for a given propagation direction and polarization of waves in the crystal), crystal length L, and the phase mismatch Δk in through the term $sinc^2(\Delta kL/2)$ (equation (3)). It is difficult to meet the phase-matching condition due to dispersion of the refractive index. If the phase-matching condition is not completely fulfilled a phase mismatch $\Delta k = k_l - k_m - k_n \neq 0$ is obtained and crystal length L is limited to the coherence length, i.e. $l_c = \pi/\Delta k$, thus reducing the conversion efficiency.

$$\eta \equiv \frac{P_{converted}}{P_{incident}} \propto d_{eff}^2 \frac{\sin^2(\Delta kL/2)}{(\Delta kL/2)^2} L^2 = d_{eff}^2 \operatorname{sinc}^2(\Delta kL/2) L^2 \tag{3}$$

As a result, it is important to operate as close as possible to the condition $\Delta k=0$. We will discuss briefly two of the most employed phase-matching techniques; Birefringent Phase-Matching (BPM) and Quasi-Phase-Matching (QPM).

3.2 Birefringent phase-matching (BPM)

As it can be seen from equation 2.b, for a targeted optical nonlinear process (i.e. a given energy conservation condition) refractive indices play an important role in achieving the phase-matching condition ($\Delta k=0$). In this respect the birefringence inherent in many nonlinear materials can be used to compensate refractive index dispersion and obtain $\Delta k=0$.

Owing to different refractive indices for the ordinary and extraordinary electric field polarization in a birefringent crystal and for specific interacting wavelengths, the optical axis of the crystal can be oriented to a particular angle to achieve phase-matching but, nevertheless, BPM cannot enable to achieve phase-matched processes in the full transparency range of the nonlinear crystal. In addition, since BPM takes advantage of both ordinary and extraordinary polarized waves, if the extraordinary wave propagates in a direction other than that of the optical axis, the propagation direction of the extraordinary wave diverges from the propagation direction of the ordinary wave. This effect is known as walk-off and reduces the effective interaction length to the distance where ordinary and extraordinary beams do not overlap thus limiting the conversion efficiency in BPM crystals.

3.3 Quasi-phase-matching (QPM)

QPM is an alternative method to achieve phase-matching in which the sign of the nonlinear susceptibility $\chi^{(2)}$ is corrected (or modulated) at regular intervals along the propagation length of the interacting waves inside the crystal. Because of dispersion of the refractive index, the interacting waves have different phase velocities and a phase-mismatch $\Delta k \neq 0$ results. From a coupled-wave equation analysis the energy transfer between the interacting waves as they propagate through the crystal coordinate z varies as $\exp(-j\Delta kz)$ and due to the periodicity of the complex exponential function the nonlinear conversion efficiency shows an oscillatory behavior with period $l_c = \pi/\Delta k$ and thus the converted power increases and diminishes periodically with (Fig. 4, curve a). In contrast, if a perfect phase-matching condition $\Delta k=0$ can be obtained, i.e. by using BPM, the converted power increases monotonically (Fig. 4, curve b). Even if there is a phase-mismatch $\Delta k \neq 0$, it is possible to invert the sign of the nonlinear susceptibility every coherence length and the converted power can be made to increase monotonically as well (Fig. 4, curve c).

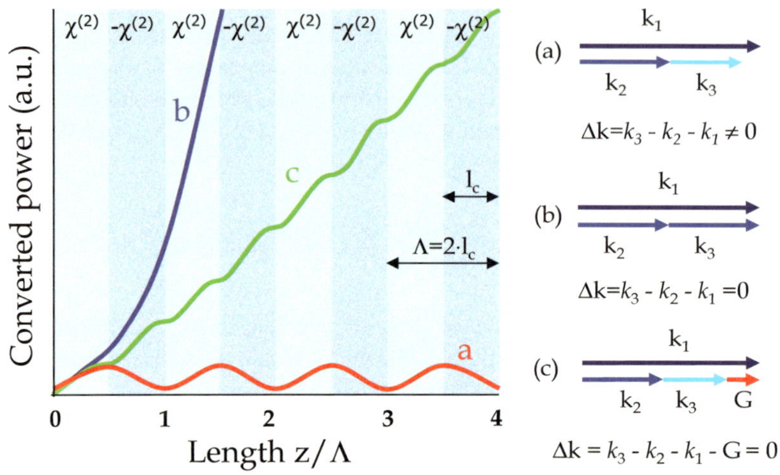

Fig. 4. Example of nonlinear converted power and wave vector mismatch for a) no phase-matching, b) BPM and c) QPM.

This is the case of QPM, where the wave vector associated with a periodic modulation of the sign of the optical susceptibility ($G=2\pi/\Lambda$) compensates for the wave vector mismatch of the interacting waves. It is important to remark that after some propagation length the converted power for QPM equals that of BPM. In contrast to BPM, QPM is not limited by walk-off (the Poynting vector is in the same direction for all interacting waves) and as a consequence longer crystals can be used (the converted crystal grows quadratically with crystal length). QPM also enables exploiting the highest non linear coefficient of a given material by choosing adequate direction of propagation and polarization in comparison to BPM where propagation direction and polarization are constrained. QPM is extensively employed in ferroelectric materials since it is a relatively easy task to reverse the sign of $\chi^{(2)}$ in such materials. The spontaneous dipolar moment can be rotated under the effect of an applied external electric field and 0°-180° alternating $\chi^{(2)}$ domains can be created and such materials are known as periodically poled crystals. Most employed periodically poled crystals are lithium niobate ($LiNbO_3$, PPLN), lithium tantalate ($LiTaO_3$, PPLT), potassium-titanyl phosphate ($KTiOPO_4$, PPKTP) and potassium niobate ($KNbO_3$, PPKN).

The alternating domains result in a periodic grating that can be modeled by means of a position dependent function (Fig. 5):

$$d(z) = d_{eff}\Pi(z) \tag{4}$$

where d_{eff} is the effective nonlinear coefficient and $\Pi(z)$ is a square function of period Λ with amplitude of ±1 (associated to opposite directions of the spontaneous polarization of ferroelectric domains) and representing the spatial distribution of domain reversals in the crystal, as it is shown in Fig. 5.

Therefore, the relative conversion efficiency associated to a collinear QPM nonlinear interaction of wavevector mismatch Δk is related to the Fourier transform $\hat{\Pi}(\Delta k)$ of the spatial distribution $\Pi(z)$:

$$\eta(\Delta k) \propto \left|\hat{\Pi}(\Delta k)\right|^2 = \left|\frac{1}{L}\int_0^L d(z)e^{-j\Delta kz}dz\right|^2 = \left|\frac{d_{eff}}{L}\int_0^L \Pi(z)e^{-j\Delta kz}dz\right|^2 \quad (5)$$

This result enables domain engineering since periodicity can be controlled and selected to provide a phase-matched interaction for any frequency conversion process within the transparency range of the material. This means an additional advantage over BPM. In this way, the design of the domain structure is reduced to find a spatial function which describes the strategic distribution of domain reversals which produces the desired QPM spectral response (Liu et al., 2001; Pousa and Capmany, 2005).

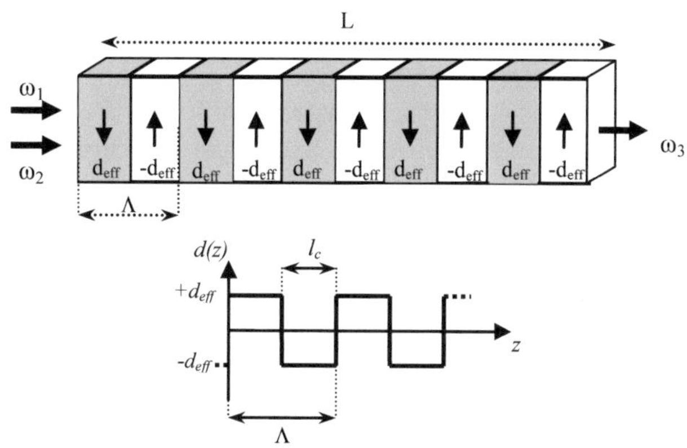

Fig. 5. Ferroelectric domain reversal and spatial distribution function in a periodically poled crystal.

Then, domain engineering permits the fabrication of ferroelectric structures which provide targeted QPM responses, consisting of a set of reciprocal vectors which participate to phase-match multiple interactions simultaneously over the same device. As QPM is sensitive to refractive index variations, the overall QPM response can be modified by altering the quasi-phase-matching condition in ferroelectric crystals through the control of the temperature (Belmonte et al., 1999) or the linear electro-optical (Pockels) effect (Xu et al., 2003). In this way, it is possible to effectively increase the number of interactions (reciprocal vectors) at the same device at no expense of nonlinearity (Torregrosa et al., 2008).

4. Intracavity nonlinear frequency conversion and external cavity lasers

As stated before, nonlinear optical frequency conversion arises only if sufficiently high optical intensities are incident to the nonlinear crystal material and becomes more pronounced as the optical intensity of the incoming waves grow higher. Usually, high optical intensities cannot be obtained from low- or moderate-power continuous-wave solid-state lasers and, as a consequence, the conversion efficiency is very low.

Since a high intensity is equivalent to a relatively high power focused to a small area, laser cavity modes are good candidates to pump nonlinear optical processes. In this sense an adequate solution is to place the nonlinear crystal within the laser resonator where the circulating power is several times higher than the output power emitted by the same laser (Fig. 6). This leads to high laser intensities (optical powers are well above 10 W and focused to areas of radius well below 100 μm) and hence the nonlinear frequency conversion is enhanced by more than two orders of magnitude (Smith, 1970).

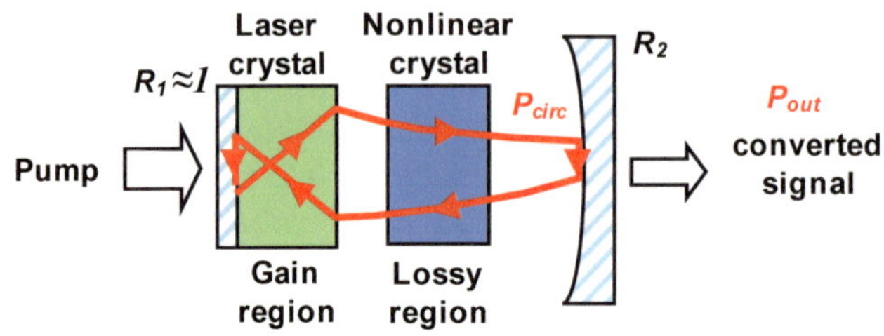

Fig. 6. Intracavity nonlinear optical conversion scheme

The power circulating inside the cavity can be approximated by $P_{circ} \approx \frac{Pout}{1-R_2}$, and, as a result, the intracavity power thus can be controlled by proper selection of the output coupler reflectivity (R_2 in Fig. 6). Careful design of the laser resonator may enhance the conversion efficiency by adjusting both size and divergence of the laser mode in the nonlinear medium (Boyd and Kleinman, 1968). Low intracavity loss due to scattering and facet reflectivity of the nonlinear crystal is desired for this type of lasers as well. If the laser emits in a multi-axial mode regime, the nonlinear medium acts as an energy exchange mechanism between them, which gives rise to noise and instabilities in the output power. In this respect, it is also interesting to operate the laser as close as possible to single-axial mode operation in order to obtain low-noise output amplitude. Furthermore, narrow-band lasers which wavelength can be tuned are highly desirable since they enable tunable optical frequency conversion owing to perturbation of the phase-matching condition.

In this regard (narrow emission linewidth and wavelength tuning capability) external-cavity solid-state lasers may play an important role. Traditionally, birefringent filters have been employed to both tune and narrow the laser emission spectrum. External-cavity solid-state lasers can be viewed as an alternative to birefringent filters because of no additional intracavity loss is produced since the tuning mechanism is not inserted within the cavity and because of their lower cost.

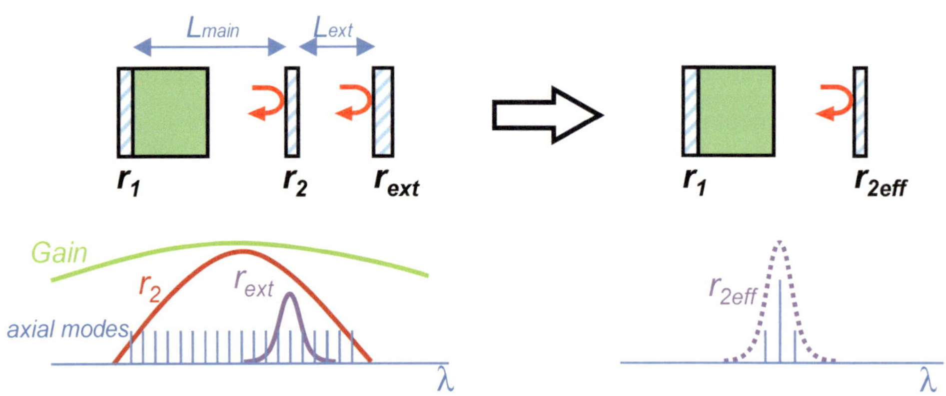

Fig. 7. External-cavity laser principle of operation.

The addition of an external reflector (r_{ext}) to a laser cavity (main cavity) modifies effectively the output electric field reflectivity of the main cavity (r_2) (Fig. 7) and then the three mirror cavity can be regarded as a two mirror cavity whit an output reflectivity r_{2eff}. This new r_{2eff} can be obtained from equation (6) (Tsunekane et al., 1998), where r_2, r_{ext} and r_{eff} are electric field reflection coefficients and τ_L round-trip time of the external cavity. As a result, the reflectivity is increased at some wavelengths.

$$r_{2eff} = \frac{r_2 + r_{ext}\exp(-j\omega\tau_L)}{1 + r_2 r_{ext}\exp(-j\omega\tau_L)} \qquad (6)$$

As a consequence of an increased effective reflectivity the laser threshold is reduced and thus the intracavity power is also enhanced. If the external reflector permits wavelength tuning of its peak reflectivity and has narrow-band reflectivity (e.g. a diffraction grating) the emission linewidth is reduced and can be tuned. For a particular relation between the length of the main cavity and the length of the external cavity, the laser can be forced to emit in a single-axial mode (Tsunekane et al., 1998).

The most widely studied intracavity nonlinear frequency conversion process in Cr-doped colquiriites has been second-harmonic generation (SHG) in Cr:LiSAF lasers for tunable blue light generation (Balembois et al., 1992; Laperle et al., 1997; Eichenholz et al. 1998; Makio et al., 2000). In the following section we will widen the utility of Cr-doped colquiriites to new intracavity frequency conversion processes such as optical parametric oscillation (OPO) and difference frequency generation (DFG).

5. Examples of intra-cavity optical frequency conversion in Cr^{3+}:LiCAF lasers

In this section several optical frequency conversion devices based on Cr^{3+}-doped $LiCaAlF_6$ colquiriite plus a nonlinear optical crystal (NLOC) inside the laser cavity will be presented. Despite the converting systems described below uses a PPSLT crystal as a NLOC to operate in the 1500 - 1700 nm spectral region, they can be generalized to any nonlinear medium presenting transparency in the operation range. In each case, the system performance will be determined by the nonlinear and dispersive properties of the material employed depending on the application.

On one hand, the use of diode-pumped laser crystals as titanium sapphire or Cr^{3+}-doped colquiriite family (Cr^{3+}:LiSAF, Cr^{3+}:LiCAF and Cr^{3+}:LisGAF) appears as attractive options to operate since they provide a broad emission around 800 nm. However, the latter group presents the advantage of being directly pumped by commercial low-cost red diodes around 665 nm. Particularly, Cr:LiCAF arise as an optimum candidate due to its spectrally broad emission and homogeneous gain profile, which allow the generation of tunable narrow linewidth pump wavelengths from 720 nm to 840 nm approximately, and whose maximum power efficiency (emission cross-section) is located around 780 nm (Payne et al., 1988).

On the other hand, periodically poled ferroelectric materials have been widely employed for QPM interactions due to their high nonlinear coefficients and wide transparent range suitable for nonlinear processes from UV to mid-IR. In last years, periodically-poled stoichiometric lithium tantalate (PPSLT) has been paid much attention due to its resistance to photorefractive damage and laser damage (Buse et al., 2007; Skvortsov et al., 1993) compared with other similar ferroelectric materials as congruent lithium tantalate (CLT) and congruent or stoichiometric lithium niobate (CLN or SLN, respectively). These properties should be specially taken into account in intracavity configurations, where powers confined in the nonlinear crystal can raise high values. However, the presence of MgO-doping in these compounds can help to mitigate optical degradation effects (Dolev et al., 2009).

In order to obtain efficient nonlinear interaction in the mentioned applications, quasi-phase matching (QPM) technique is employed in microstructured second order nonlinear crystals to correct the accumulated phase mismatching of the participating waves along propagation length ($\Delta k \neq 0$)). This way, the reciprocal vector $G=2\pi/\Lambda$ introduced by a periodic structure with a period Λ compensates the wavevector mismatch $\Delta k=k_p-k_i-k_s-G=0$ experimented by interacting waves as a consequence of the dispersion in nonlinear materials. Although QPM efficiency is lower than the efficiency obtained by perfect phase matching, the conversion efficiency can be compensated by accessing to larger nonlinear coefficients.

The emission spectrum of Cr:LiCAF is particularly well suited for the above mentioned processes. An optical parametric oscillator (OPO) operating near degeneracy can be constructed using the Cr:LiCAF laser radiation as the OPO pump wave, obtaining, therefore, two tunable wavelengths (signal and idler) in the 1550 nm optical communications band. Similarly, it tailors to the pump wave characteristics participating in wavelength conversion processes based on DFG in the vicinity of the degeneration in the same spectral band.

The present section is not aimed to obtain high lasing and frequency conversion efficiencies but to show the potential of these devices and techniques in several applications and address future improvements.

5.1 Optical parametric oscillation in a Cr^{3+}:LiCAF+PPSLT laser

Tunable dual-wavelength lasers are of particular interest for generation of tuneable terahertz (THz) radiation based on heterodyne mixing in photo-mixers, and for WDM channel conversion based on four wave mixing (FWM) in semiconductor optical amplifiers (SOAs). In this regard optical parametric oscillators are, by their nature, highly coherent dual-wavelength tunable optical sources (Nabors et al. 1990; Lee and Wong, 1993). In this section we present a doubly resonant intra-cavity pumped optical parametric oscillator (OPO) consisting of a Cr:LiCAF tunable laser and a periodically poled stoichiometric lithium tantalate (PPSLT) nonlinear crystal. The OPO radiation (signal + idler) can be tuned by means of any perturbation of the phase-matching condition (i.e. by either changing temperature or pump wavelength). Moreover, the laser cavity design permits tuning of the Cr:LiCAF radiation, which is used as pump wave for the parametric oscillation process.

The experimental set-up is shown schematically in Fig. 8. A 665 nm fiber coupled laser diode (LD) is used as optical pump for the Cr:LiCAF. Light from the laser diode is collimated (CL) and then focused (FL) into a 3% at. doped 1.5 mm long 3 mm diameter Cr:LiCAF crystal. The laser crystal was a-cut for polarized oscillation as required for efficient quasi-phase matching (d_{33}). One side of the Cr:LiCAF (M1) has dielectric coatings directly deposited for broadband high reflectivity for both 793±20 nm and 1586±100 nm spectral bands, and on the other side it is antireflection coated (AR) for both 800 nm and

1580 nm spectral bands, in an attempt to minimize intracavity loss. A periodically poled stoichiometric lithium tantalate (PPSLT) crystal is used as the non linear material to downconvert the pump wave (Cr:LiCAF laser) into two parametric waves (signal and idler). The crystal is 20 mm long with a 2x2 mm^2 cross sectional area, has anti-reflection (AR) coatings on both sides (AR@750-850 nm and AR@1500-1700 nm) and a grating period of 22.1 μm for type-0 QPM interaction. It is mounted in an oven for temperature control of the QPM condition. M1 and the output mirror (M2) form the main cavity for laser (Cr:LiCAF) and parametric (PPSLT) oscillation and it is 10 cm radius of curvature concave mirror and partially reflective coated for R≈98.5%@770-830 nm and R≈96.5%@1450-1650 nm.

The output from the main cavity is collimated (CL) and PPSLT OPO radiation is separated from Cr:LiCAF emission by means of a dichroic beam splitter (BS). Cr:LiCAF laser beam is then expanded using an anamorphic prism pair (4x) (APP) and fed back to the main cavity by means of a diffraction grating (G) (1200 lines/mm, blazing@750 nm) in a Littrow configuration. Beam expansion prior to diffraction grating improves spectral resolution of the diffraction grating. OPO pump wave tuning is therefore achieved through tilting of the diffraction grating. This allows OPO pump tuning, narrows the laser linewidth and increases the intracavity laser power (due to an increased effective reflectivity). Under the experimental conditions of Fig. 8, the OPO pump was tunable from 775 to 830 nm (Fig. 9.a).

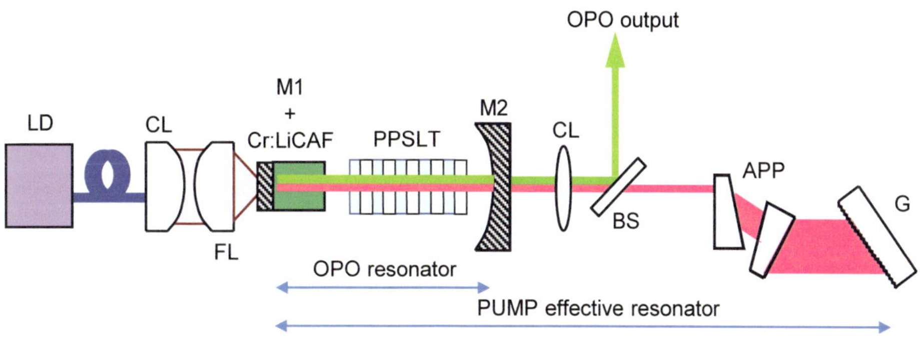

Fig. 8. Schematic of the intracavity pumped OPO experimental setup. LD: pump laser diode, CL: collimating lens, FL: focusing lens, M1: dielectric mirror directly deposited on the laser material input face, M2: concave mirror, BS: beam splitter, APP: anamorphic prism pair and G: ruled diffraction grating.

It is worth noting that there are two different cavities for Cr:LiCAF laser and OPO emission respectively. Parametric waves are resonated between M1 and M2 whereas Cr:LiCAF radiation is resonated between M1 and the effective reflectivity provided by M2 and the diffraction grating. If no external feedback is applied to the main cavity, Cr:LiCAF laser oscillates free-running with a relatively broad spectrum centered at 820 nm and more than 4 nm width, 650 mW of absorbed 665 nm power threshold and around 12% slope efficiency. This rather low slope efficiency is thought to be due to the high M^2 (beam quality factor) of the 665 nm pump system yielding a poor overlap between 665 nm pump beam and Cr:LiCAF laser cavity mode.

When optical feedback is applied, the Cr:LiCAF emission can be tuned and its spectrum is narrowed below 30 pm (optical spectrum analyzer limited). Fig. 9.b shows the relative output power of the Cr:LiCAF laser at 792 nm for different operation conditions. If PPSLT is not inserted in the main cavity, the absorbed 665 nm pump power threshold is approximately 400 mW (Fig. 9.b squares) and a slope efficiency lower than 1% is obtained. This additional reduction of the slope efficiency is a consequence of an increased effective reflectivity of the output coupling when external feedback is added.

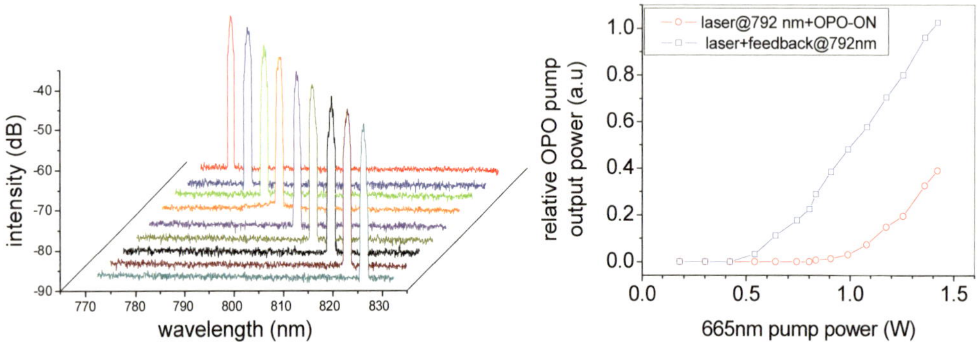

Fig. 9. Emission of the Cr:LiCAF laser: a) Wavelength tuning spectrum and b) output power under different operation conditions.

If the PPSLT crystal is placed with its extraordinary polarization parallel to the polarization of the laser (OPO ON, Fig. 9.b circles) and after adjusting the temperature of the non linear crystal, parametric generation takes place. As it can be observed, the 665 nm absorbed pump threshold power amounts to 750 mW and the slope efficiency is further reduced owing to the effect of residual reflectivity on faces of the PPSLT crystal and due to parametric generation. When parametric oscillation starts, it behaves as a wavelength dependant loss at the OPO pump wavelength (792 nm). Under this operation conditions the circulating Cr:LiCAF radiation is nearly 2 W but because of such OPO pump wavelength dependant loss, Cr:LiCAF 792 nm radiation coexists with 820 nm free-running radiation and, as the 665 nm power increases, both the fed back (792 nm) and the free-running (820 nm) radiation reach similar peak values. Thus, the available intra-cavity 792 nm power is sensitively below 2 W. As a result of this, a maximum OPO signal+idler output power of 10 mW is achieved. The OPO output power can be further enhanced by careful design of the mirror reflectivity around 800 nm, in order to fully suppress the free-running Cr:LiCAF oscillation (Stothard et al. 2009), and improving the beam quality factor (M^2) in order to improve the available intra-cavity OPO pump power.

After achieving parametric oscillation and starting from a degenerate OPO emission ($\omega_s \approx \omega_i$), the signal and idler waves were moved away from degeneracy by either temperature (Fig. 10. a) or pump wavelength (Fig. 10. b) tuning. Two examples of OPO emission spectrum for two different tuning conditions are shown in Fig. 10.c and in Fig. 10.d. Both tuning parameters allowed variable signal and idler wavelength separations between 0-250 nm (0-30 THz), limited by the reflectivity of the main cavity output mirror (M2) and the quasi-phase-matching condition. As it can be observed it follows the well-known parabolic relation between OPO output wavelength (signal and idler) and the tuning parameter (temperature or wavelength).

Because of operation near degeneracy (i.e. near the vertex of the parabola) the slope relating the change in OPO output wavelength to any tuning parameter is high and thus small changes in any tuning parameter produces a noticeable change in the OPO output wavelength (Lindsay et al., 1998; Wang et al. 2002), in contrast to OPOs working far from degeneracy. Therefore, if a fast signal-idler wavelength difference tuning is desired, pump wavelength tuning should be used instead of temperature tuning and, moreover, the faster the method employed for OPO pump wavelength tuning the faster will be the wavelength spacing between OPO signal an idler waves.

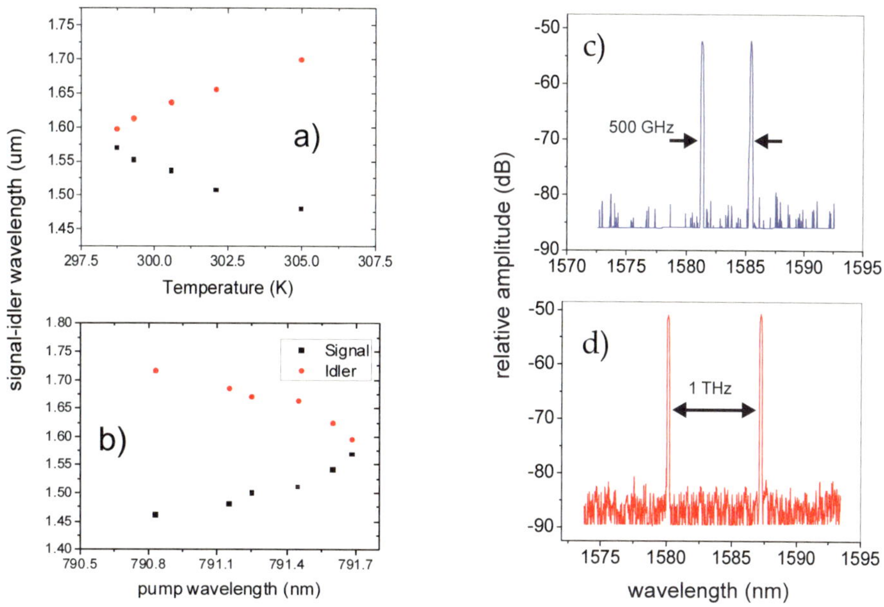

Fig. 10. Tuning of the dual-wavelength (signal+idler) OPO emission: a) temperature tuning, b) Cr:LiCAF laser tuning, c) and d) are examples of OPO emission spectrum with signal-idler frequency spacing of 0.5 THz and 1 THz respectively.

5.2 Difference frequency generation (DFG)

In recent years, the combination of the development of coherent sources based on nonlinear difference frequency generation (DFG) and the advances in the fabrication of nonlinear optical structures have led to cover new spectral regions and enhance spectral bands not efficiently developed in the infrared (IR) spectral region. In this context, continuous wave (CW) single-pass wavelength conversion of near and mid IR solid state lasers based on intracavity DFG in nonlinear media appears as an attractive approach to satisfy the present needs in optical communications (Yoo, 1996), sensing (Wysocki et al, 1994) or spectroscopy applications (Chen et al, 2006).

Difference frequency generation is produced when two waves, referred to as signal (at a frequency ω_s) and pump (an intense wave at a higher frequency ω_p), combine in a nonlinear medium to produce a new wave called idler at the difference frequency $\omega_i = \omega_p - \omega_s$ (after energy conservation fulfillment). The wave-mixing process results from the nonlinear interaction among the optical waves and the nonlinear material characterized by its susceptibility $\chi^{(2)}$, as it is schematically represented in Fig. 11.

Wavelength conversion processes are commonly characterized by the conversion efficiency defined in equation (3). But in practice, interacting waves have some finite spatial extent in the plane transverse to the direction of propagation, so the spatial confinement is taken into account through the Boyd-Kleinman factor h in such equation. This factor collects the effects produced by the interaction of focused gaussian beams in nonlinear media (Boyd and Kleinman, 1968) depending on the material properties as walk-off and absorption, the phase mismatching, and the focusing conditions experimented by the participating waves. A trade-off between the confinement and the interaction length is required to guarantee the optimal overlapping between modes along the nonlinear, and thus enhancing the conversion efficiency. In this sense, type-0 QPM interactions, in which all participating waves have identical polarization (extraordinary), avoids the presence of walk-off and enables the access to the largest coefficient d_{33} in the second-order nonlinear tensor, leading to more efficient processes.

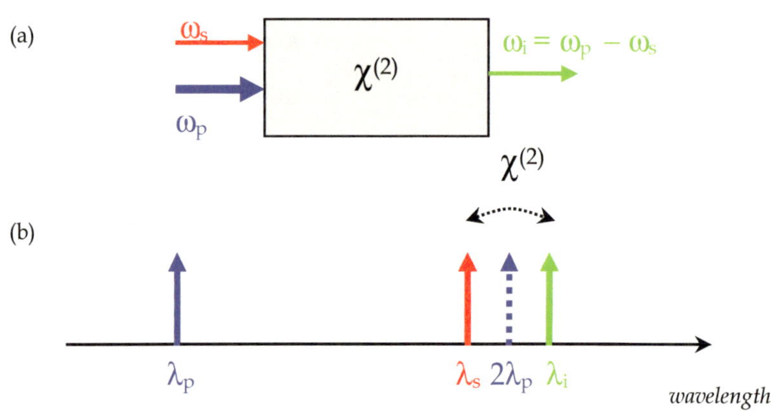

Fig. 11. Difference frequency generation (a) Diagram of the wave generation ω_i from the mixing of an intense pump wave ω_p and a weak wave ω_s in a nonlinear medium $\chi^{(2)}$.
(b) Spectral allocation of components involved in the wavelength conversion process.

Since the nonlinear mixing is proportional to the pump intensity, the efficiency of continuous-wave (CW) optical frequency generation can be increased by placing the nonlinear crystal inside the resonant cavity where the pump wave is generated. Then, the power level of the converted wave can be improved some magnitude orders in comparison with the provided by external processes. This way, the efficiency of DFG processes based on crystals with low nonlinear coefficient can be also compensated by the presence of high intracavity pump power. On the other hand, the conversion efficiency also describes sum and second harmonic generation similarly to that obtained in DFG. However, unlike the formers it is to be noted that when DFG occurs, not only idler photons are generated, but also new signal photons are created from the decomposition of pump photons as a consequence of the energy conservation fulfilment, and thus providing the amplification at both signal and idler wavelengths.

Unlike other conversion techniques, DFG preserves both phase and amplitude features of the incoming signal, providing a strict degree of transparency. In addition to this, DFG enables the conversion of multiple input wavelengths simultaneously what makes it an ideal technology to be exploited in the next generation of optical converters in communication applications (Yoo, 1996). This performance is achieved thanks to the wide spectral bandwidth experimented by the signal and idler waves in QPM devices (Fejer et al, 1992), which also can be extended for broadband conversion as it is seen in the next section. In order to show the potential of this technique, a flexible and versatile wavelength converting system for optical telecom signals based on single-pass DFG in a PPSLT crystal placed inside a CW tuneable solid-state laser is described in the following lines. In such applications, wavelength conversion usually covers the spectral region ranging from 1500 to 1700 nm, so the generation of a pump wave around 800 nm is required.

On the left of Fig.12 it is represented the theoretical QPM tuning curves in 1500-1700 nm band (for both signal and idler waves) as a function of the PPSLT temperature and for different pump wavelengths (792, 793 and 794 nm) from Sellmeier equations (Bruner et al., 2003). Furthermore, conversion efficiency curves for the operation temperatures (28, 41 and 53°C) are shown in detail on the right side from extracted slides. So, given a fixed external signal, it can be seen how idler wave is shift to higher wavelengths (up to 8 nm) as pump wavelength is increased.

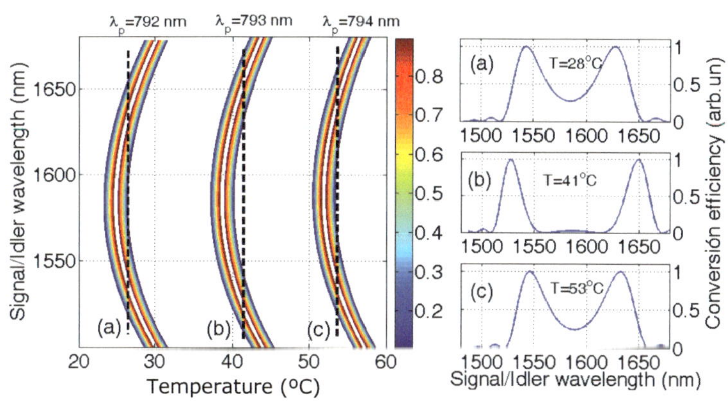

Fig. 12. QPM tuning curves for the PPSLT crystal (L=20 mm and Λ=22.1μm) for different operation temperatures: (a) T=28°C, (b) T=41°C, (c) T=53°C.

The experimental set-up of the wavelength converting system is schematically shown in Fig. 13. The laser cavity consists of a two-arm V-folded main cavity defined by three mirrors (M1, M2 and M3). A 3 mm long a-cut end-pumped Cr^{3+}:LiCAF crystal rod doped with 3 at.% of chromium, and with a 3 mm of diameter is employed as the gain medium. Mirror M1 is coated directly on the input facet of Cr^{3+}:LiCAF crystal whereas mirrors M2 and M3 are plano-concave spherical mirrors arranged in a confocal configuration to produce a beam waist in the middle of the second arm, where the center of the PPSLT is placed. Thus, tightly confined signal and pump beams can be easily made to overlap in a collinear interaction thus improving the conversion efficiency. The PPSLT is 20 mm long with a 2 × 2 mm^2 cross sectional area and a grating period of 22.1 μm, which is mounted in an oven for temperature control of QPM conditions and to reduce potential photorefractive effects. Self-injection-locking is provided by an external coupled cavity based on a Littrow-mounted ruled diffraction grating which allows for tuneable oscillation around 792±8 nm with a spectral width narrower than 1 nm. Then, the pump wave is collinearly mixed with a signal generated from a tunable multi-wavelength erbium doped fiber laser whose output, comprised between 1530 and 1570 nm, is amplified and partially polarized prior to its introduction into the laser cavity in order to enhance the conversion efficiency.

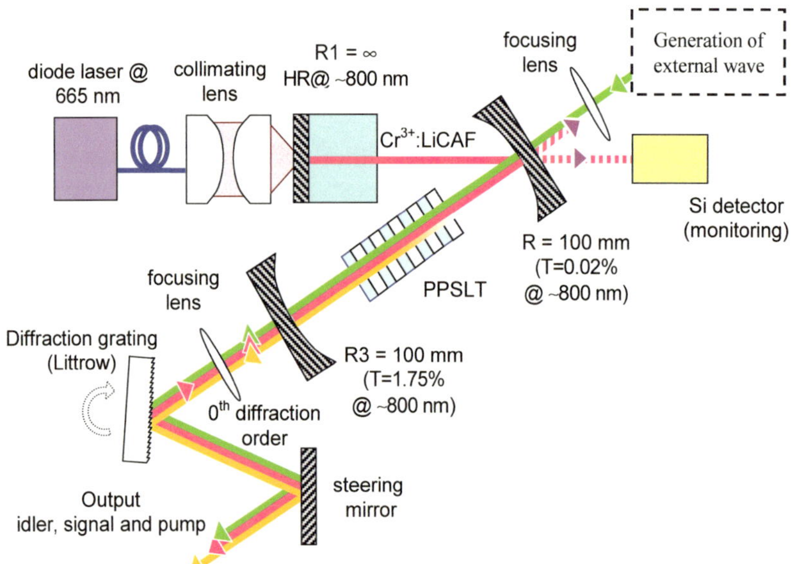

Fig. 13. Experimental setup for DFG processes.

Simultaneous multi-wavelength conversion is represented in Figure 14 for several arbitrary wavelengths (channels) and combinations at 1547, 1551 and 1555 nm. A pump wavelength at 794 nm (represented in Fig. 14 as OSA second order as a spectral reference) is used in the mixing process when the PPSLT crystal is heated up to 53°C to satisfy QPM condition. Idler waves generated at ~1631, 1627 and 1623 nm, comprising a spectral range of 7.8 nm where no more than a 4% of penalty with regard to the maximum QPM theoretical conversion efficiency is experimented, so in this case QPM condition does not suppose any limiting factor. The spectral allocation of the converted spectrum can be tuned by tilting the grating thanks to the characteristic broadband gain of Cr^{3+}:LiCAF (Payne et al., 1988) and by temperature adjustment of the PPSLT crystal to satisfy the required QPM condition.

Although the experimental conditions available restricted the laser operation to only 1.5 times above threshold with a modest intra-cavity power around 2 W, improving the laser design to approach the typical performances achievable with diode-pumped Cr3+:LiCAF lasers (Demirbas et al., 2009) should lead to high conversion efficiencies, even in excess of 100% if the laser is carefully designed to achieve net parametric gain at reasonably feasible values of intra-cavity power. Due to the nature of the DFG process, the conversion efficiency becomes independent of the input signal power and it is therefore suitable for weak input signals.

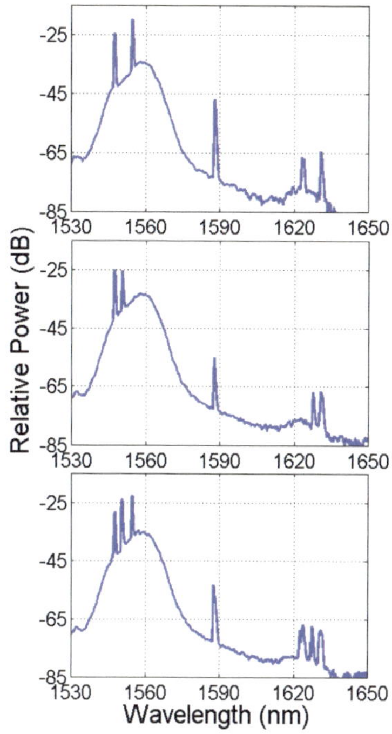

Fig. 14. Simultaneous wavelength conversion.

5.3 Broadband conversion

The emergence of applications which require the use of broad spectral regions where there are not optical sources or they have not been efficiently developed have made grow the interest in broadband wavelength conversion techniques recently. The availability of broadband optical sources is interesting in a variety of applications which ranges from wavelength division multiplexing (WDM), passive optical networks (PON) based on spectral slicing techniques (Jung et al., 1998), interferometric sensors (Wysocki et al., 1991), and biomedicine such as optical coherence tomography (Schmitt, 1999).

For this aim, the intracavity broadband conversion based on DFG processes in QPM devices can be an attractive and effective alternative to those techniques based on the generation of nonlinear processes such as cross-gain or cross-phase modulation (XPG, XPM) and four wave mixing (FWM) in semiconductor optical amplifiers (SOAs) following interferometric arrangements and highly nonlinear fiber configurations (Bilenca et al., 2003). This scheme exploits not only the advantages of active conversion, but also the wide spectral acceptance of the signal/idler waves in periodically poled ferroelectric crystals. Furthermore, due to the DFG nature, the spectrum of the converted wave undergoes the spectral inversion with regard to the pump wave while the spectral fidelity is preserved. This opens up a way for wavelength conversion of weak broadband sources since no threshold is required, as it is derived from equation (3).

Given a broadband input signal with a bandwidth of B and centered at ω_s, the allocation of the targeted idler spectrum depends on the spectral position of the pump wave ω_p. In this way, a proper choice of the pump wavelength is required to satisfy the quasi-phase matching condition imposed by the nonlinear crystal features, and then, different performances can be obtained; a broadband signal can be displaced to a spectral region where no coherent source is available (Fig. 15.a), a spectrally variable bandwidth up to 2B can be achieved by tuning the pump wave so that $\omega_p/2$ falls in the vicinity of the spectral components delimiting the signal band (Fig. 15.b), and spectral shaping from total or partial overlap of involved spectra provided by QPM structures (Fig. 15.c). In this case, domain engineering in nonlinear ferroelectric crystals could provide particular QPM tuning curves that lead to dynamic and flexible resultant responses depending on the overlapping between the signal and idler spectra.

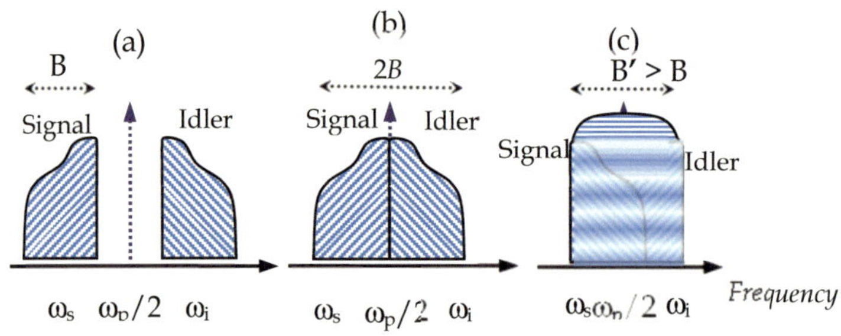

Fig. 15. Broadband conversion operation modes based on DFG from a pump wave at ωp and a broadband signal wave centered at ωs with a spectral bandwidth of *B*.

Intra-cavity assisted broadband wavelength conversion for an incoherent continuous-wave input source (ASE, amplified spontaneous emission source) has been recently demonstrated in the spectral region around 1535-1635 nm (Torregrosa et al., 2011). Following an experimental setup similar to that proposed in the previous section, Fig. 16 shows the broadband conversion results when an ASE source based on a diode-pumped Erbium-doped fiber is used. The output of the ASE source, which is spectrally extended from 1530-1565 nm, is collinearly mixed with a pump wave tuned at 794 nm (represented as OSA second order as a spectral reference) in a PPSLT crystal at 53°C. As a result of the mixing process, a broadband idler ranging from 1605 to 1635 nm is generated, in which spectral inversion is revealed clearly. Furthermore, a shift up to 8 nm of the broadband spectrum is achieved by tuning the pump wavelength accordingly with the crystal temperature to satisfy QPM condition, as it is shown in the inset of Fig. 16.

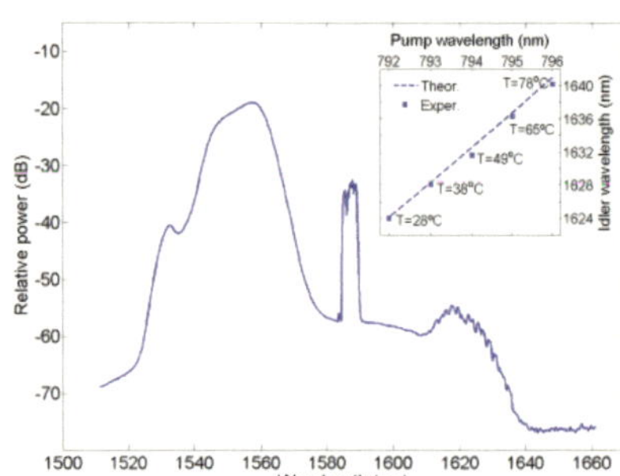

Fig. 16. Power spectra of the measured broadband conversion process

Despite the very limited experimental conditions available with a modest intra-cavity power laser operating only 1.5 times above threshold, it is shown that improving the laser design should lead to a high conversion efficiency (even in excess of 100%). Fortunately, the efficiency can be improved by increasing the intracavity pump power since the efficiency is proportional to the pump wave intensity according to equation (3). Thus, the converting scheme here described emerges as an attractive and versatile tool for extending the availability of outstanding ASE sources to spectral regions other than their natural emission spectrum using a simple periodically poled crystal.

6. Conclusions and future work

Spectral caracteristics of solid-state lasers based on Cr-doped colquiriites allow to select the emission wavelength from a wide gain spectrum, and can be pumped by comercially available laser diodes. Owing to its spectral emission in the vecinity of 800 nm, they can be employed in intracavity nonlinear conversion to generate radiation in the 1500-1700 nm spectral band where DFG and OPO have not been studied in detail so far, in contrast to the SHG case.

Intracavity wavelength conversion is characterized by presenting conversion efficiencies typically two orders of magnitude greater than those offered by external processes. However, the results obtained present reduced conversion efficiencies due to the restrictions imposed by experimental conditions available, such as the limited 665 nm pump power, and mainly, the inefficient adaptation between the 665 nm pump mode and laser oscillation mode as a result of the poor quality of the former (M^2>>1). From the spectral point of view, the performance obtained reveals the advantages and versatility to operate in a huge variety of applications.

Since QPM is associated to the use of ferroelectric crystals, such as lithium or tantalate niobate, by strategic domain distributions it is possible to provide QPM tuning responses associated to particular applications in addition to the ability to operate in regimes not accessible to conventional birefringently phasematched media.

Finally, although the experimental setups of the converting systems here presented appear as bulky schemes (as a result of the space required for handling, measuring and characterizing the system performance), the underlying technology to diode pumped solid-state lasers enables the implementation of compact devices by taking advantage of the microchip laser technology (Zayhowski, 1997). A further compactness could be achieved by doping with Cr^{3+} ions into the nonlinear medium, as it has been previously demonstrated in periodically poled ferroelectric crystals (Capmany et al., 2000). This way, the laser oscillator and the nonlinear optical converter would be integrated in the same crystal, leading to promising laser schemes as self-frequency converters.

7. References

Balembois, F.; Georges, P.; Salin, F.; Roger, G. and Brun, A. (1992), Tunable blue light source by intracavity frequency doubling of a Cr-doped LiSrAlF6 laser. *Applied Physics Letters.*, Vol. 61, No. 20, pp. 2381–2383. ISSN: 0946- 2171 (Print) ISSN: 1432-0649 (Online).

Belmonte M., Skettrup T.; Pedersen C.; (1999), Frequency doubling in LiNbO3 using temperature-dependent QPM, *Journal of Optics A-Pure And Applied Optics*, Vol. 1, No. 1, pp. 60-63, ISSN (printed): 1464-4258. ISSN (electronic): 1741-3567.

Bilenca A.; Alizon R.; Mikhelashhvili V.; Dahan D.; Eisenstein G.; Rchwertberger R.; Gold D.; Reithmaier J. P. and Forchel A., (2003), Broad-Band Wavelength Conversion Based on Cross-Gain Modulation and Four-Wave Mixing in InAs-InP Quantum-Dash Semiconductor Optical Amplifiers Operating at 1550 nm, *IEEE Photonics Technology Letters*, Vol. 15, No. 4, pp. 563-565, ISSN: 1041-1135.

Boyd G. D. and Kleinman, D. A., (1968), Parametric Interaction of Focused Gaussian Light Beams, *Journal of Applied Physics*, Vol. 39, No. 8, pp. 3597-3641, Print: ISSN 0021-8979, Online: ISSN 1089-7550.

Bruner A.; Eger D.; Oron M. B.; Blau P.; Katz M. and Ruschin S., (2003), Temperature-dependent Sellmeier equation for the refractive index of stoichiometric lithium tantalate, *Optics Letters*, Vol. 28, pp. 194-196, ISSN: 0146-9592 (print), ISSN: 1539-4794 (online).

Buse K.; Imbrock J.; Krätzig E. and Peithmann K., (2007), Photorefractive Effects in LiNbO3 and LiTaO3, In: *Photorefractive Materials and Their Applications 2*, P. Günter, J. P. Huignard, pp. 83-126, Ed. Springer, ISBN 978-0387-33924-5, USA.

Capmany, J.; Montoya, E.; Bermúdez, V.; Callejo, D.; Diéguez, E. and Bausá, L. E., (2000), Self-frequency doubling in Yb3+ doped periodically poled LiNbO3:MgO bulk crystal, *Applied Physics Letters*, Vol. 76, No. 11, pp. 1374-1376, ISSN 0003-6951 (print), ISSN 1077-3118 (online)

Chen W.; Cousin J.; Poullet E.; Boucher D.; Xiaoming G.; Sigrist M.W. and Tittel F.K., (2006), Laser Difference-Frequency Generation in the Mid-Infrared and Applications to High-Resolution Molecular Spectroscopy and Trace Gas Detection, *Joint 31st International Conference on Infrared Millimeter Waves and 14th International Conference on Terahertz Electronics*, 2006. pp. 583.

Demirbas U.; Li D.; Birge J. R.; Sennaroglu A.; Petrich G. S.; Kolodziejski L. A.; Kärtner F. X. and Fujimoto J. G., (2009), Low-cost, singlemode diode-pumped Cr:colquiriite lasers, *Optics Express*, Vol. 17, No. 16, pp. 14374–14388, ISSN: 1094-4087.

Dolev I.; Ganany-Padowicz A.; Gayer O.; Arie A.; Mangin J. and Gadret G. (2009). Linear and nonlinear optical properties of MgO:LiTaO3, *Applied Physics B*, Vol. 96, No 2-3, pp. 423-432. ISSN: 0946- 2171 (Print) ISSN: 1432-0649 (Online).

Eichenholz, J. M. and Richardson, M. (1998). Measurement of Thermal Lensing in Cr-Doped Colquiriites, *IEEE Journal of Quantum Electronics*, Vol. 34, No. 5, pp. 910-919, ISSN: 0018-9197.

Eichenholz, J. M.; Richardson, M. and Mizell, G. (1998), Diode-pumped, frequency doubled LiSAF microlaser. *Optics Communications*, Vol. 153, No. 5, pp. 263–266. ISSN: 0030-4018.

Fejer M. M.; Magel G. A.; Jundt D. H and Byer R. L., (1992) Quasi-phase-matched second harmonic generation: tuning and tolerances, *IEEE Journal of Quantum Electronics*, Vol. 28, No. 11, pp. 2631–2654 (1992), ISSN: 0018-9197.

Fernandez-Pousa, C. R. and Capmany, J., (2005), Dammann grating design of domain-engineered lithium niobate for equalized wavelength conversion grids, *IEEE Photonics Technology Letters*, Vol. 17, No. 5, pp. 1037-1039, ISSN: 1041-1135.

Jung D. K.; Shin S. K.; Lee C. H. and Chung Y. C., (1998), Wavelength-division-multiplexed passive optical network based on spectrum-slicing techniques, *IEEE Photonics Technology Letters*, Vol. 10, No. 9, pp. 1334-1336, ISSN: 1041-1135.

Laperle, P.; Snell, K. J.; Chandonnet, A.; Galarneau, P. and Vallée, R. (1997), Tunable diode-pumped and frequency-doubled Cr:LiSAF lasers. *Applied Optics*, Vol. 36, No. 21, 5053–5057. ISSN: 1559-128X (print), ISSN: 2155-3165 (online).

Lee, D. and Wong, N. C. (1993), Stabilization and tuning of a doubly resonant optical parametric oscillator, *Journal of the Optical Society of America B*, Vol. 10, No. 9, pp. 1659-1667, ISSN: 0740-3224 (print), ISSN: 1520-8540 (online).

Lindsay, I. D.; Turnbull, G. A.; Dunn, M. H. and Ebrahimzadeh, M. (1998). Doubly resonant continuous-wave optical parametric oscillator pumped by a single-mode diode laser, *Optics Letters*, Vol. 23, No. 24, pp. 1889-1891, ISSN: 0146-9592 (print), ISSN: 1539-4794 (online).

Liu, H; Zhu, YY; Zhu, SN; Zhang, C; Ming, NB, (2001), Aperiodic optical superlattices engineered for optical frequency conversion, *Applied Physics Letters*, Vol. 79, No. 6, pp. 728-730, ISSN 0003-6951 (print), ISSN 1077-3118 (online).

Nabors, C. D.; Yang, S. T.; Day, T.; Byer, R. L. (1990), Coherence properties of a doubly resonant monolithic optical parametric oscillator, *Journal of the Optical Society of America B*, Vol. 7, No. 5, pp. 815-820, ISSN: 0740-3224 (print), ISSN: 1520-8540 (online).

Maiman, T. H. (1960). Stimulated Optical Radiation in Ruby. *Nature*, Vol. 187, pp. 493-494. ISSN: 0028-0836

Makio, S.; Miyai, T.; Sato, M. and Sasaki, T. (2000) 67 mW Continuous-wave blue light generation by intracavity frequency doubling of a diode pumped Cr:LiSrAlF6 laser. *Japanese Journal of Applied Physics*, Vol. 39, pp. 6539–6541, ISSN: 0021-4922 (print), ISSN: 1347-4065.

Payne, S. E.; Chase, L. L.; Newkirk, H. W.; Smith, L. K. and Kupke W. F., (1988), LiCaAlF6:Cr3+: A promising New Solid Satate Laser Material, *IEEE Journal of Quantum Electronics*, Vol. 24, No. 11, pp. 2243-2252, ISSN: 0018-9197.

Payne, S.; Chase, L.; Smith, L.; Kway, W. and Newkirk, H. (1989). Laser performance of LiSrAlF6:Cr3+, *Journal of Applied Physics*, Vol. 66, No. 66, pp. 1051-1056. , Print: ISSN 0021-8979, Online: ISSN 1089-7550.

Schmitt J. M., (1999). Optical Coherence Tomography (OCT): A Review, *IEEE Journal of Selected Topics in Quantum Electronics*, Vol. 5, No. 5, pp. 1205-1215, ISSN: 1077-260X.

Skvortsov, L. A. and Stepantsov, E. S. (1993), Laser damage resistance of a lithium niobate-tantalate bicrystal system, *Quantum Electronics*, Vol. 23, No. 11, pp. 981-982. ISSN 1063-7818 (Print). ISSN 1468-4799 (Online).

Smith, R. G., (1970), Theory of Intra-cavity Optical Second-Harmonic Generation, *IEEE Journal of Quantum Electronics*, Vol. 6, No. 4, pp. 215-223, ISSN: 0018-9197.

Smith, L.; Payne, S.; Kway, W.; Chase, L.; Chai, B. (1992). Investigation of the laser properties of Cr3+:LiSrGaF6, *IEEE Journal of Quantum Electronics*, Vol. 28, No. 11, pp. 2612 - 2618, ISSN: 1077-260X.

Stothard, D. J.; Hopkins, J-M.; Burns, D. and Dunn, M. H. (2009). Stable, continuous-wave, intracavity, optical parametric oscillator pumped by a semiconductor disk laser (VECSEL), *Optics Express*, Vol. 17, No. 13, pp. 10648-10658, ISSN: 1094-4087.

Torregrosa, A. J. ; Maestre, H. ; Fernández-Pousa, C. R. and Capmany J., (2008) Electro-Optic Reconfiguration of Quasi-Phase Matching in a Dammann Domain Grating for WDM Applications, *Conference on Lasers and Electro-Optics/Quantum Electronics and Laser Science Conference and Photonic Applications Systems Technologies*, paper CWC1.

Torregrosa, A. J. ; Maestre, H. and Capmany, J., (2011) Wavelength conversion of a broadband ASE source based on intra-cavity difference-frequency-mixing in a Cr3+:LiCAF-PPSLT laser, *The European Conference on Lasers and Electro-Optics (CLEO/Europe)*, paper CD_P23.

Tsunekane, M.; Ihara, M.; Taguchi, N. and Inaba, H. (1998). Analysis and Design of Widely Tunable Diode-Pumped Cr:LiSAF Lasers with External Grating Feedback, *IEEE Journal of Quantum Electronics*, Vol. 34, No. 7, pp. 1288-1296, ISSN: 1077-260X.

Wang, H.; Ma, Y.; Zhai, Z.; Gao, J.; Xie, C. and Peng, K. (2002). Tunable Continuous-Wave Doubly Resonant Optical Parametric Oscillator by Use of a Semimonolithic KTP Crystal, *Applied Optics*, Vol. 41, No. 6, pp. 1124-1127. ISSN (printed): 1464-4258. ISSN (electronic): 1741-3567.

Wysocki P. F. ; Digonnet M.J.F. ; Kim B. Y. ; Shaw H. J. (1994), Characteristics of Erbium-doped Superfluorescent Fiber Sources for Interferometric Sensor Applications, *IEEE Journal of Ligthwave Technology*, Vol. 12, No. 3, pp.550-567, ISSN: 0733-8724.

Xu F. ; Liao J. ; Wang Q. ; Du J. ; Xu Q. ; Liu S. ; He J.L. ; Wang H.T. ; Ming N.B. (2003), Electro-optically controlled efficiencies in a QPM coupled parametric process, *Applied Physics B: Lasers and Optics*, Vol. 76, No. 7, pp. 797-800, ISSN: 0946- 2171 (print version) ISSN: 1432-0649 (electronic version).

Yoo S.J.B., (1996) Wavelength conversion technologies for WDM network applications, *IEEE Journal of Ligthwave Technology*, Vol.14, No. 6, pp. 955-966, ISSN: 0733-8724.

Zayhowski, J. J. (1997), Microchip optical parametric oscillators, *IEEE Photonics Technology Letters*, Nol.9, No.7, pp.925-927, ISSN: 1041-1135.

7

Frequency Upconversion in Rare Earth Ions

Vineet Kumar Rai
Laser and Spectroscopy Laboratory, Department of Applied Physics
Indian School of Mines, Dhanbad, Jharkhand,
India

1. Introduction

A large number of solid host materials have been considered to study the consequence of host materials on the lasing properties of the active ions since the development of the solid state lasers. With increasing interest in the photonic devices the appeal for employing photonic glasses has grown much interest because they may be chemically and mechanically stable hence being ideal for substituting the crystalline systems in many conditions. Rare earth doped solid materials are of great importance due to their wide range of applications in lasers, temperature sensors, optical amplifiers, etc. among other devices [Savage, 1987; Digonnet, 1993; Jackson, 2003; Rai, 2007]. Study of the spectroscopic properties of rare earth doped systems is very important from several points of view. The effect of environment on the energy levels, variations in their emission and absorption characteristics, upconversions observed in near infrared (NIR), visible pumping etc. can easily be studied from these. The longer lifetime of the energy levels is one of the prime requirements for lasing. Lasing in these rare earth ions occur due to the transition between the spin-orbit splitted components of the same state or between the states arising from the unfilled (4f) electronic configuration, which are usually forbidden as per electric dipole selection rules. The electric dipole transitions between states of the 4f configuration for free ions are strictly forbidden by the parity rule. The transitions can take only due to the mixing of states of opposite parity in the wave functions. The optical properties of rare earth ions doped solid host materials depend on the chemical composition of the host materials [Reisfield, 1973; Babu et al., 2000; Dumbaugh, 1992; Rai, 2006; Dieke et al., 1968; Ebendroff, 1996]. Rare earth ions are found to be very much sensitive to small changes in chemical surroundings. On the other hand network modifiers affect the local environment around the fluorescent ions and thereby its optical properties. Therefore, it becomes essential to get the information about the symmetry and bonding of the probe ion and how they change their optical properties with chemical composition of the solid host materials and also from site to site variations. In the solid host materials with higher phonon frequencies, it is not easy to get efficient infrared and visible emission even in the Er^{3+}, Nd^{3+}, Ho^{3+}, Tm^{3+}, etc. The effect of host materials on the optical properties of rare earth ions introduced as doping along with compositional changes of modifiers have been studied by different workers [Babu et al., 2000; Rai et al., 2006; Nageno et al., 1994; Tripathi et al., 2006; Rai, 2010; Rai et al., 2008; Sai Sudha et al., 1996; Lin et al., 2002; Hussain et al., 2000; Nachimuthu et al., 1997; Tripathi et al., 2006]. They concluded that due to their improved radiative transition probabilities, these glasses are promising materials for different technological applications.

For getting the high quantum efficiency, concentration of the rare earth ions should be high, but it may cause certain problems like concentration quenching due to the interaction between the excited and its unexcited neighbours. Therefore, in order to make the devices with high optical characteristics, the concentration of the rare earth ions has to be kept low so that the luminescence quenching is minimized. Specifically, the systems of main interest are those having low cut-off phonon frequencies, high refractive index etc. because in such systems the quantum efficiency for the rare earth ions introduced as doping is improved significantly [Yamane et al., 2000]. It is also possible to prevent the quenching effect by modifying the environment felt by the luminescent ions [Rai, 2010; Snoeks et al., 1995; Kumar et al., 2005]. Therefore, the glasses containing metallic nanoparticles doped with low concentration of rare earth ions are of particular interest because the large local field acting around the rare earth ions positioned near the nanoparticles may increase the luminescence efficiency when the frequency of the excitation beam and or / the luminescence frequency are near resonance with the surface plasmon frequency of the nanoparticles [Kumar et al., 2005]. There is a large demand in generation of visible light sources for different photonics applications. One effective way for generating visible light is the frequency upconversion in which the absorption of two or more photons is followed by the emission of high energy photon. Rare earth ions are the appropriate candidates for the frequency upconversion (UC) processes owing to their large number of energy levels, narrow emission spectral lines, long lifetime of excited states and they can be easily populated by the near-infrared radiation.

The study of frequency upconversion luminescence in lanthanides doped solid materials is mainly due to their potential applications in different areas such as color display, two photon imaging in confocal microscopy, high density optical data storage, upconvertors, under sea communications, IR lasers, indicators, temperature sensors, biomedical diagnostics etc. The main interest in recent years is concentrated in the identification of new upconversion mechanisms and their characterization in rare earth doped materials. Generally, the frequency upconversion may arise by any of the following mechanisms (a) A multiphoton absorption (b) Energy transfer interaction (c) Excited state absorption (d) Avalanche energy transfer.

Excited state absorption (ESA) involves single ion, which depends on multistep absorption in individual ion, while the frequency upconversion through energy transfer between groups of rare earth ions can be very efficient and therefore can be used in making upconversion lasers [Yamane et al.,2000; Kassab et al., 2007; Silva et al., 2007]. The upconversion phenomenon are motivated not only by attempts to enrich the fundamental knowledge for understanding the mechanism of interaction between the rare earth ions doped into a variety of host materials but also the search for new light sources emitting in the visible region. Keeping in mind the attention has been paid to solid host materials doped with rare earth ions that show strong and broad absorption bands matching with the emission wavelength of commercial diodes [Yamane et al., 2000; Kaminskii, 1990; Kaplyanskii, 1987; Kenyon, 2002; Lezama et al., 1985].

Solid host materials based on heavy metal oxides are of particular interest for different photonics applications due to their low cut -off phonon energy and large refractive index

[Rai et al., 2008; Yamane et al., 2000]. In hosts with low cut-off phonon energies, the luminescence intensity of rare earth ions is enhanced by several times [Digonnet 1993; Rai et al., 2006; Kenyon, 2002]. This enhancement is neither due to any external field nor a Maxwell field, which was explained in detail very early on by Lorentz, but due to the local field which depends on the presence of all the other atoms in the dielectric [Born et al., 1999]. Inorganic glasses have been used as optical materials for a long time due to their isotropy, hence to make the large size and high optical homogeneity more easily, and high transparency over a wide spectral range from ultraviolet to infrared, as well as linear functional properties, high mechanical strength, high chemical resistance and simple fabrication procedures for obtaining good optical quality samples. The host materials of main interest are those for which the quantum efficiency of rare earth ions luminescence is enhanced, allowing the development of more efficient lasers, upconvertors, optical amplifiers, etc. [Yamane et al., 2000; El-Mallawany, 2001]. The present chapter deals with the frequency upconversion in some of the lanthanides, specially Pr^{3+} and Nd^{3+} doped solid host materials and different excitation processes, mainly, excited state absorption (ESA) and the energy transfer (ET) processes responsible for the luminescence in the rare earth ions.

2. Frequency upconversion in triply ionized lanthanides doped in solid host materials

The frequency upconversion 'conversion of low energy photon into higher energy photon' emission is receiving significant attention due to a variety of photonic applications. Glassy materials doped with rare earth ions are capable of high peak power generation due to their high saturation frequencies, broad emission bandwidths etc. The oxide glasses have compatibility with waveguide fabrication process and their optical damage threshold is quite high. They have high chemical and thermal stability and are thus suitable materials for technological applications. However, the high phonon energy corresponding to the stretching vibrations of the oxide glass network modifiers creates difficulty in generation of upconversions. The frequency upconversion in rare earth ions doped fluoride, ZBLAN and ZBS based glasses has also been observed, but due to its hygroscopic nature their applications are limited [Oliveria et al., 1998; Parker, 1989]. The frequency upconversion in lanthanides doped solid host materials has been studied by several workers [Pacheco et al., 1988; Pelle et al., 1993; Mujaji et al., 1993; Victor et al., 2002; Rai et al., 2006; Rai et al., 2007; Rai et al., 2008; Rai et al., 2008; Rai, 2009; Tripathi et al., 2007; Mohanty et al., 2011; Balda et al., 1999; Ganem et al., 1992; Smart et al., 1991; Luis et al., 1994].

2.1 Frequency upconversion in Pr^{3+} doped glasses

Triply ionized praseodymium doped glasses and crystals are one of the most extensively studied systems because of their wide applications in many optical devices. The Pr^{3+} contains several metastable states that provides the possibility of simultaneous emissions in the blue, green, orange, red and infrared (IR) regions [Kaminskii, 1990]. Upconversion (UC) emission has been observed in Pr^{3+} doped different glasses as well as crystal lattices [Rai et al., 2006; Rai et al., 2007; Rai et al., 2008; Rai et al., 2008; Rai, 2009; Balda et al., 1999; Ganem et al., 1992; Smart et al., 1991; Luis et al., 1994]. Among the heavy metal oxide (HMO) glasses the ones based on tellurium oxide (TeO_2) are important candidates to be used in practical devices because they present high mechanical strength, high chemical durability, high optical damage threshold, and small absorption coefficient over the visible and near infrared regions of the electromagnetic spectrum. Furthermore, in general these glasses are nonhygroscopic and can be fibered. Frequency upconversion in Pr^{3+} doped tellurite based glasses has been studied by different workers [Rai et al., 2006; Rai et al., 2007; Rai et al., 2008; Rai et al., 2008; Rai, 2009; Balda et al., 1999]. For excitation of the samples the NIR radiation from a Ti-sapphire laser pumped by second harmonic of Nd:YVO_4 (yttrium vanadate) (CW) laser [Rai et al., 2006; Rai et al., 2007] as well as a Nd:YAG (yttrium aluminum garnet) laser pumped dye laser (8 ns pulses, tunable from 570 to 600 nm) was used [Rai et al., 2007; Rai et al., 2008; Rai, 2009; Balda et al., 1999] as an excitation source. The excitation beam was focused onto the sample with a lens and the fluorescence was collected in a direction perpendicular to the incident beam; the signal was analyzed by a monochromator attached with a photomultiplier tube. All the measurements were made at room temperature.

2.2 (a) Excited state absorption (ESA)

The upconversion emission in triply ionized praseodymium doped TeO_2-Li_2O glass upon CW excitation from a Ti-sapphire laser has been reported [Rai et al., 2007]. The absorption spectrum of 0.5 mol% of Pr^{3+} doped tellurite glass shows four absorption bands in the 400–800 nm region (Fig. 1). These absorptions bands are mainly due to the transitions from the ground ($^3H_4$) state to the $^3P_2$, $^3P_1$, $^3P_0$ and $^1D_2$ excited states. Upconversion emission in the Pr^{3+} doped TeO_2-Li_2O glasses upon excitation at ~816nm from a Ti-sapphire laser has been observed in the blue, orange and red regions corresponding to the $^3P_2 \rightarrow {}^3H_4$, $^3P_0 \rightarrow {}^3H_4$, $^3P_1 \rightarrow {}^3H_5$, $^3P_0 \rightarrow {}^3H_5$, $^1D_2 \rightarrow {}^3H_4$ and $^3P_0 \rightarrow {}^3F_2$ transitions. Thus upconversion at several wavelengths can be obtained by pumping with single NIR wavelength. It is well known that the upconversion luminescence intensity (I_0) is proportional to the nth power of the pump intensity (I_i), where 'n' is the number of pump photons required to populate the emitting level. To ascertain the number of photons involved in the upconversion process the intensity of the emitted band was measured as a function of NIR radiation power. The slope of the graph $\log I_0$ versus $\log I_i$ gives a value ~1.93 for $^3P_0 \rightarrow {}^3F_2$ transition, indicating that two photons are involved in the upconversion process.

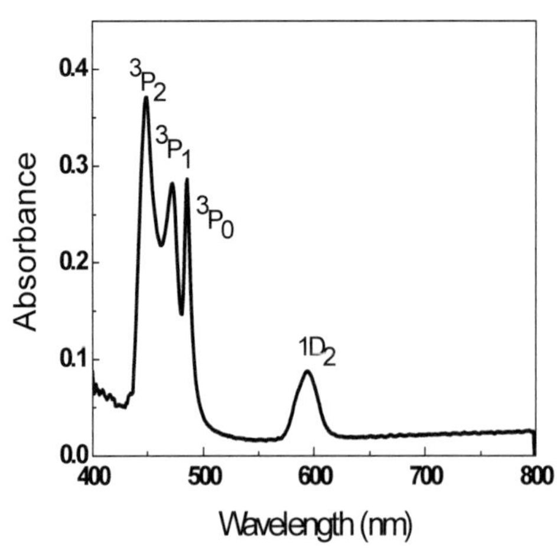

Fig. 1. Absorption spectrum of 0.5 mol% Pr^{3+} doped TeO_2-Li_2O glass. Reprinted with permission from ref. (Rai, 2009), Copyright 2009, Springer.

In the present case, ^3P_J as well as $^1D_2$ levels are populated as a result of two photon absorption resulting upconversion transition to various lower lying levels. The upconversion luminescence may arise by one of the following mechanisms: (a) Sequential absorption of two photons, (b) excited state absorption (ESA), (c) photon avalanche and (d) energy transfer (ET) processes. In this case the upconversion emission is observed even at lower pump power of the laser, the upconversion emission by sequential absorption of two photons is ruled out. Now, if we consider the energy level diagram of Pr^{3+} (Fig. 2), $^1G_4$ level of it lies at ~9850 cm^{-1}. The energy corresponding to NIR radiation is ~12252 cm^{-1}. The $^1G_4$ level is therefore excited through phonon-assisted excitation. The phonon energy of TeO_2 lattice is ~650 cm^{-1} which requires four phonons to be involved in the process. If upconversion emission is due to energy transfer process between two Pr^{3+} ions, the excited Pr^{3+} ions in $^1G_4$ state transfer its excitation energy to the next neighboring Pr^{3+} ion in this excited state. During this process one ion is excited to the upper level and other one return to the ground state. But the emission is observed from the levels lying at energy ~22000 cm^{-1} also, therefore ET is not responsible at least for all upconversion emissions. May be that a part of intensity for the transitions from the levels lying at energy <19700 cm^{-1} is contributed to this channel. But the upconversion emission bands do not show any significant concentration dependence. This also supports ET a less probable channel for upconversion emission. Since the dependence of the upconversion emission intensity versus pump power does not show any inflection, which confirms that the avalanche energy transfer is also not a reasonable process for the upconversion. Therefore the dominant mechanism may be most probably ESA. In this mechanism the excited Pr^{3+} ion in $^1G_4$ level reabsorbs another incident photon of the same energy and is promoted to ^3P_J level. The excited ions from this state relax to different lower-lying excited states and give fluorescence corresponding to the $^3P_2 \rightarrow {}^3H_4$, $^3P_0 \rightarrow {}^3H_4$, $^3P_1 \rightarrow {}^3H_5$, $^3P_0 \rightarrow {}^3H_5$, $^1D_2 \rightarrow {}^3H_4$ ($^3P_0 \rightarrow {}^3H_6$) and $^3P_0 \rightarrow {}^3F_2$ transitions. Normally, upconversion emission is not observed from $^3P_2$ level; however in this case a weak upconversion emission from this state has also been detected. This may be due to exact energy matching at the second step. The upconversion emission was measured at different temperatures of the glass. It has been found that the upconversion emission decreases with increasing the temperature, which reveals that the upconversion is not due to any thermally

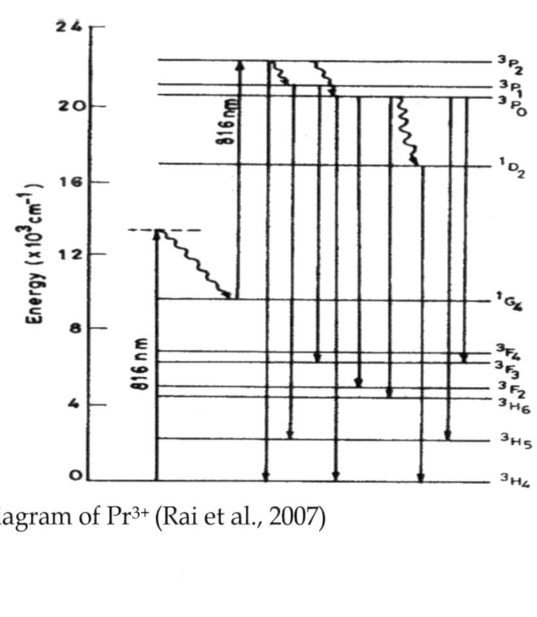

Fig. 2. Energy level diagram of Pr^{3+} (Rai et al., 2007)

activated process. The upconversion efficiency for the ${}^3P_0 \rightarrow {}^3F_2$ transition at 0.5 mol% concentration of Pr^{3+} doped in tellurite glass at room temperature (~30 ^{0}C) is found to be ~1.7%.

A broadband upconversion emission in Pr^{3+} doped aluminum, barium, calcium fluoride glass has been observed at different wavelengths throughout the visible region upon non resonant excitation (~810 nm) corresponding to the ${}^3H_4 \rightarrow {}^1G_4$ transition from a Ti-sapphire laser [Rai et al., 2006]. Upconversion emissions are observed for nearly all bands for which one photon fluorescence is seen. The peaks observed in orange-red region due to the ${}^1D_2 \rightarrow {}^3H_4$ and ${}^3P_0 \rightarrow {}^3H_6$ transitions are also observed in the fluorescence spectrum. One sharp upconversion peak observed near 641 nm is due to the ${}^3P_o \rightarrow {}^3F_2$ transition. Weak upconversion is also observed at ~676 and ~731 nm and they are due to the ${}^3P_1 \rightarrow {}^3F_3$ and ${}^3P_o \rightarrow {}^3F_4$ transitions. Thus, using NIR radiation, one can get upconversion emissions throughout the visible region. The Pr^{3+} ions on absorption of ~12,300 cm^{-1} radiation are promoted to the ${}^1G_4$ level through phonon assisted excitation. The upconversion emission from the ${}^3P_J$ levels is observed, it is possible only when excited Pr^{3+} ion in the ${}^1G_4$ level reabsorb the second incident photon of the energy ~12,300 cm^{-1} (i.e. the ESA process). Energy transfer (ET) amongst the excited ions will not be sufficient to populate the ${}^3P_J$ levels. Therefore, the upconversion emissions observed in blue region are due to the excited state absorption (ESA). The ${}^3P_J$ levels also give upconversion emissions in the orange and red regions. The ${}^1D_2$ level is populated by the relaxation from ${}^3P_J$ levels. The other possibility of population to the ${}^1D_2$ level is via energy transfer (ET) among the Pr^{3+} ions, i.e.

$$ {}^1G_4 + {}^1G_4 \rightarrow {}^1D_2 + {}^3H_4 + \Delta E \text{ (phonon energy)} $$

For this, the intensity of upconversion emission band corresponding to the ${}^1D_2 \rightarrow {}^3H_4$ transition should be large enough and show a quadratic behavior with the concentration of rare earth ions, whereas the upconversion emission corresponding to the ${}^1D_2 \rightarrow {}^3H_4$ transition appears weak. Therefore, the most probable process of the ${}^1D_2$ level population is via relaxation from ${}^3P_J$ levels. The mechanism involved in the upconversion emissions observed in the visible region is due to excited state absorption (ESA).

2.3 (b) Energy transfer (ET)

Upconversion luminescence in triply ionized praseodymium-doped TeO_2-Li_2O glass using direct excitation into the ${}^1D_2$ level from a Nd:YAG (yttrium aluminum Garnet) laser pumped dye laser (~8 ns pulses, tunable from 570 to 600 nm; peak power: ~20 kW, linewidth: ~0.5 cm^{-1}) has been reported [Rai, 2009]. The method for preparation of the doped glasses is reported elsewhere in detail [Rai et al., 2007]. The absorption spectrum of a 0.5 mol% praseodymium doped glass sample in the 400-800nm is shown in Fig. 1. The band positions observed in the absorption spectrum are in good agreement with the values reported for the other glasses. The variations in their bandwidths and relative intensities are only due to the site-to-site variations of the crystal field strengths. The spectra observed for the other samples are similar, except for the band intensities and their line widths, which are concentration dependent. A simplified energy level scheme of the Pr^{3+} electronic levels is shown in Fig. 3. An intense luminescence at ~480 nm and at ~680 nm respectively corresponding to the ${}^3P_0 \rightarrow {}^3H_4$ and ${}^1D_2 \rightarrow {}^3H_5$ transitions was seen under the excitation at ~590 nm (Fig. 4). Similar features were observed for all the doped samples with their relative intensities. The upconversion intensity shows a quadratic dependence with the laser intensity as well as with the concentration of the Pr^{3+} ions. Thus, two incident laser photons and two triply ionized praseodymium ions are responsible for this upconversion emission to occur, while for the emission at ~680 nm ascribed to the $^1D_2 \rightarrow {}^3H_5$ transition a linear dependence on the laser intensity as well as on the concentration of the rare-earth ions with the fluorescence intensity was observed. This confirms that the energy transfer is affecting the upconversion luminescence.

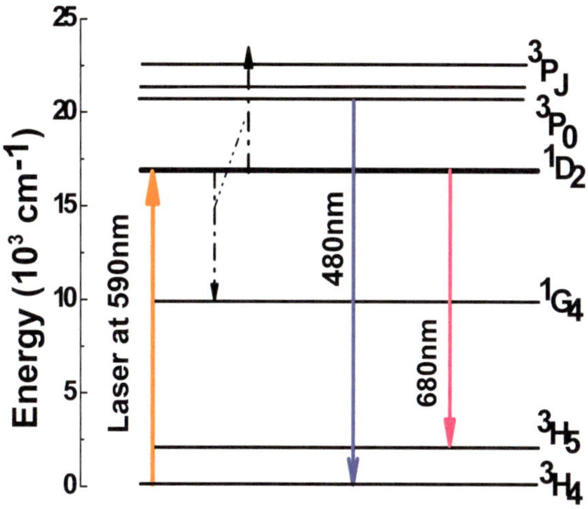

Fig. 3. Energy level diagram of Pr^{3+}. Reprinted with permission from ref. (Rai, 2009), Copyright 2009, Springer.

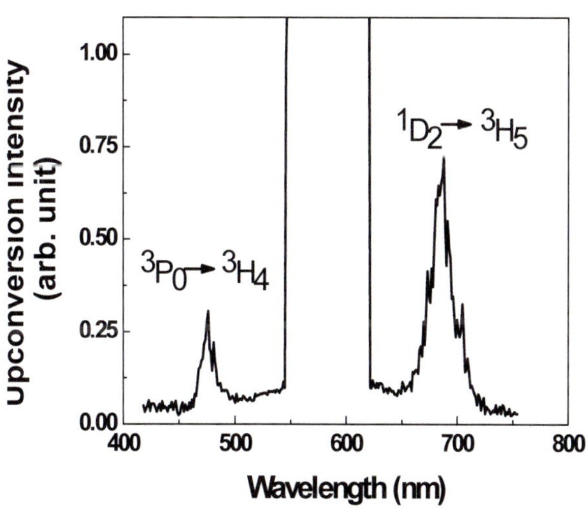

Fig. 4. Upconversion spectrum of 0.5 mol% Pr^{3+} doped TeO_2-Li_2O glass. Reprinted with permission from ref. (Rai, 2009), Copyright 2009, Springer.

In fact, a pair of neighboring Pr^{3+} ions in the ground state is excited by the two incident laser photons (at ~590 nm) directly into the $^1D_2$ state. One of the excited ion in the $^1D_2$ state transfers its excitation energy to its neighboring second excited ion in the same state and returns to the ground state. The second excited ion after receiving the additional energy is excited upward to the ^3P_J state; the radiative relaxation from the $^3P_0$ level gives an intense blue emission at ~480nm attributed to the $^3P_0 \rightarrow ^3H_4$ transition. The phenomenon of the energy-transfer process containing a pair of Pr^{3+} ions can be given as,

2 photons (2hυ) + 2 ions in the $^3H_4$ (ground) state $\rightarrow ^1D_2 + ^1D_2 \rightarrow ^3P_J + ^1G_4 \rightarrow ^3P_0 + ^3H_4$ +phonons.

The temporal evolution of the emission at ~480 nm shows rise and decay times which are illustrated in Fig. 5 for all the doped samples. From this figure, it is clear that the rise and decay times depend upon the concentration of the praseodymium ions. The time dependence of the upconversion intensity can be expressed as,

$$I_{UP} \alpha [\exp(-t/\tau_r) - \exp(-t/\tau_d)] \tag{1}$$

where ´τ_r´ and ´ τ_d´ are the rise and decay times of the upconverted fluorescence respectively. The rise and decay times is observed to decrease as the concentration of the rare-earth ions increases from 0.2 mol% to 1.0 mol%, which also confirms that energy transfer is the dominant mechanism among the excited and its neighboring unexcited ions.

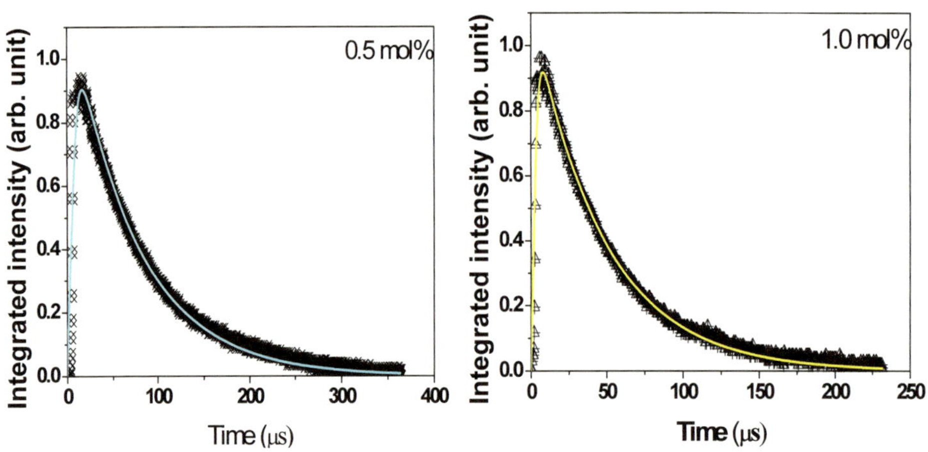

Fig. 5. Temporal evolution of upconversion emission at ~480nm. Reprinted with permission from ref. (Rai, 2009), Copyright 2009, Springer.

Energy transfer process between donors and acceptors

Energy transfer among the donors and acceptors modifies the fluorescence decay and the dynamical behavior of the emission at ~ 680nm and can be described by Inokuti- Hirayama (I-H) model [Inokuti et al., 1965]. According to the I - H model,

$$I(t) = I_0 \exp\{-t/\tau_0 - \Gamma(1-3/S)C/C_0(t/\tau_0)^{3/S}\} \tag{2}$$

or

$$I(t) = I_0 \exp\{-t/\tau_0 - P(t)\}$$

where

$$P(t) = \Gamma(1-3/S)C/C_0(t/\tau_0)^{3/S}$$

and ´τ_0´ is the lifetime of the isolated ions in absence of the energy transfer. C and C_0 are the concentration of the rare earth ions and the critical concentration respectively. 'Γ' is a gamma function. $P(t)$, the transfer function that assumes different forms at different concentrations of rare earth ions, origin of the interaction between the ions and time range [Inokuti et al., 1965]. P(t) involves the contributions due to the energy transfer (ET) processes taking place for the frequency upconversion and ET between donors and acceptors that contributes to the fluorescence emission. The temporal behavior of the $^1D_2 \rightarrow {}^3H_5$ transition peaking at ~680nm as a function of praseodymium ion concentrations was monitored. The radiative lifetime observed for fluorescence emission at ~ 680nm is ~250μs at ~0.01mol% for which no upconversion emission is seen. It is found that the decay curve deviates from its pure exponential shape and lifetime of the $^1D_2$ level decreases with increasing concentration. This indicates that the fast diffusion process should be ignored. Therefore, the static transfer process is the dominant mechanism for energy transfer within the concentration range of our samples. The decay time (τ_D) for the $^1D_2$ level can be written as,

$$1/\tau_D = W_{Rad} + W_{NR}, \text{ where radiative decay rate } (W_{Rad}) = 1/\tau_0 \quad (3)$$

W_{NR}, the nonradiative relaxation rate (which involves the multiphonon emission and energy transfer among the rare earth ions) and its value determined for all doped samples using equation (3) are ~1.5 x 10^3, ~5.7x10^3 and ~22.5x10^3 Hz respectively. A plot for W_{NR} versus concentration shows a quadratic behavior with a slope ~ 1.7±0.1 < 2. This difference is due to the emission of multiple phonons through nonradiative relaxation, employing that two rare earth ions are involved in the energy transfer process.

At longer times the acceptors at large distances are taken into account so according to I - H Model [Inokuti et al., 1965],

$$P(t) = \Gamma(1-3/S)C/C_0(t/\tau_0)^{3/S} = \beta\tau^{3/S} \quad (4)$$

where $\beta = \Gamma(1-3/S)C/C_0(1/\tau_0)^{3/S}$, and S= 6, 8 and 10 for dipole-dipole, dipole-quadrupole and quadrupole-quadrupole type interactions respectively. 'β' is function of radiative lifetime of isolated ions, critical concentration (at which the energy transfer rate for an isolated ion pair is equal to the spontaneous decay rate of the donor). Moreover, in order to get the information about the type of the interactions involved, a log-log plot for P(t) versus time (t) by extracting the exponential factor exp(-τ/τ_0) from equation (2) was made and found best fit with slope 3/S ≈ 0.51±0.01 (i.e. S ≈ 6) for all the doped samples, thereby indicating the interaction to be of dipole-dipole type (Fig 6).

Let N_1, N_2 be the populations in the pair states |1> = |$^1D_2$, $^1D_2$> and |2> = |$^3P_0$, $^3H_4$> respectively and γ_1, γ_2 represent respectively the relaxation rates of state |1> due to all possible mechanism except the transfer to state |2> and the total radiative relaxation rate from state |2>. If W_{12} is the energy transfer rate from state |1> to state |2>, the temporal evolution after excitation of the initial state of the pair and of its final fluorescent state may be given as [Pelle et al., 1993],

$$\frac{dN_1}{dt} = -[\gamma_1 + W_{12}]N_1 \tag{5}$$

$$\frac{dN_2}{dt} = W_{12}N_1 - \gamma_2 N_2 \tag{6}$$

The solutions of the above equations are

$$N_1(t) = N_1(0)e^{-(\gamma_1 + W_{12})t} \tag{7}$$

$$N_2(t) = \frac{N_1(0)W_{12}}{\gamma_1 - \gamma_2 + W_{12}}(e^{-\gamma_2 t} - e^{-(\gamma_1 + W_{12})t}) \tag{8}$$

In order to compare the observed decay and rise times, it is assumed that the two rates (γ_1 + W_{12}) and γ_2 are smaller than τ_r^{-1}, while τ_d^{-1} corresponds to the smaller rate.

Here the observed decay times are longer than the actual lifetime of the $^3P_0$ level (~8.9μs). On comparing the equation (1) with equation (8), it can be seen that τ_r^{-1} and τ_d^{-1} corresponds to γ_2 and (γ_1 + W_{12}) respectively. Therefore, the value of W_{12} can be determined using the relation, $\tau_d^{-1} = W_{12} + \gamma_1$ and observed to be large (≈ 1075 Hz) compared to the other fluoride based host matrices.

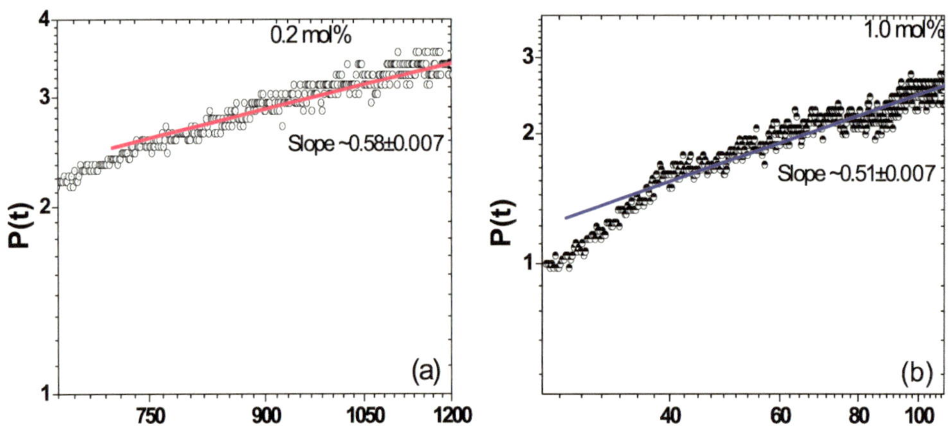

Fig. 6. log – log plot of P(t) versus t. Reprinted with permission from ref. (Rai, 2009), Copyright 2009, Springer.

Orange to blue frequency upconversion in Pr^{3+} doped TeO_2-ZnO and chalcogenide ($Ga_{10}Ge_{25}S_{65}$) glasses containing silver nanoparticles under excitation with a nanosecond laser operating at ~590 nm, in resonance with the $^3H_4 \rightarrow {}^1D_2$ transition has been studied [Rai et al., 2008; Rai et al., 2008]. The absorption spectra of the GGSPr sample with and without metallic nanoparticles annealed for different time intervals have been shown in Fig 7. The absorption bands associated with the Pr^{3+} transitions from the ground state ($^3H_4$) to different excited states viz. $^3P_2$, $^3P_1$, $^3P_0$ and $^1D_2$ and a feature at ~472 nm attributed to surface plasmon resonance (SPR) of the NPs are observed. The SPR wavelength calculated using Mie theory

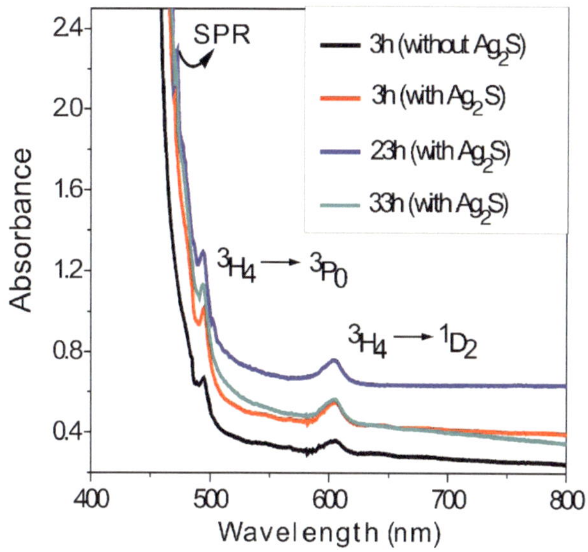

Fig. 7. Absorption spectra of the GGSPr sample (with and without metallic nanoparticles) for different annealing times as indicated in the legend. Reprinted with permission from ref. (Rai et al., 2008), Copyright 2008, American Institute of Physics.

[Prasad, 2004], considering spherical particles, the dielectric function of bulk silver [Palik, 1985] and the glass refractive index, is ~485 nm. This result shows a reasonable agreement with the experimental value since the effect of the metal glass interface is not considered in the calculations neither the nanoparticles shape. The amplitude of the band at ~472 nm increases with increasing the concentration of nanoparticles which grows up for heat-treatment time of ~23 hour (h). For a heat treatment time longer than 23 h, a reverse trend is observed for the concentration. No shift of the band peaking at ~472 nm was observed for different annealing times. The excited Pr^{3+} ions exchange energy in presence of the nanoparticles, originating efficient conversion from orange to blue. The enhancement in the intensity of the upconversion luminescence corresponding to the $^3P_0 \rightarrow ^3H_4$ transition of Pr^{3+} ions, is found to be due to the proximity between the surface plasmon band and the [^3P_J (J=0, 1, 2); $^1I_6$) manifold. The intensity of the emission corresponding to the $^1D_2 \rightarrow ^3H_5$ transition was not affected by the presence of the metallic nanoparticles because the $^1D_2$ multiplet is located far from the surface plasmon band (Fig. 8).

Pr^{3+} doped samples without silver were heat treated for different time intervals, but no variation in the luminescence intensity was observed, indicating that the presence of nanoparticles is essential to obtain enhanced emission. No evidence of change in the glass structure neither nucleation of nanoparticles due to the influence of the laser was observed. This is understood considering that the laser pulses have low energy and the pulse repetition rate (~5 Hz) is small. This result confirms that the enhancement of the upconversion intensity in other samples is attributed to the large local field acting on the emitting ions due to presence of the metallic nanoparticles and not due to changes in the glass structure. On the other hand, quenching was also observed for samples heat treated for longer times, which indicates that ET occurs from the Pr^{3+} ions to the metallic nanoparticles. This result is important in the sense that it demonstrates that there is an optimum heat-treatment time of samples to observe enhanced frequency UC due to nucleation of metallic nanoparticles.

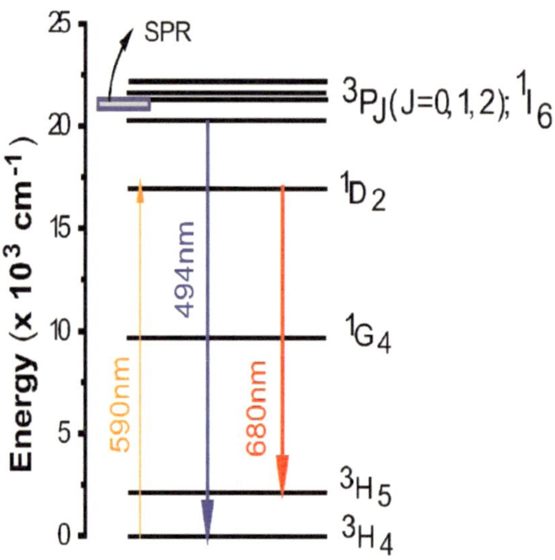

Fig. 8. Simplified energy level diagram of Pr^{3+}. Reprinted with permission from ref. (Rai et al., 2008), Copyright 2008, American Institute of Physics.

2.4 Frequency upconversion in Nd^{3+} doped glasses

Optical spectroscopy and upconversion luminescence in Nd^{3+} doped $Ga_{10}Ge_{25}S_{65}$ glass has been reported [Rai et al., 2009]. The mechanisms leading to the upconversion emissions upon nonresonant excitation at ~1.064mm in the Nd^{3+} doped chalcogenide glass corresponding to the green, orange, and red regions as well as the dynamics of the process have been investigated. The samples were prepared by the melt-quenching method. The absorption spectrum of the Nd^{3+} (0.25 mol %) doped chalcogenide glass sample contains several absorption bands at ~884, ~806, ~795, ~750, ~682, ~594, and ~532 nm corresponding to the transitions from the ground state ($^4I_{9/2}$) to the $^4F_{3/2}$, $^4F_{5/2}$, $^2H_{9/2}$, $^4F_{7/2}$, $^4F_{9/2}$, $^4F_{5/2}$, and $^4G_{7/2}$ excited states respectively. The spectra of other concentrated samples are similar but the intensity of the absorption bands depends linearly on the concentration of the Nd^{3+} ions.

The upconversion spectrum observed under infrared excitation at ~1.064μm is shown in Fig. 9. Three luminescence bands are observed at ~535, ~600, and ~670 nm corresponding to the $^4G_{7/2}\rightarrow^4I_{9/2}$, [$^4G_{7/2}\rightarrow^4I_{11/2}$; $^4G_{5/2}\rightarrow^4I_{9/2}$], and [$^4G_{7/2}\rightarrow^4I_{13/2}$; $^4G_{5/2}\rightarrow^4I_{11/2}$] transitions respectively. The upconversion intensity versus pump power as well as concentration of the Nd^{3+} has been observed to follow a quadratic dependence, indicating that two laser photons and two Nd^{3+} ions contribute to the emission of each upconversion photon. To obtain additional information about the upconversion process the temporal evolution of the $^4G_{7/2}\rightarrow^4I_{9/2}$ transition peaking at ~535 nm was studied. The luminescence signal shows decay times of ~53.4±0.4, 32.6±0.3 and 18.0±0.1μs for the Nd^{3+} concentrations of 0.05, 0.10, and 0.25 mol %, respectively. Since the radiative lifetime τ_R of the $^4G_{7/2}$ level is ~81.23μs. The measured decay times indicate a strong interaction among the Nd^{3+} ions. Since the energy difference $\left[E\left(^4G_{7/2}\right)-E\left(^4G_{5/2}\right)\right]$ has about the same value as $\left[E\left(^4I_{11/2}\right)-E\left(^4I_{9/2}\right)\right]$ (Fig. 10). Then the decrease in the lifetime of level $^4G_{7/2}$ is attributed to the cross-relaxation process ($^4G_{7/2}$; $^4I_{9/2}$) → ($^4G_{5/2}$ → $^4I_{11/2}$). The actual lifetime is related to the crossrelaxation rate by $\tau = \tau_R/(1+W_{CR}\tau_R)$ with W_{CR} being equal to ~6.4 x 10^3 Hz (0.05 mol% Nd^{3+}), 18 x 10^3 Hz (0.1 mol% Nd^{3+}), and 43 x 10^3 Hz (0.25 mol% Nd^{3+}).

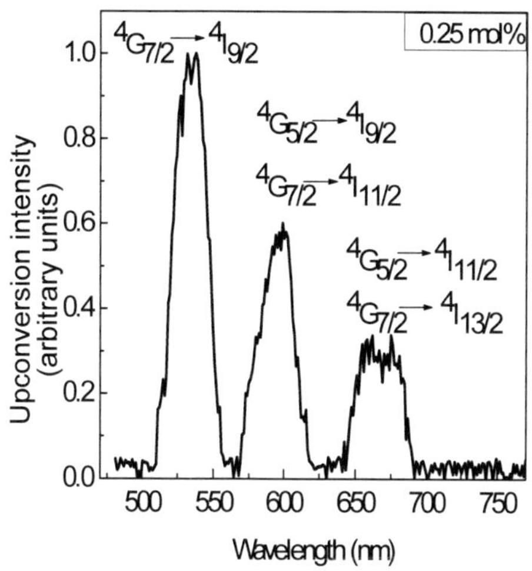

Fig. 9. Frequency upconversion spectrum for excitation using a laser operating at 1.064 μm. Reprinted with permission from ref. (Rai et al., 2009), Copyright 2009, American Institute of Physics.

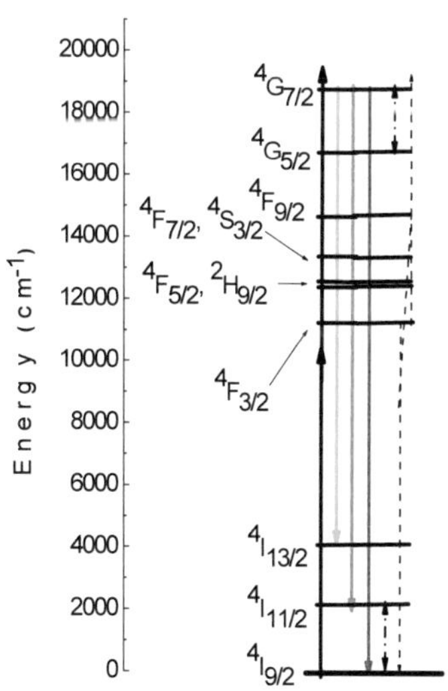

Fig. 10. Simplified energy level scheme for the Nd^{3+} ions. Reprinted with permission from ref. (Rai et al., 2009), Copyright 2009, American Institute of Physics.

To understand the origin of the upconversion luminescence there are three possible excitation pathways. As the laser frequency ' ω_L ' is off - resonance with the Nd^{3+} transitions starting from the ground state ($^4I_{9/2}$). However, two-photon absorption (TPA) is resonant for a transition from ground state to the $^4G_{7/2}$ state. The upconversion luminescence intensity in this case would be quadratic with the laser intensity and would vary linearly with the Nd^{3+} concentration.

Another possibility would be due to a one-photon transition to the $^4F_{3/2}$ level followed by energy transfer (ET) between pairs of excited Nd^{3+} ions. As a result of this process two ions excited to the $^4F_{3/2}$ level may interact and exchange energy in such way that one decay to the ground state and the other is promoted to the $^4G_{7/2}$ level from where it decays radiatively to lower lying levels. However, although this process of energy transfer has been observed in other glasses [Rai, 2009; Rai et al., 2008]. In the present case this process is not expected to be more efficient than the TPA process discussed above because the one-photon frequency detuning for the electronic transition $^4I_{9/2} \rightarrow {}^4F_{3/2}$ is $\approx$ 1500 cm^{-1}. The absorption has to be phonon-assisted and the corresponding probability is small. Another possibility to excite the $^4F_{3/2}$ level would be a resonant one-photon transition originating from the $^4I_{11/2}$ level. However the population in the $^4I_{11/2}$ level is small at ~300 K. In both cases the UC luminescence intensity would present a quadratic dependence with the laser intensity and with the Nd^{3+} concentration.

3. Acknowledgements

Author is grateful to University Grants Commission (UGC) and Department of Science & Technology (DST), New Delhi, India for the financial assistance.

4. References

Babu, P.; Jayasanker C. K., 'Optical spectroscopy of Eu^{3+} ions in lithium borate and lithium fluoroborate glasses', Physica B, 279 (2000) 262-281.

Balda, R.; Fernández, J.; Adam, J. L.; Mendioroz, A; Arriandiaga, M. A., Energy transfer and frequency upconversion in Pr^{3+}-doped fluorophosphate glass', J. Non-Crystalline Solids, 256-257 (1999) 299-303.

Born, M.; Wolf, E., Principles of optics (Cambridge University Press, Cambridge, 1999).

Dieke, G. H.; Crosswhite, H. M., Spectra and energy levels of rare earth ions in crystals, Inter Science, New York, 1968.

Digonnet, M.J.F., 'Rare earth doped fiber lasers and amplifier', (Dekker, New York, 1993).

Dumbaugh, W. H.; Lapp, J. C., Heavy-Metal Oxide Glasses', J. Am. Ceramic Soc., 75 (1992) 2315–2326.

Edward D. Palik, Handbook of Optical Constants of Solids (Academic, New York, 1985).

Ebendroff, H.; Priems, H.; Ehert, D., 'Spectroscopic properties of Eu^{3+} and Tb^{3+} ions for local structure investigations of fluoride phosphate and phosphate glasses', J. Non-Cryst. Solids, 208 (1996) 205-216.

El-Mallawany, R. A. H., Tellurite glasses handbook-Physical properties and data (CRC, Boca Raton, FL, 2001).

Ganem, J.; Dennis, W.M.; Yen, W.M., 'One-color sequential pumping of the 4f5d bands in Pr-doped yttrium aluminum garnet', J. Lumin., 54 (1992) 79-87.

Hussain, N. S.; Buddhudu, V. S., 'Absorption and photoluminescence spectra of Sm^{3+}:TeO_2–B_2O_3–P_2O_5–Li_2O glass', Mater. Res. Bull., 35 (2000) 703 -709.

Inokuti, M.; Hirayama, F., 'Influence of energy transfer by the exchange mechanism on donor luminescence', J. Chem. Phys., 43 (1965) 1978 (1-12).

Jackson, S. D., 'Continuous wave 2.9 µm dysprosium doped fluoride fiber laser', Appl. Phys. Lett., 83 (2003) 1316-1318.

Kaminskii, A.A., Laser crystals, second ed., Springer-Verlag, Berlin, 1990.

Kaplyanskii, A. A.; Macfarlane, R. M., 'Spectroscopy of Solids containing Rare Earth Ions', (North-Holland, New York, 1987).

Kassab, L. R. P.; Araújo C. B. de; Kobayashi, R. A.; Pinto R. A.; Silva, D. M. da, 'Influence of silver nanoparticles in the luminescence efficiency of Pr^{3+}-doped tellurite glasses', J. Appl. Phys., 102 (2007) 103515 (1-4).

Kenyon, A. J., 'Recent developments in rare-earth doped materials for optoelectronics', Progr. Quantum Electron., 26 (2002) 225–284.

Kumar, G. M.; Rao, D. N.; Agarwal, G. S., 'Experimental studies of spontaneous emission from dopants in an absorbing dielectric', Opt. Lett., 30 (2005) 732-734.

Lezama, A.; Leite, J. R. Rios; Araújo, C. B. de, 'Triad spectroscopy via ultraviolet up-conversion in Pr^{3+}:LaF_3', Phys. Rev. B, 32 (1985) 7139–7142.

Lin, H.; Pun, E. Y. B.; Huang, L. H.; Liu, X. R., 'Optical and luminescence properties of Sm^{3+}-doped cadmium–aluminum–silicate glasses', Appl. Phys. Lett., 80 (2002) 2642 (1-3).

Luis E.E. de Araujo, Gomes, A.S.L.; Araujo, Cid B. de; Messaddeq, Y.; Florez, A.; Aegerter, M. A., 'Frequency upconversion of orange light into blue light in Pr^{3+}-doped fluoroindate glasses', Phys. Rev. B, 50 (1994) 16219-16223.

Mohanty, D. K.; Rai, V. K.; Dwivedi, Y.; Rai, S. B., Enhancement of upconversion intensity in Er^{3+} doped tellurite glass in presence of Yb^{3+}. Appl. Phys. B, 104 (2011) 233-236.

Mujaji, M.; Jones, G.D.; Syme, R.W.G., 'Site-Selective Spectroscopy of Ho^{3+} Ions in $CsCdBr_3$ Crystals' Phys. Rev. B, 48 (1993) 710-725.

Nachimuthu, P.; Jagnnathan, R.; Nirmal Kumar, V.; Rao, D. N., 'Absorption and emission spectral studies of Sm^{3+} and Dy^{3+} ions in PbO-PbF_2 glasses', J. Non-Cryst. Solids, 217 (1997) 215-223.

Nageno, Y.; Takabe, H.; Moringama, K.; Izumitani, T., 'Effect of modifier ions on fluorescence and absorption of Eu^{3+} in alkali and alkaline earth silicate glasses', J. Non-Cryst. Solids, 169 (1994) 288-294.

Oliveria, A.S.; Araujo, M.T. de; Gouveia-Neto, A.S.; Medeiros Neto, J.A.; Sombra, A.S.B.,' Frequency upconversion in Er^{3+}/Yb^{3+}-codoped chalcogenide glass', Appl. Phys. Lett., 72 (1998) 753 (1-3).

Pacheco, Elio M.; Araujo, Cid B. De, 'Frequency up-conversion in a borate glass doped with Pr^{3+}', Chemical Physics Letters., 148 (1988) 334-336.

Parker, J. M., 'Fluoride Glasses', Annu. Rev. of Mater. Sci., 19 (1989) 21-41.

Pelle, F.; Glodner, Ph., 'Pair processes in $CsCdBr_3$:Er^{3+}: A study by up-conversion and excited-state absorption of efficient UV and blue anti-Stokes emissions', Phys. Rev. B, 48 (1993) 9995–10010

Prasad, P. N., Nanophotonics (Wiley, New York, 2004).

Rai, A.; Rai, V.K., 'Optical properties and upconversion in Pr^{3+} doped in Aluminium, Barium, Calcium Fluoride glass-I', Spectrochim. Acta A, 63 (2006) 27-31.

Rai, V. K., 'Local field effect of the host matrix on the decay time of doped emitters', Appl. Phys. B, 100 (2010) 871-874.

Rai, V. K., 'Temperature sensors and optical sensors', Appl. Phys. B, 88 (2007) 297-303.

Rai, V. K.; Araújo, C. B. de; Ledemi, Y.; Bureau, B.; Poulain, M.; Zhang, X. H.; Messaddeq, Y., ' Frequency upconversion in a Pr^{3+}doped chalcogenide glass containing silver nanoparticles', J. Appl. Phys., 103 (2008) 103526 (1-4).

Rai, V. K.; Araújo, Cid B. de; Ledemi, Y.; Bureau, B.; Poulain, M.; Messaddeq, Y.,' Optical spectroscopy and upconversion luminescence in Nd^{3+} doped $Ga_{10}Ge_{25}S_{65}$ glass', J. Appl. Phys., 106 (2009) 103512 (1-5).

Rai, V. K.; Araujo, Cid B. de; Menezes, L. de, 'Two photon absorption in TeO_2-PbO glasses excited at 532 and 590nm', Appl. Phys. A, 91 (2008) 441-443.

Rai, V. K.; Kumar, K.; Rai, S.B., 'Upconversion in Pr^{3+} doped tellurite glass', Optical Materials, 29 (2007) 873–878.

Rai, V. K.; Menezes, L. de S.; Araújo, C. B. de,' Infrared-to-green and blue upconversion in Tm^{3+}-doped TeO_2–PbO glass', J. Appl. Phys., 103 (2008) 053514 (1-4).

Rai, V. K.; Menezes, L. de S.; Araujo, C. B. de; Kassab, L. R. P.; Silva, D. M. da; Kobayashi, R. A., 'Surface-plasmon-enhanced frequency upconversion in Pr^{3+} doped tellurium-oxide glasses containing silver nanoparticles', J. Appl. Phys., 103 (2008) 93526.

Rai, V. K.; Rai, S. B.; Rai, D.K., 'Optical studies of Dy^{3+} doped tellurite glass: Observation of yellow-green upconversion', Optics Communications, 257 (2006) 112-119.

Rai, V.K., 'Upconversion due to energy transfer involving Pr^{3+} ions in pairs', Appl. Phys. B, 95 (2009) 329–333.

Reisfield, R., 'Spectra and energy transfer of rare earths in inorganic glasses', Structure and Bonding, 13 (1973) 53-98.

Sai Sudha, M. B.; Ramakrishna, J., 'Effect of host glass on the optical absorption properties of Nd^{3+}, Sm^{3+}, and Dy^{3+} in lead borate glasses', Phys. Rev. B, 53 (1996) 6186–6196.

Savage, J. A., 'Materials for infrared fiber optics', Mater. Sci. Rep., 2 (1987) 99-138.

Silva, D. M. da; Kassab, L. R. P.; Lüthi, S. R.; Araújo, C. B. de; Gomes, A. S. L.; Bell, M. J. V., 'Frequency upconversion in Er^{3+} doped PbO–GeO_2 glasses containing metallic nanoparticles', Appl. Phys. Lett., 90 (2007) 81913 (1-3).

Smart, R.G.; Hanna, D.C.; Tropper, A.C.; Davey, S.T.; Carter, S.F.; Szebesta, D., 'CW room temperature upconversion lasing at blue, green and red wavelengths in infrared-pumped Pr^{3+}-doped fluoride fibre', Electron. Lett., 127 (1991) 1307-1309.

Snoeks, E.; Lagendijk, A.; Polman, A., 'Measuring and modifying the spontaneous emission rate of erbium near an interface', Phys. Rev. Lett., 74 (1995) 2459–2462.

Tripathi, G.; Rai, V. K.; Rai, D.K; Rai, S.B., 'Upconversion in Er^{3+} doped Bi_2O_3–Li_2O–BaO–PbO tertiary glass' Spectrochimica Acta Part A, 66 (2007) 1307–1311.

Tripathi, G.; Rai, V. K.; Rai, S. B., 'Optical properties of Sm^{3+}: CaO-Li_2O-B_2O_3-BaO glass and codoped Sm^{3+}:Eu^{3+}'. Appl. Phys. B, 84 (2006) 459-464.

Tripathi, G.; Rai, V. K.; Rai, S. B., 'Spectroscopic studies of Eu^{3+} doped calibo glass: Effect of the addition of barium carbonate, energy transfer in the presence of Sm^{3+}', Optics Communications, 264 (2006) 116-122.

Victor, T.K.; Raffaella, R.; Maurizio, M.; Maurizio, F., in: S. Jiang (Ed.), Rare earth doped materials and devices VI, Proc. SPIE, 4645 (2002) 174–182.

Yamane, M.; Y. Ashara, Glasses for Photonics (Cambridge University Press, Cambridge, England, 2000).

Part 4

Semiconductor Quantum-Dot Nanostructure Lasers

8

Parameters Controlling Optical Feedback of Quantum-Dot Semiconductor Lasers

Basim Abdullattif Ghalib[1], Sabri J. Al-Obaidi[2] and Amin H. Al-Khursan[3,*]
[1]Laser Physics Department, Science College for Women, Babylon University, Hilla,
[2]Physics Department, Science College, Al-Mustansiriyah University, Baghdad,
[3]Nassiriya Nanotechnology Research Laboratory (NNRL),
Science College, Thi-Qar University, Nassiriya,
Iraq

1. Introduction

A simple model to describe the dynamics of a single mode semiconductor laser subject to a coherent optical feedback is proposed in 1980 by Lang and Kobayashi (LK). Feedback loop depends on external mirror and creates a passive external cavity, which is explicitly taken into account via the complex delayed electric field variable $E(t-\tau)$ fed back into the laser. The round trip time is the main feature of the LK model of the laser beam. The LK model has open the door to a very complex dynamics since the system phase space has infinite dimensions and sustain a chaotic regime [1]. Optical feedback consist of two subjects, coherent and incoherent feedback, depending on whether the coherence time of the laser light is larger or smaller than the delay time (τ) respectively [2]. There are five distinct regimes that are defined by the level of the feedback power ratio, this is discusses in section 2. The great importance for dynamics of semiconductor lasers with optical feedback is due to the potential applications of such lasers for secure communications by means of chaotic synchronization. External perturbations such as injected signal, feedback, or pump current modulation are required to achieve a chaotic output. From a practical point of view, optical feedback provided by a back reflecting mirror is one of the simplest ways to achieve chaotic oscillations from a semiconductor laser, even weak optical feedback leads to complex dynamics. In particular, it can sustain a chaotic regime of low-frequency fluctuations with sudden irregular intensity dropouts followed by a gradual intensity recovery [6].

To improve semiconductor laser performance, a nanoscale active region, in the form of two-dimensional quantum wells (one degree of freedom), one-dimensional quantum wires (two degrees of freedom), or zero-dimensional quantum dots (three degrees of freedom) are used [3]. Since quantum dot (QD) semiconductor materials have discrete energy subbands, one could expect symmetric emission lines, then the subject of great current interest is a sensitivity of QD semiconductor lasers to optical feedback [4]. QD lasers acquired more

* Corresponding Author

importance after significant progress in nanostructure growth by self-assembling technique. The first demonstration of a QD laser with low threshold current density was reported in 1994 [5]. A QD laser emits at wavelengths determined by the energy levels of the dots, rather than the bandgap energy. Thus, they offer the possibility of improved device performance and increased flexibility to adjust the wavelength. They have the maximum material and differential gain, at least 2-3 orders higher than QW lasers [6]. A QD laser is a semiconductor laser that uses QDs as the active laser medium in its light emitting region. Due to the tight confinement of charge carriers in QDs, they exhibit an electronic structure similar to atoms. Lasers fabricated from such an active media exhibit higher device performance compared to traditional semiconductor lasers based on bulk or quantum well active medium. Improvements in performance can appear in wide modulation bandwidth, while both lasing threshold, relative intensity noise, linewidth enhancement factor and temperature sensitivity are reduced. QD semiconductor lasers displays an interesting hybrid of atomic laser and standard quantum well semiconductor laser properties. Optical feedback containing very commonly in a wide variety of fields including biology, ecology and physics. In biology they occur in regulation and stabilization processes, e.g. blood cell-production, neural control and respiratory physiology and control of physiological systems (heart rate, blood pressure, motor activity) [7,8].

This chapter covers a review and a study of optical feedback in QD lasers. Section 2 reviews the characteristics of optical feedback regimes, while in section 3, the new rate equations for laser dynamics to describe the active region and Parameters used in the calculations. Section 4 includes a study of time delay effect on optical feedback at threshold current, phase and time delay.

2. Diode lasers with optical feedback

Optical feedback depends on several parameters and effects on the operating characteristics of a diode laser. One of these effects is the re-injection of a fraction of light into the laser diode after a time (τ) later delayed optical feedback. The optical feedback regimes consists of five distinct regimes defined by the level of the feedback power ratio. These regimes are depend on the internal parameters of the solitary diode laser, such as the linewidth enhancement factor, the diode dimensions and the facet coatings. They are:

Regime I corresponds to low feedback level were broadening or narrowing of the optical linewidth is observed depending on the feedback phase, the importance of this regime lies not in the manipulation of linewidths achievable, as greater control can be achieved in higher regimes. In regime II, two modes are observed do not simultaneously exist. As the feedback is increased towards regime III the mode hopping frequency and the mode splitting frequency increases. The transition to this regime from regime I is characterized by an observed line broadening. This regime overlaps regime I. The properties of regime III are single-mode operation and stability arises. The minimum linewidth mode has the best phase stability for this reason regime is inappropriate for most applications. This regime occupies only a very small value of feedback power ratios. Regime IV, which is observed for higher feedback levels, is associated with the coherence collapse this regime is useless for coherent communications. However, applications such as imaging or secure data transmission require highly incoherent sources. In regime V, a stable emission with a narrow linewidth at high feedback levels. This regime is characterized by very narrow-linewidth stable single-mode low intensity noise operation. The coherence of the laser is regained. It operates as a long cavity laser with a short active region. Experimentally it is usually required to antireflection coat the diode laser front facet in order to reach this regime. Due to the strong feedback in this regime the system is also much less sensitive to additional reflections. The system operating in this regime is often referred to as an external cavity diode laser (ECDL) [12].

3. The rate equations for laser dynamics

In the QD laser, we considers a separate system for electrons and holes in the QD ground state (GS) and exited state (ES) which typically applies for the self-organized QDs in the $InN/In_{0.8}Al_{0.12}N/In_{0.25}Al_{0.75}N$ material system. The model used here is plotted in Fig. 1, where the various mechanisms that occurs in the laser cavity are abstracted. Our model recognizes between lifetimes according to carrier type (electron and hole) although their values are taken here the same for simplicity. First the carriers are injected in the wetting layer with rate I/q and relax in the dot. The carriers are captured in the ES with a rate $1/\tau_{c,w}^{e}, 1/\tau_{c,w}^{h}$ and from the ES to the GS with rate $1/\tau_{c,E}^{e}$, $1/\tau_{c,E}^{h}$. The carriers escape also from the GS back to the ES with rate $1/\tau_{e,G}^{e}, 1/\tau_{e,G}^{h}$ or from the ES back to the WL with rate $1/\tau_{e,E}^{e}$, $1/\tau_{e,E}^{h}$. The recombination processes of carriers from the WL and the QD confined states with rates $\tau_{r,w}^{e}$ $\tau_{r,w}^{h}$ $\tau_{r,E}^{e}$ $\tau_{r,E}^{h}$ $\tau_{r,G}^{e}$ $\tau_{r,G}^{h}$, respectively. It is assumed that the stimulated emission can take place only due to recombination between the electrons and holes in the ES and GS. Then the rate equation system becomes:

$$\frac{dE_G}{dt} = -\frac{E_G}{2\tau_s} + \frac{1}{2}\Gamma g_G \upsilon_g(\rho_G^e + \rho_G^h - 1)E_G + \frac{1}{2}i\alpha(\rho_G^e + \rho_G^h)E_G + \frac{\gamma}{2}E_G(t-\tau) + R_{sp,GS} \quad (1)$$

$$\frac{dE_E}{dt} = -\frac{E_E}{2\tau_s} + \frac{1}{2}\Gamma g_E \upsilon_g(\rho_E^e + \rho_E^h - 1)E_E + \frac{1}{2}i\alpha(\rho_E^e + \rho_E^h)E_E + \frac{\gamma}{2}E_E(t-\tau) + R_{sp,ES} \quad (2)$$

$$\frac{d\rho_E^e}{dt} = -\frac{\rho_E^e}{\tau_{r,E}^e} - \frac{\Gamma g_E \upsilon_g}{N_{QD}}(\rho_E^e + \rho_E^h - 1)|E_E|^2 - \frac{\rho_E^e(1-\rho_G^e)}{\tau_{c,E}^e} + \frac{\rho_G^e(1-\rho_E^e)}{\tau_{e,G}^e} - \frac{\rho_E^e}{\tau_{e,E}^e} + \frac{N_{w_e}}{N_{QD}\tau_{c,w}^e} \quad (3)$$

(3)

$$\frac{d\rho_E^h}{dt} = -\frac{\rho_E^h}{\tau_{r,E}^h} - \frac{\Gamma g_E \upsilon_g}{N_{QD}}(\rho_E^e + \rho_E^h - 1)|E_E|^2 - \frac{\rho_E^h(1-\rho_G^h)}{\tau_{c,E}^h} + \frac{\rho_G^h(1-\rho_E^h)}{\tau_{e,G}^h} - \frac{\rho_E^h}{\tau_{e,E}^h} + \frac{N_{w_h}}{N_{QD}\tau_{c,w}^h} \quad (4)$$

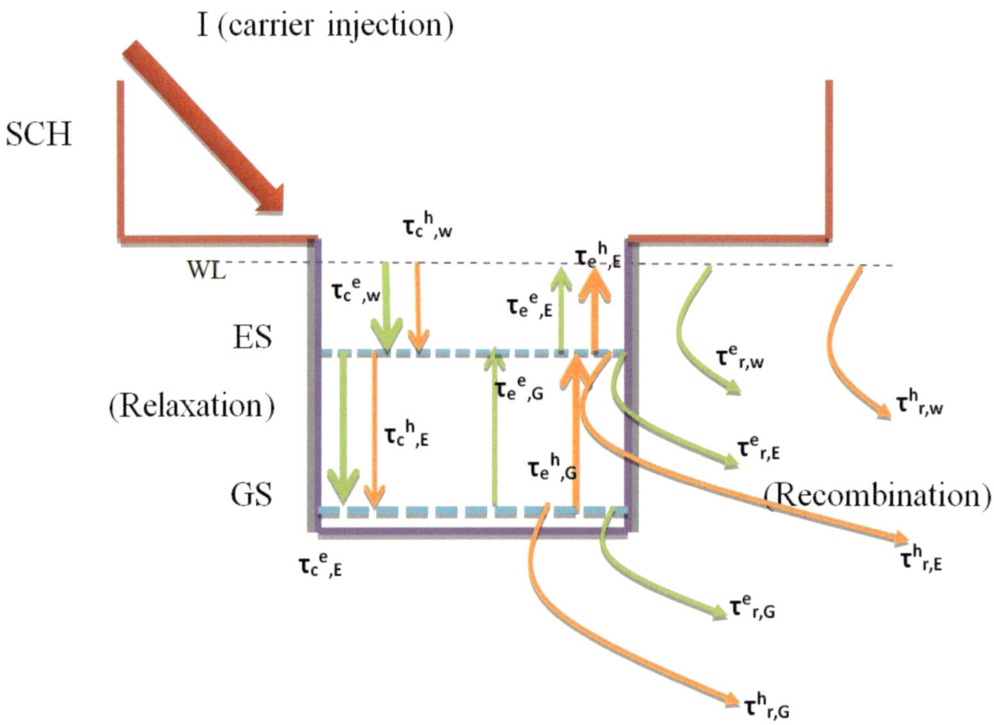

Fig. 1. Energy diagram of the active layer of the QD laser

$$\frac{d\rho_G^e}{dt} = -\frac{\rho_G^e}{\tau_{r,G}^e} - \frac{\Gamma g_G \upsilon_g}{N_{QD}}(\rho_G^e + \rho_G^h - 1)|E_G|^2 + \frac{\rho_E^e(1-\rho_G^e)}{\tau_{c,E}^e} - \frac{\rho_G^e(1-\rho_E^e)}{\tau_{e,G}^e} \tag{5}$$

$$\frac{d\rho_G^h}{dt} = -\frac{\rho_G^h}{\tau_{r,G}^h} - \frac{\Gamma g_G \upsilon_g}{N_{QD}}(\rho_G^e + \rho_G^h - 1)|E_G|^2 + \frac{\rho_E^h(1-\rho_G^h)}{\tau_{c,E}^h} - \frac{\rho_G^h(1-\rho_E^h)}{\tau_{e,G}^h} \tag{6}$$

$$\frac{dNw_e}{dt} = \frac{J_{c1}}{q} + \frac{\rho_E^e}{\tau_{e,E}^e} - \frac{N_{w_e}}{N_{QD}\tau_{c,w}^e} - \frac{N_{w_e}}{N_{QD}\tau_{r,w}^e} \tag{7}$$

$$\frac{dNw_h}{dt} = \frac{J_{c2}}{q} + \frac{\rho_E^h}{\tau_{e,E}^h} - \frac{N_{w_h}}{N_{QD}\tau_{c,w}^h} - \frac{N_{w_h}}{N_{QD}\tau_{r,w}^h} \tag{8}$$

Where E_G, E_E are complex amplitudes of electric field in the QD GS and ES, respectively, g_G, g_E are gain in GS and ES, respectively, Γ is the optical confinement factor, υ_g is the group velocity, α is the linewidth enhancement factor, γ is the feedback level, $\rho_G^e, \rho_G^h, \rho_E^e, \rho_E^h$ are occupation probabilities in GS, ES for electrons (e) and holes (h) respectively, τ is the time delay, $R_{sp,GS}$, $R_{sp,ES}$ are the spontaneous emission rates in GS and ES, respectively. τ_S is the photon lifetime, N_{w_e}, N_{w_h} are the electron and hole carrier densities in the wetting layer, N_{QD} is the QD density. J_{c1}, J_{c2} are the current densities of electrons and holes, respectively. Equations (1-8) are solved numerically to describe the dynamics of the carrier densities in wetting layer for electrons and holes and the occupation probability in ground and exited states for electrons and holes.

4. Calculations and discussion

4.1 Effect of time delay on optical feedback

Figure 2 shows the time series of photon density in the GS. It shows a negligible value of the GS field. This can be results since the laser works at ES. The time series of the photon density in ES at three values of time delay is illustrated in Fig. 3(a)-(b) at the two values of threshold current (1.5J_{th} and 4.5 J_{th}) respectively. From these figures one can show that the amplitude of the electric field in the ES increases with increasing current density. Approximately, as the current density doubles, ES photon density four times increases. On the other hand, the round trip delay time in the external cavity, τ, has a different effect. At shorter delay times, ES photon density four time increases, while further increment in the delay time doubles it. The laser delays by (3.5-4.5 ns) before the population inversion built then the relaxation oscillations built. The electric field amplitude is increased at longer round-trip external cavity delay time, i.e. longer external cavity length. The periodic oscillations are completely removed at shorter τ. This can be attributed to the observation [14] that these instabilities exhibited by QDs are related to long external cavity. Fig. 4 shows the time series of occupation probability of electrons in ES at different values of external cavity delay time τ. Fig. 5 shows the time evolution of electron and hole densities in WL. The same curve is obtained for all the three delay times. The symmetric rise of both densities results from taking the same parameters (at most) for both carriers, although the hole density somewhat higher than electron WL. The relaxation oscillations appears in the behavior of ES occupation while it is completely removed in the WL carrier behavior. This can be reasoned to the faster relaxation time from WL to ES and longer escape time to WL. Fig. 6 shows the three-dimensional (3D) plot of the ES photon density and GS occupation probability vs. time. It shows that the feedback oscillations of ES field raises when the GS occupation probability goes to unity.

4.2 Coherent and non-coherent optical feedback

The coherence and non-coherence are depends on the phase between incident and reflected waves by external cavity.

Figure 7 shows the ES photon density at three phases. The longer turn-on delay time is shown to increases with ϕ. When $\phi = 0$ the ES photon density takes a high value and the feedback oscillations are shown. When $\phi = \pi / 2$ the ES field amplitude is reduced and the oscillations appear at longer time. This reduction results from interference between forward

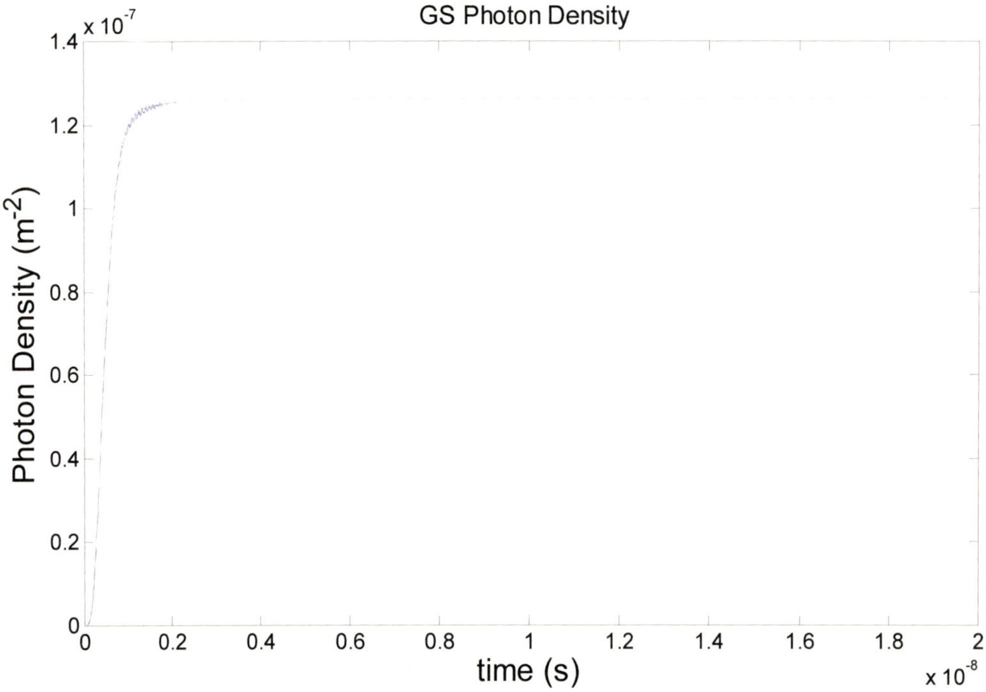

Fig. 2. Time variation of Photon density in GS when $J_{c1}=1.5\ J_{th}$,fi=0, α=2 andγ=0.025e12.

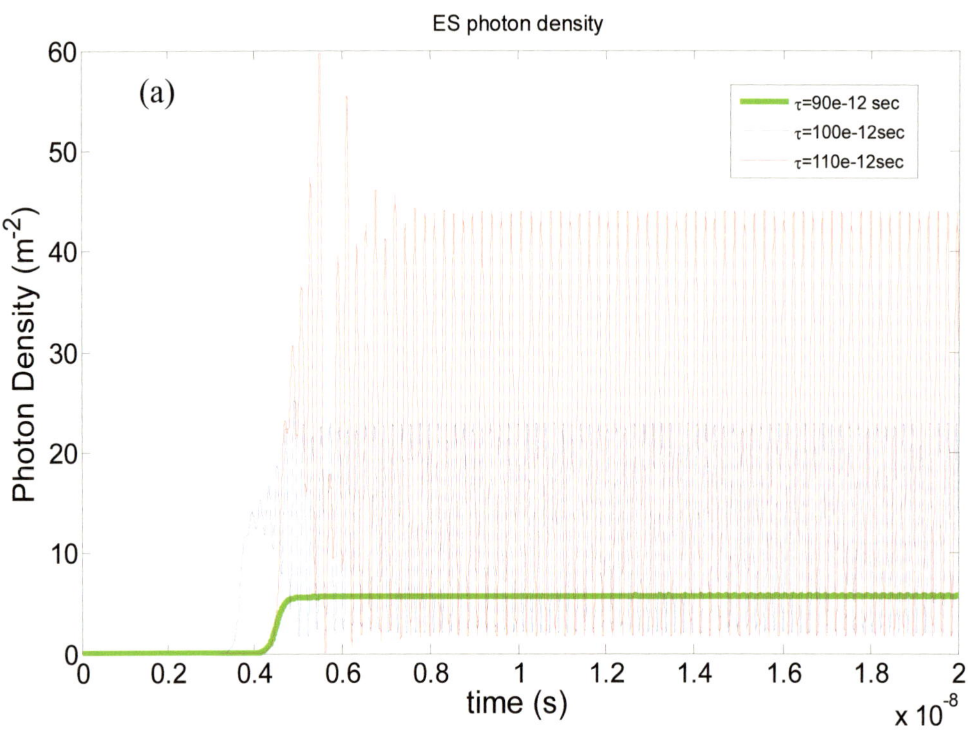

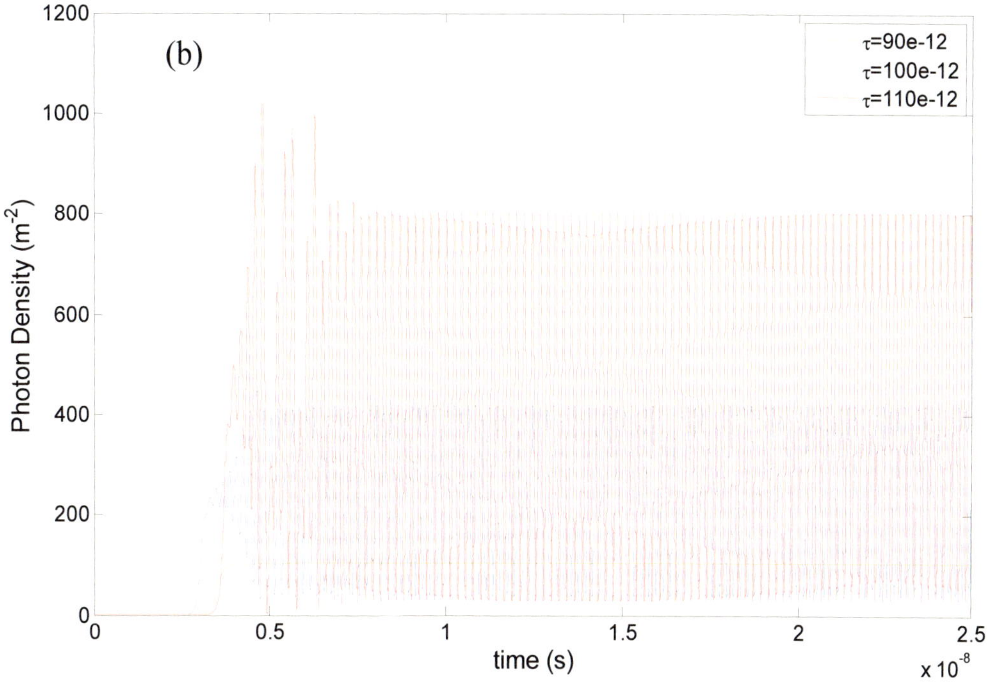

Fig. 3. Time variation of Photon density at three values of time delay when (a): J_{c1}=1.5 and J_{c1}=4.5 J_{th}. The simulation is done at the following parameters ϕ =0, α=2 and γ=0.025e12.

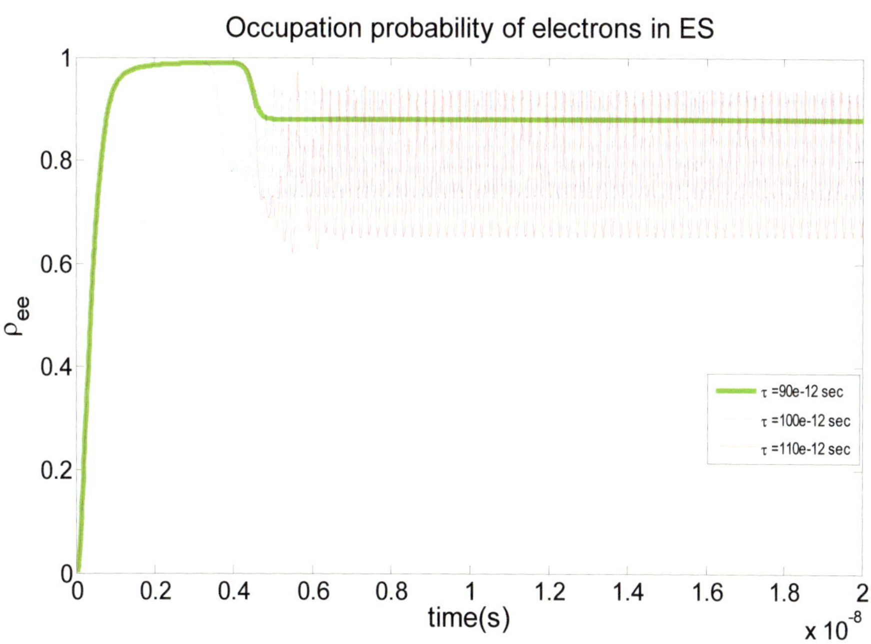

Fig. 4. Time series of electron occupation probability in ES at three values of time delay when J_{c1}=1.5 J_{th}, ϕ =0, α=2 and γ=0.025e12.

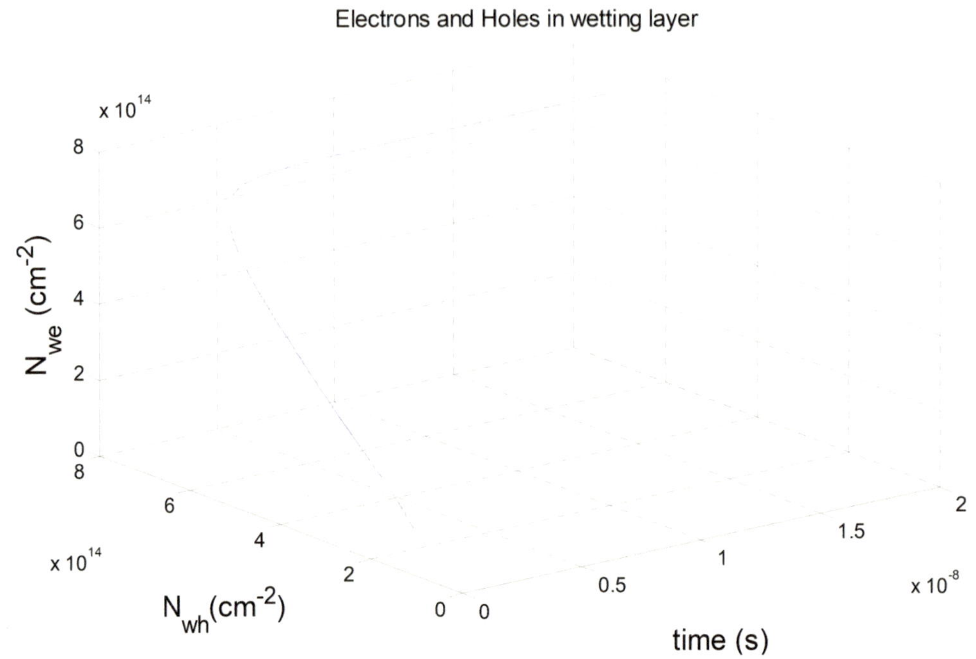

Fig. 5. 3D plot of WL electron and hole densities at $J_{c1}=1.5\ J_{th}$, $\varphi=0$, $\alpha=2$ and $\gamma=0.025e12$

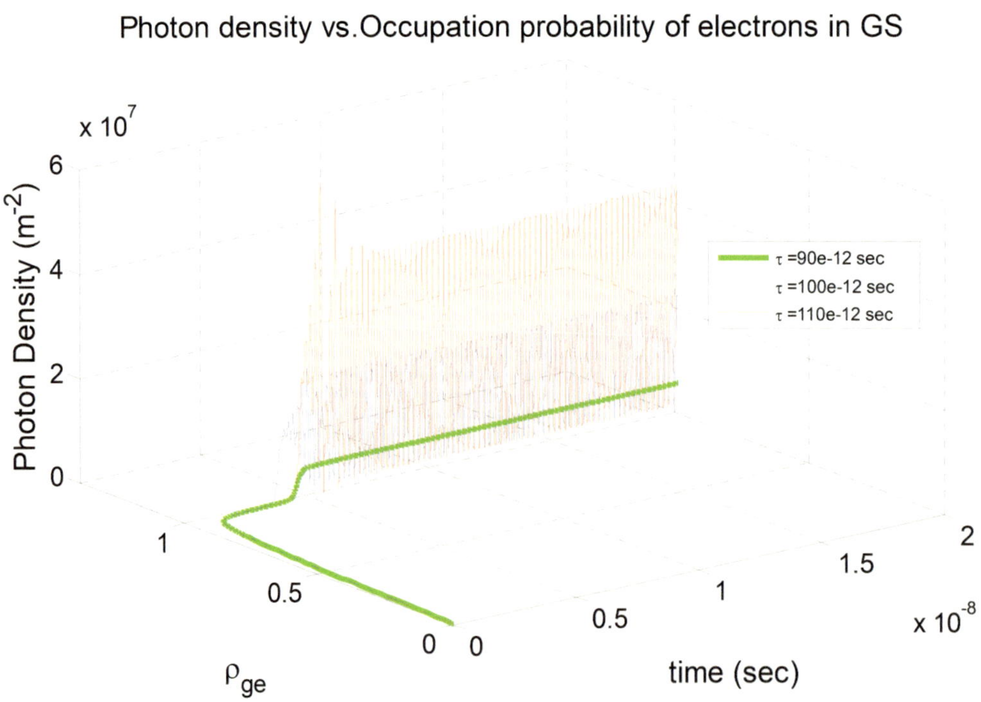

Fig. 6. 3D plot of photon density in ES, occupation probability in ES vs. time at three values of time delay when $J_{c1}=1.5\ J_{th}$, $\phi=0$, $\alpha=2$ and $\gamma=0.025e12$.

and backward waves. When $\phi = \pi$ electric field is completely damped. This results from the destructive interference between the laser and the delay fields. To discuss this case let us study ES occupation probability at these phases. This is shown in Fig. 8. A point one must refer to here is the time spent before feedback oscillations appears. When the fields are constructively interfere ($\phi = 0$) the oscillations are appear earlier (~ after 5ns) then electron occupation is reduced. When $\phi = \pi / 2$, the interference have small effect where the oscillations appear at time (>1ns) and the reduction in electron occupation is small. In Fig. 9 a 3D plot of ES photon density vs. occupation probability of electrons and holes in ES. Although the occupation probability of electrons in ES goes to unity, the interference results in zero field at π-phase. This is also stressed by Fig. 9. Fig. 10 shows the 3D plot of ES photon density vs. occupation probability of (a): electrons and (b) holes in ES. Fig. 11 shows ES photon density vs. (a): electrons and (b): holes in wetting layer. One can discusses the case of $\phi = \pi$ by comparing Figs. 7 and 8. With increasing ϕ the carrier density increases while the photon density reduces. For small ϕ, the relaxation oscillations are high and depletes the carrier density where the laser gain attains. For $\phi = \pi$ case the laser turns off, but due to current injection the carrier density increases until the gain is achieved again then, the laser turns on and undergoes relaxation oscillations. The memory of similar earlier events is retained [15] within the external cavity and reinjected into the laser cavity. Finally an equilibrium state is achieved. Depending on this, one can also relates the longer turn-on delay time with increasing ϕ shown in Fig. 7 to the carrier depletion occurs with increasing incoherence.

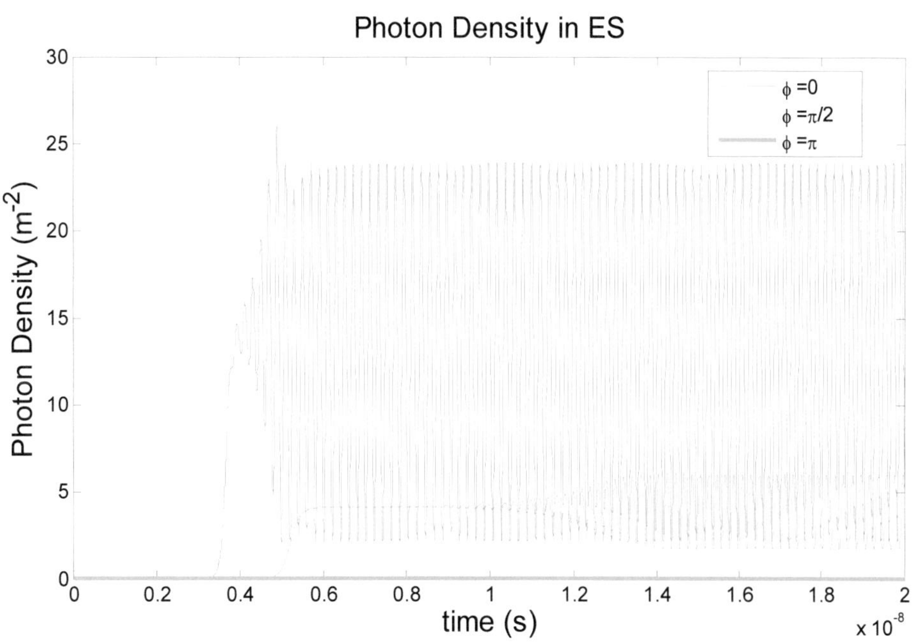

Fig. 7. Time series of ES photon density at three phase values when J_{c1}=1.5 J_{th}, $\tau = 100ps$, α=2 and γ=0.025e12.

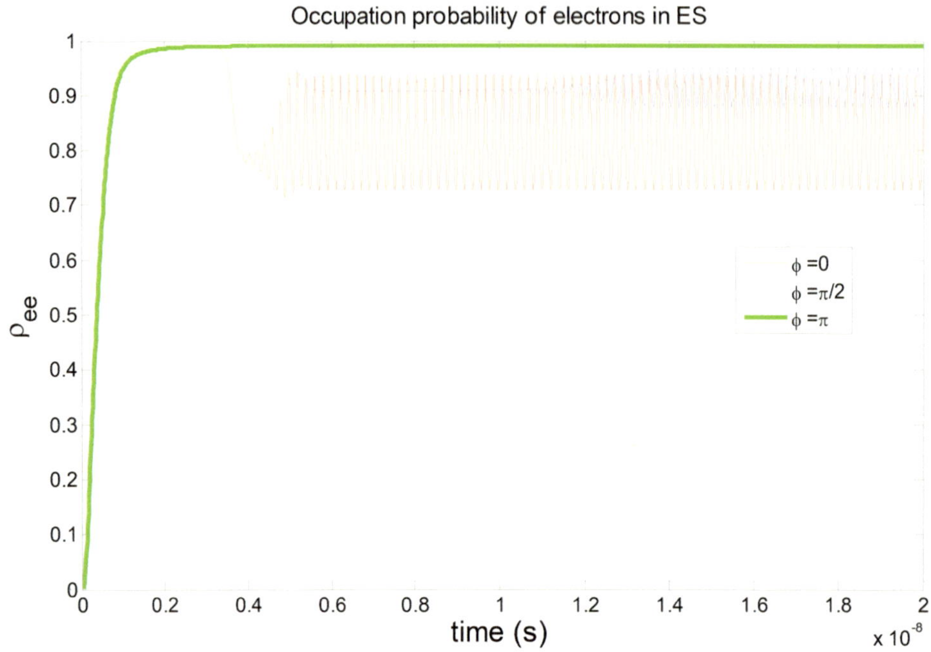

Fig. 8. Time series of occupation probability of ES electrons at three values of ϕ when J_{c1}=1.5 J_{th}, $\tau = 100ps$, α=2 and γ=0.025e12.

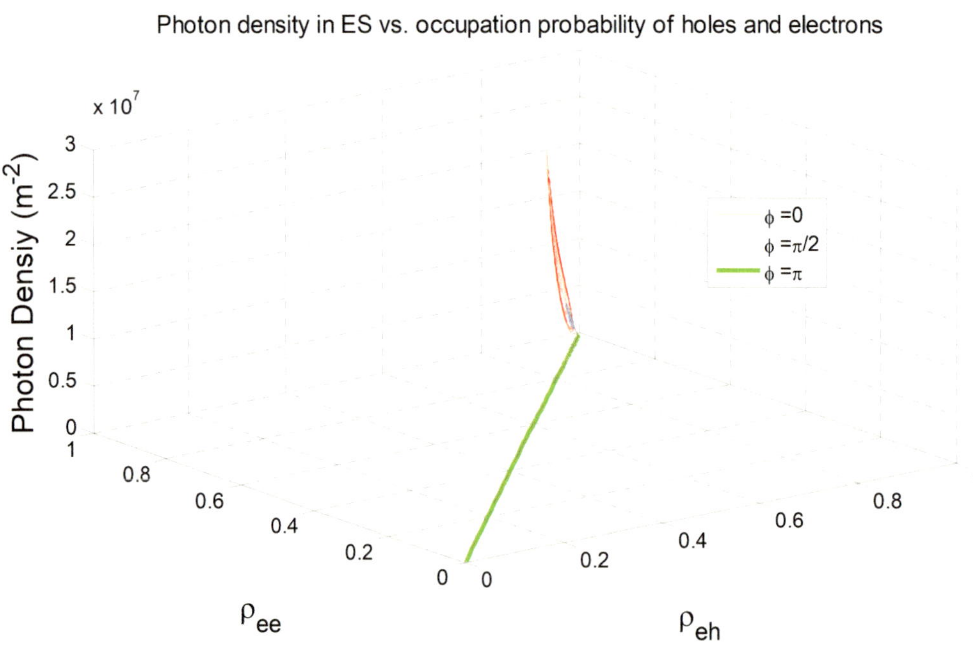

Fig. 9. 3D plot of ES photon density vs. occupation probability of electrons and holes in ES at three values of ϕ when J_{c1}=1.5 J_{th}, $\tau = 100ps$, α=2 and γ=0.025e12.

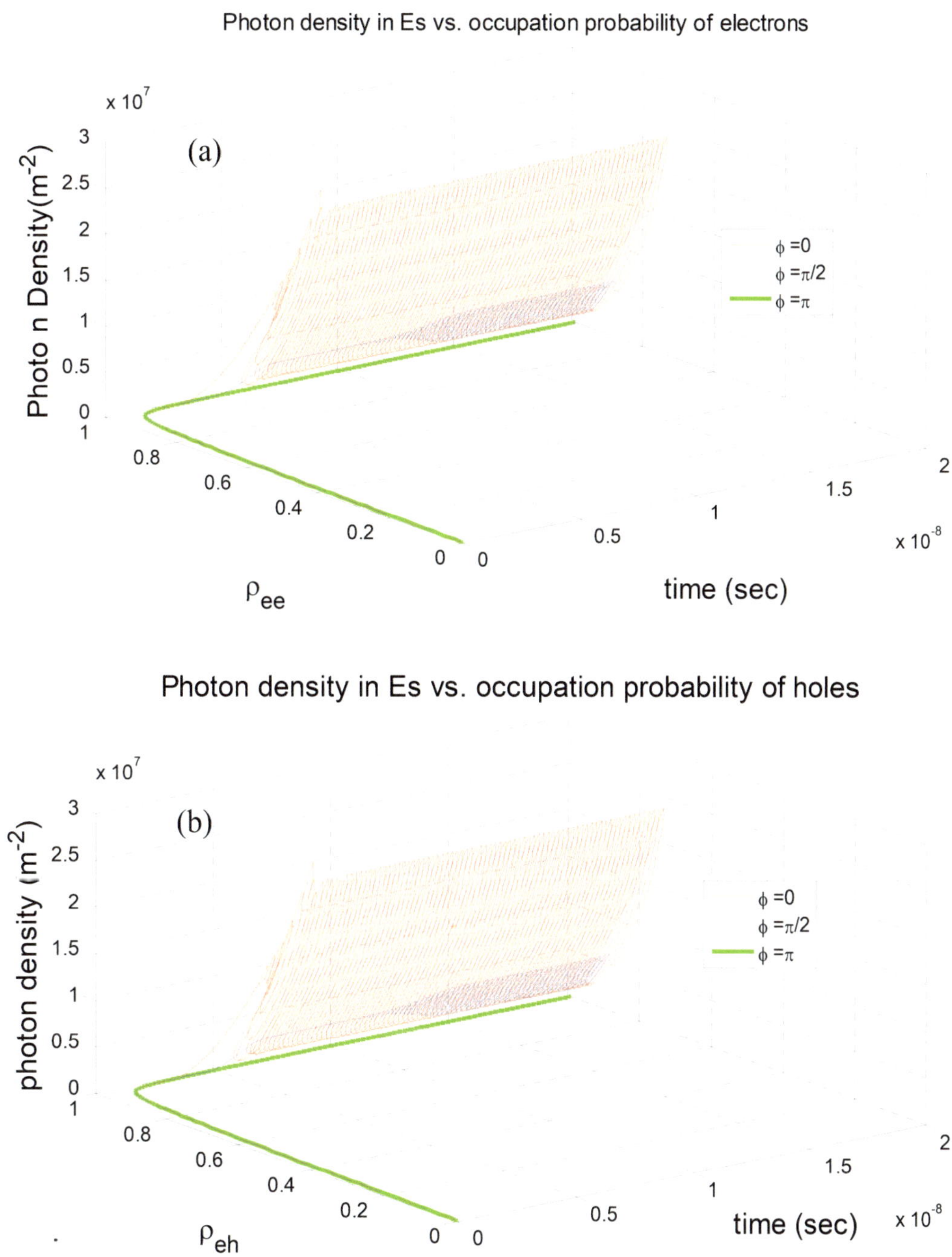

Fig. 10. 3D plot of electric field in ES vs. occupation probability of (a): electrons and (b) holes in ES at three values of ϕ when J_{c1}=1.5 J_{th}, $\tau = 100ps$, α=2 and γ=0.025e12.

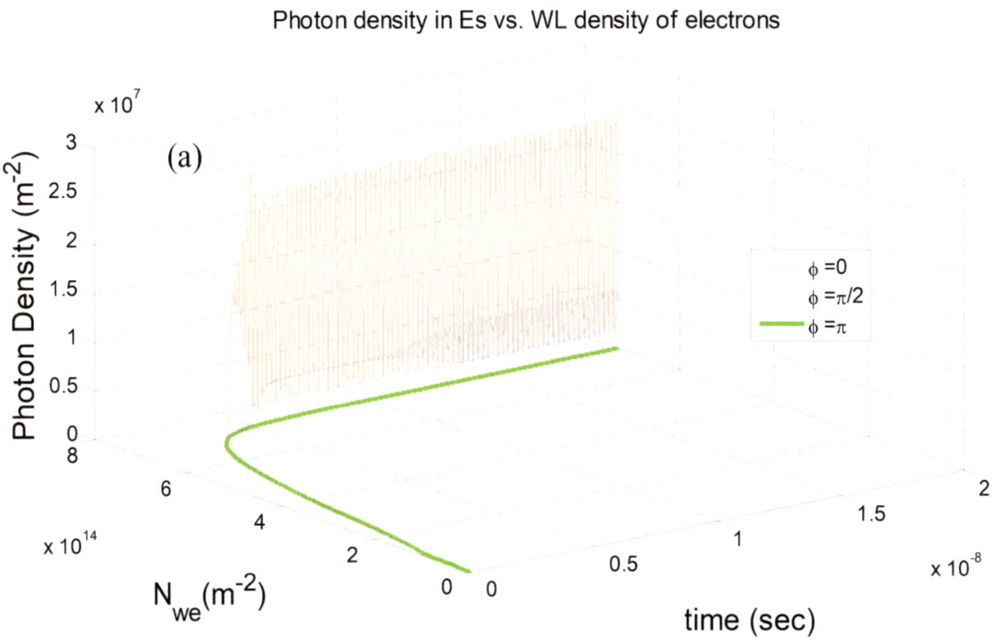

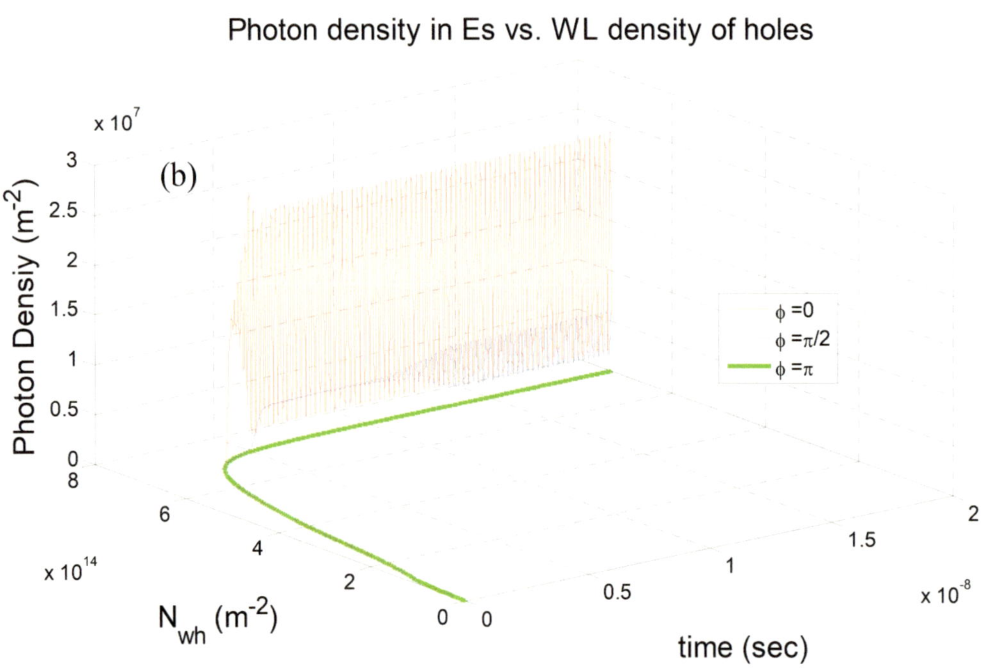

Fig. 11. 3D plot of electric field in ES vs. (a): electrons and (b): holes in wetting layer at three values of φ when $Jc1=1.5\ Jth, \tau = 100ps$, α=2 and γ=0.025e12.

Parameter	Symbol	Value	Unit
Recombination lifetime of electrons in GS	$\tau^{e}_{r,G}$	0.4	ns
Recombination lifetime of holes in GS	$\tau^{h}_{r,G}$	0.4	ns
Recombination lifetime of electrons in ES	$\tau^{e}_{r,E}$	0.4	ns
Recombination lifetime of holes in ES	$\tau^{e}_{r,E}$	0.4	ns
Recombination lifetime of electronsin WL	$\tau^{e}_{r,w}$	1000	Ps
Recombination lifetime of holes in WL	$\tau^{h}_{r,w}$	1000	Ps
Carrier capture time of electronsin WL	$\tau^{e}_{c,w}$	3	ns
Carrier capture time of holesin WL	$\tau^{h}_{c,w}$	3	ns
Carrier escape time of electrons from ES to WL	$\tau^{e}_{e,ES}$	1000	ps
Carrier escape time of holes from ES to WL	$\tau^{h}_{e,ES}$	1000	ps
Carrier relaxation time of electrons from GS to ES	$\tau^{e}_{e,GS}$	1.2	Ps
Carrier relaxation time of holes from GS to ES	$\tau^{h}_{e,GS}$	1.2	ps
Carrier relaxation time of electrons from ES to GS	$\tau^{e}_{c,ES}$	0.16	Ps
Carrier relaxation time of holes from ES to GS	$\tau^{h}_{c,ES}$	0.16	Ps
Optical confinement factor	Γ	$7*10^{-3}$	
Density of QDs	N_{QD}	$5*10^{14}$	cm^{-2}
Laser length	L	$2*10^{-3}$	cm
Effective thickness of the active Layer	L_w	$0.2*10^{-6}$	cm

Table 1. Parameters used in calculations.

5. Conclusions

The feedback in quantum dot lasers is discussed. The rate equations model using the delay differential equations is stated and solved numerically to elucidate the behavior of different states in the quantum dot laser. excited states in quantum dot is shown to have an important effect on the feedback. Effect of decoherence is studied and is shown to delays the laser field due to carrier depletion.

6. References

[1] D. Pieroux and P. Mandel, "Low-frequency fluctuations in the Lang- Kobayashi equation", Physical Review E 68, (2003): 036204.

[2] J. M. Pol, " Semiconductor Laser Dynamics Compound-cavity", Ph.D. University of Balears (2002).

[3] E. Kapon, D. M. Hwang and R. Bhat, "Stimulated Emission in Semiconductor Quantum Wire Heterostructures," Phys. Rev. Lett., 63, No.4 (1989): 430.

[4] D. O`Brien, S. P. Hegarty and G. Huyet and A. V. Uskov, "Sensitivity of quantum-dot semiconductor lasers to optical feedback," Optics Letters, 29, No.10 (2004): 1072.

[5] M.Grundmann, D. Bimberg, V. M. Ustinov, S. S. Ruvimov, M. V. Maximov, P. S. Kop'ev, Zh. I. Alferov, U. Richter, P. Werner, U. Gosele, H. J. Kirstaedter, N. N. Ledentsov. "Low threshold, large to injection laser emission from (InGa)As quantum dots," Electron. Lett., 30 (1994): 1416.

[6] M. Datta and Z. Wasiczko Dilli, "Quantum Dot Lasers", (CambridgeUniversity Press, Cambridge, 1997) 11.

[7] D. Bimberg, M. Grundmann, and N. N. Ledentsov, "Quantum Dot Heterostructures", (John Wiley & Sons Ltd., New York, 1999).

[8] M. Kuntz, N. N. Ledentsov, D. Bimberg, A. R. Kovsh, V. M. Ustinov, A. E. Zhukov and Y. M. Shernyakov, "Spectrotemporal response of 1.3 mm quantum-dot lasers," Appl. Phys. Lett. 81, No.20 (2002): 3846.

[9] I. V. Koryukin and P. Mandel, "Dynamics of semiconductor lasers with optical feedback: Comparison of multimode models in the low-frequency fluctuation regime," Physical Review A 70 (2004): 053819.

[10] S. B. Kuntze, B. Zhang, L. Pavel, "Impact of feedback delay on closed- loop stability in semiconductor optical amplifier control circuits," J. Lightwave Technology, 27, No.9 (2009): 1095.

[11] J. L. Chern, K. Otsuka, and F. Ishiyama, "Coexistence of Two Attractors in Lasers with Delayed Incoherent Optical Feedback," Opt. Comm. 96, (1993): 259.

[12] R. W. Tkach and A. R. Chraplyvy, "Regimes of feedback effects in 1.5μm distributed feedback lasers," J. Lightwave Technology, LT-4, (1986): 1655.

[13] A. V. Uskov. D. O'Brien, S. P. Hegarty, and G. Huyet, "Sensitivity Of quantum-dot semiconductor lasers to optical feedback," Optics Letters, 29, No.10 (2004): 1027.

[14] E. A. Viktorov, P. Mandel and G. Huyet, "Long-cavity quantum dot laser," Optics Letters, 32, (2007): 1268.

[15] R. Ju and P. Spencer, "Dynamic Regimes in Semiconductor Lasers Subject to Incoherent Optical Feedback," J. Lightwave Technology, 23 (2005): 2513.